THE COLLABORATIVE FIGHT

STUDIES IN **CMR**
CIVIL-MILITARY RELATIONS

William A. Taylor, *Series Editor*

THE COLLABORATIVE FIGHT

Pursuing Jointness in the US Military

Paul R. Birch and Lina M. Svedin

University Press of Kansas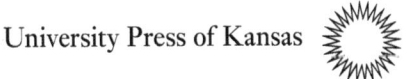

Published by the University Press of Kansas (Lawrence, Kansas 66045), which was organized by the Kansas Board of Regents and is operated and funded by Emporia State University, Fort Hays State University, Kansas State University, Pittsburg State University, the University of Kansas, and Wichita State University.

The views expressed in this publication are those of the author and do not necessarily reflect the official policy or position of the Department of Defense or the US government. The public release clearance of this publication by the Department of Defense does not imply Department of Defense endorsement or factual accuracy of the material.

Library of Congress Cataloging-in-Publication Data

Names: Birch, Paul R., author. | Svedin, Lina M., author.
Title: The collaborative fight : pursuing jointness in the US military / Paul R. Birch and Lina M. Svedin.
Other titles: Pursuing jointness in the US military
Description: Lawrence, Kansas : University Press of Kansas, [2024] | Series: Studies in civil-military relations | Includes bibliographical references and index.
Identifiers: LCCN 2023019475 (print) | LCCN 2023019476 (ebook)
  ISBN 9780700636211 (cloth)
  ISBN 9780700636228 (ebook)
Subjects: LCSH: Unified operations (Military science) | United States—Armed Forces—Organization.
Classification: LCC U260 .B53 2024 (print) | LCC U260 (ebook) | DDC 355.4/6—dc23/eng/20230912
LC record available at https://lccn.loc.gov/2023019475.
LC ebook record available at https://lccn.loc.gov/2023019476.

British Library Cataloguing-in-Publication Data is available.

Printed in the United States of America
10 9 8 7 6 5 4 3 2 1

The paper used in this publication is acid free and meets the minimum requirements of the American National Standard for Permanence of Paper for Printed Library Materials Z39.48-1992.

We dedicate this book to Stephen "Chef" Chiabotti,
whose shenanigans got us into this mess and
whose warmth and passion made us want to stay in it.

Contents

Tables _____

Series Editor's Foreword

William A. Taylor

The ways in which individual military services work together toward national security objectives and the extent to which they succeed at doing so are perennial themes of military history and contemporary operations. Jointness has been an aspirational goal of many modern militaries, including the US defense establishment since World War II. In this fine volume, Paul Birch and Lina Svedin explore the complex realities and myriad difficulties associated with such lofty ambitions. Combining thorough research with sweeping interviews of senior leaders and experienced practitioners, the authors produce a comprehensive account of both the challenges and the opportunities inherent in jointness. In doing so, they explore prevailing theories from a range of disciplines, including general ones on organizational cooperation and others specifically related to the military services, and apply them to relevant case studies—AirLand Battle from 1973 to 1991, the Joint Primary Aviation Training System from 1988 to the present, and air support in counterinsurgency from 2001 to 2012—to share important insights on such contemporary endeavors as Joint All-Domain Command and Control (JADC2). As a result, the authors elucidate essential factors that enable enhanced coordination and identify others that diminish it over time or prevent it from ever occurring in the first place.

In this thought-provoking study, Birch and Svedin challenge popular conceptions about the degree to which the separate military services achieve jointness, or even consistently seek it. The probing questions and innovative answers that the authors provide are both serious and substantial. Not only is jointness a central objective for policymakers, but its prevalence—or lack thereof—can have profound consequences on the battlefield. Deficiencies in the ability of the distinct services to collaborate diminish US military power, in turn limiting strategic options. The authors thereby pierce the comfortable veneer that jointness is self-perpetuating by examining uncomfortable truths lurking beneath the surface. Indeed, competition among the services is often more prevalent than cooperation between them. Throughout this book, the authors remind us that multiple forms of jointness exist, permeating both institutional prerogatives and operational capabilities. This novel survey identifies factors that detract from cooperation and proposes adjustments that can improve group effort in the long term, thereby benefiting the efficiency of the defense establishment in peace and buttressing military effectiveness during war.

This superb analysis advances our understanding of civil-military relations in meaningful ways. Three key themes stand out. First, this book showcases the multidisciplinary nature of civil-military relations as a field, as well as the diverse perspectives necessary to address many of its most pressing issues. By incorporating a broad range of theories and works from different fields, the authors lay useful foundations, both theoretical and empirical, for grasping the prevailing contours of contemporary civil-military relations. Second, this work provides unique understandings about civilian control of the military by examining the underlying complexities of military coordination. The authors address a vexing conundrum: If interservice rivalry is the norm, then how can policymakers ensure that the various military services achieve national security objectives for the common good rather than pursuing their own parochial interests? This study illuminates the value of ensuring oversight of the military in a way that privileges cooperation over rivalry, while not trying to completely eliminate the latter either. Third, this examination affords cogent discernments about a vital premise of civil-military relations—collaboration. The authors challenge readers to question the extent to which the military services strive for meaningful partnership of their own accord and consider ways in which civilian policymakers can enhance teamwork moving forward.

The admirable result is a welcome addition that contributes to a nuanced understanding of civil-military relations as a field of inquiry and to one practical arena within it—jointness—that if pursued over a long-term horizon has substantial rewards in terms of both institutional and strategic outcomes. The ensuing story is one that explores how specific leaders, military organizations, and the defense establishment writ large often pursue divergent interests, interrogates why that dynamic persists, ponders how policymakers can enhance collaboration, and reinforces why it all matters to US national security during both peace and war. These consequential topics are what this original and valuable book is all about.

Acknowledgments

We are grateful for the many people and institutions that have supported our research and time working together.

A host of research librarians, archivists, historians, officers, enlisted members, and others assisted us at the Library of Congress, the National Archives, the US Army Center for Military History, the Pentagon Library, the US Strategic Command, the US Air Force Air Education and Training Command, the US Naval History and Heritage Command, the US Marine Corps School of Advanced Warfighting, the US Central Command, the US Air Forces Central, and the Air Force Historical Research Agency. They all have our lasting gratitude.

In addition, nearly sixty professionals who are serving or have served in the US defense establishment agreed to interviews for this project, and many of them indulged several rounds of questions. They include active and retired military members of all ranks, congressional staffers, college professors, distinguished authors, and corporate leaders. The collective generosity of time and insight that they demonstrated have been invaluable. These professionals proved to be passionate about military cooperation, and many provided an extraordinary amount of assistance. Their wisdom inspired the best of what this book has to offer and we alone stand for any shortcomings or errors reflected here.

We would like to give a special thanks to the Library of Congress for providing us with some real estate inside that researcher's paradise and thought-inspiring setting for a year of our work. Finally, we would like to thank those close to us who gave up time and attention, so that we could finish this collaborative operation.

Acronyms and Abbreviations

ACCE	air component coordination element
ALB	AirLand Battle
ATC	Air Training Command
BAI	battlefield air interdiction
BRAC	Defense Base Closure and Realignment Commission
CAOC	combined aerospace operations center
CAP	combat air patrol
CAS	close air support
CENTCOM	US Central Command
COIN	counterinsurgency
DoD	Department of Defense
FM	Field Manual, *also* frequency modulation
GAO	Government Accounting Office
GPS	Global Positioning System
ISR	intelligence, surveillance, and reconnaissance
JCS	Joint Chiefs of Staff
JFDG	Joint Force Development Group
JPATS	Joint Primary Aviation Training System
JSTARS	Joint Surveillance and Target Attack Radar System
MCO	major combat operations
NATO	North Atlantic Treaty Organization
NDAA	National Defense Authorization Act
OEF	Operation Enduring Freedom
OIF	Operation Iraqi Freedom
OSD	Office of the Secretary of Defense
PATS	Primary Aviation Training System
SAC	Strategic Air Command
SASC	Senate Armed Services Committee
SOF	special operations forces
SUPT	Specialized Undergraduate Pilot Training
TAC	Tactical Air Command
TACP	tactical air control party

TFX	Tactical Fighter Experimental
TRADOC	Training and Doctrine Command
UAV	unmanned aerial vehicle

Chapter One

The Collaborative Fight

US Armed Services' Ambivalent Relationship to Jointness

The Armed Forces of the United States have embraced "jointness"
as their fundamental organizing construct at all echelons.
Jointness implies cross-service combination wherein the capability
of the joint force is understood to be synergistic, with the sum
greater than its parts (the capability of individual components).
—Joint Publication 1, *Doctrine for the Armed Forces of the*
United States, March 2013

Young man, the Soviet Union is our adversary.
The Navy is our enemy.
—General Curtis E. LeMay, US Air Force
Attributed by George Will, October 2013

The Collaborative Imperative

The beautiful picture of brothers in arms vanquishing a tyrant. The power
of a well-orchestrated army and navy winning historic battles. Overwhelming military might and ability through teamwork. That's the image the US
military services portray to the public and tell themselves throughout their
official doctrine. But perhaps there is a fatal flaw in that armor. Often, the
services think more about their own turf than about the overarching objective of national security and maintaining an advantage against the United
States' external enemies. The issue is not new. The Joint Staff was established in 1942, aiming to achieve joint military action, both strategically in
Washington and tactically on the battlefield. The Pentagon (the Joint Chiefs
in particular) proclaims "jointness" in military affairs. But if we dig deeper,
we find that the house is not quite in order—or that the joint barracks rarely
passes inspection. Interservice rivalry is real, the competition for budget
dollars fierce, and adjusting functions or sharing resources between services
seems nearly impossible. For all the talk of jointness, there seems to be more
competition among services than there is cooperation. And jointness without
cooperation is dead.

So what do we do? There is a real need for the ability to fight jointly on the ground. The need for operational jointness is acutely experienced by the troops and commanders involved in US security engagements around the globe. They attempt to solve the need for operational collaboration with duct tape and bailing wire—that is to say, with service tools not designed to fit with each other, with minimal or questionable authority, and with very little functioning infrastructural support. We can do better than that, can we not? One, we wish to support our troops in the field and, two, it is a security imperative to have real joint thinking and unified action in military affairs. We are entering into an era of "joint all domain command and control" (JADC2), where the reliance on joint communication and coordination is pivotal. The troubles of accomplishing jointness should give us grave cause of concern: If the Army, Air Force, and Navy habitually have trouble buying radios that can talk with each other, why do we think we will build Army weapons that can be directed to targets discovered by Air Force Aircraft via information networks constructed by the Navy? In order to have a better chance at making JADC2 work we need to figure out what facilitates and inhibits interservice cooperation in a broader sense, so we can specifically target the structural and cultural mechanism needed to make institutional jointness work in reality.

We take on the challenge in this book by establishing a theoretical framework for understanding interservice jointness—a framing theory of institutional jointness—based on prior research on collaboration and on military behavior in areas other than joint force and capacity development. We examine cases in depth where the US military has had some success in pursuing joint institutional ventures or actions and we use these bodies of knowledge and data to draw conclusions about what it takes to make interservice jointness real, and what these conclusions imply for JADC2 and future iterations of joint endeavors in the US military.

The Development of an Ambiguous Relationship: Mixed Signals about Jointness

The fundamental problem is this: the US military sends mixed signals about the extent to which it is a joint organization. "Jointness," borrowing from the Joint-doctrine epigraph that opened this chapter, is the notion that the six US military services (Army, Navy, Marines, Air Force, Space Force, and the Coast Guard) cooperate in their efforts to ensure national security, and it is a de facto point of departure for discussions about US military policy. Jointness is an assumption formally entrenched in public law, propagated by those who steer and supervise the defense establishment, encouraged by the military services themselves, and embraced—perhaps subconsciously—by the general population. In the area of public law, two major American defense reorganization bills explicitly implore joint-force effectiveness and management. They are the National Security Act of 1947 (US Congress 1947) and the Goldwater-Nichols Department of Defense Reorganization Act of 1986 (US Congress 1986). There have also been formal executive structures put in place since 1903 to encourage components of the US military to pursue institutional jointness and work in concert with one another. The exemplar of this type of cooperation, prior to the establishment of the Joint Chiefs, was the War and Navy Departments' Joint Board orders to the services to collaborate on and develop joint war plans. These orders resulted in the "color-coded" war plans promulgated during the 1920s and 1930s, the most prominent of which was War Plan Orange (devised for an anticipated conflict with Japan).

Evidence that outside observers assume jointness to be the normal state of affairs is provided by the thousands of media reports that refer monolithically to "the military" or "the Pentagon," implying a unified approach to military affairs and efforts (e.g., Shanker 2010; Nissenbaum 2013; Dorell 2013). Reading any of the Pentagon's joint doctrine—consisting of hundreds of pamphlets that cover everything from general approaches to theater war to detailed logistics procedures—might lead a casual observer to the conclusion that close cooperation between the military services is the norm. The American public, represented in Gallup Polls, answers clearly and unambiguously that we have a unified military when asked to compare the armed forces to other national institutions.[1]

Other evidence reveals considerable skepticism and institutional contradictions with respect to jointness, though. Vague presumptions of harmonious cooperation do not hold up to detailed scrutiny. Even if the United States has won its largest battles, campaigns, and wars through joint cooperation on the battlefield, a mantle of jointness rarely adorns the broader scope of military affairs. Nor is it always in view when major security issues arise. When President Theodore Roosevelt established the Joint Board in 1903, he did so realizing that cooperation between the American Army and Navy had grown difficult and elusive.[2] The major legislative initiatives of 1947 and 1986 likewise derived their legislative impetus from less-than-outstanding examples of joint cooperation. For the media that covers Pentagon politics, "interservice rivalry" is a real and well-documented phenomenon that plays out in venues well beyond service-academy sporting events. Interservice rivalry in fact manifests at every interface between the services from appropriations bills to command of forces in harm's way. Members of the defense establishment, some of whom commissioned those glossy brochures on jointness, see the problem firsthand. Anyone who has observed weapons acquisition or the adoption of operational plans, or of even simple administrative rules that affect more than one service, can attest to the fact that intense, parochial responses from the various military institutions accompany every initiative.

Second to defense insiders, academics seem to have the clearest view of just how great the divisions are among the US armed services. For centuries, scholars have realized the difficulties of effective combined-arms efforts, which are the historic intra- and interservice predecessors to jointness.[3] Treaties enabling joint or combined action and the politics that accompany them are notoriously difficult. Accounts as disparate as Hebrew scriptures, Sun-Tzu, Thucydides, Livy, and Niccolò Machiavelli demonstrate that military cooperation—when it happens—is fleeting, given to the whim of individual and organizational self-interest, and often marked by breeches of trust.[4]

For decades, scholars have explicitly described, even in an American context of military unity, the pulling and hauling of bureaucratic politics that characterize interactions among the services, the Department of Defense, and elected officials (Huntington 1957, 428–455).[5] Huntington's (1957) concept of "inter-branch rivalry" is an innate part of the way he models pluralistic American defense strategies.[6] Writing immediately prior to the Goldwater-Nichols reforms of 1986, Thomas MacKubin Owens offered a pessimistic prognosis for joint institutions (1985, 106–107). He recognized the difficulties of overcoming the internal defense bureaucracy in light of the underlying strategic dilemma: an inability "to predict and control the actions of possible enemies" (Huntington 1957, 418).[7] The idea is that where one or more services benefit from a guess at where the next war might arise—in today's era, the driving example is an anticipated military conflict with China in and around the Straits of Taiwan, which makes the Air Force and Navy more relevant than the Army—any service on the outside of that assumption can make a counterpoint argument that a *different* war is more likely to arise. Other scholars such as Carl Builder have lamented the inability or unwillingness of the Joint Chiefs of Staff to fulfill this role. (It is the organization ostensibly best suited to make informed judgments about needed tradeoffs among military forces, objectives, and risks—it's in the name after all, right?) Instead, he cautioned that when the nation needs full accountings of military options, associated costs, and inherent risks, "the JCS is, regrettably, not the place to get them" (Builder 1989, 151).

Other scholars such as Ian Horwood (2006, 2) have remarked on the longevity and persistence of interservice rivalry and the tendency of the services to measure their success in terms of resources they deny their sister services. Congressional staffers and defense civil servants have also had an unencumbered view of this Schadenfreude. James Locher's (2002, 15) portrayal of "service supremacists" in the Pentagon described influence peddlers who put preference and independence over warfighting capability.

Thomas Ehrhard (2000, 58) described the fierce rivalries that rage on among the services, debunking any simplistic notion that the defense establishment takes "unified positions." Finally, David Johnson (2007, 49) has indicated that joint doctrine sometimes exists to heighten interservice tensions, rather than providing the unity it purports.

Yet even in those circles where the obstacles to jointness are most apparent, there is still a desire to see more of the cooperative ethic it represents. Samuel Huntington, in his foundational treatise on civil-military affairs, called on the civilian masters of the US military to allow the "military spirit" to live on, because "military life subordinates man to duty for society's purposes" (1957, 465).[8] Huntington's optimism was indeed founded on a monolithic view of the military, but his sentiment reflects a larger aspiration and assumption: that the institution as a whole has a self-denying ability that will allow the people in it to work for the best possible national outcomes. Other scholars and practitioners alike call this "a public service ethos," and view it as foundational to what it means to work in public organizations (Perry and Wise 1990; Lawton, Rayner, and Lasthuizen 2013, 49–70).

The Case for Jointness

Writing about cooperation between air and ground forces in the Second World War, David Spires lifted up "mutual trust, respect, and a common mission-directed interest" (1998, 316). The sentiment arises from the most basic reason for seeking jointness: it leads, ostensibly, to increased operational efficiency and defense capability. Unity of command, a benefit of jointness extolled by Clausewitz (1984) and other scholars of military strategy (Gray 1999, 212), cannot be extricated from victory in modern war.[9] Ehrhard (2000, 635) expressed a desire for a "catalyst" who understood the various services' innovation paradigms well enough to enable more efficient weapons-system integration. Monte Cannon, who wrote about the difficulty of achieving battlefield synergy among the services, nonetheless espoused "crafting a unifying vision of victory as a foundation for joint command and control" (2012, 299). Jeffrey Donnithorne (2013, 517) examined the challenges to jointness executed within the military bureaucracy, but deemed the pursuit of cooperation worthy, putting the onus on civilian masters to craft coherent policy and legislation that facilitate jointness and prevent uncooperative shirking. Unless the epigraph that opens this chapter is a cynical ruse designed to fool outsiders who read military doctrine, the military services themselves seem eager to behave in a cooperative manner—even if their members might not agree with all of the definitions used in and principles espoused by joint doctrine or feel that one service's bias skews the doctrine (e.g., Wilkerson 1997,

66).[10] A snide but popular aphorism used to express this skepticism is, "You cannot spell Joint without the letters A, R, M, and Y."

The reasons for institutional and operational jointness seem obvious from a first-order, rational view of US military history.[11] Joint operations have enabled the most visible and significant military successes—how could the D-Day landings of World War II have happened without coordination among the military branches and services?[12] Robert Scales argued, "Only 100 ground combat hours were necessary for the Army to re-establish itself convincingly as a successful land combat force" (1997, 5) in Desert Storm. To those who felt that Desert Storm marked the end of interservice infighting and validated the wisdom of Goldwater-Nichols by proving jointness a fait accompli, success of this magnitude was impossible without all services working together.[13] Cannon argued that Desert Storm in no way served as an impeccable example of joint cooperation, however. Instead, from his point of view, the far superior US military juggernaut was able to roll over a feeble Iraqi resistance even though the American services suffered "significant incongruity of often disparate operations and the presence of somewhat familiar tensions at the seams between components" (Cannon 2012, 116), suggesting the lack of institutional jointness among the services.

When jointness fails, as it has visibly several times since the World War II experience that led President Dwight Eisenhower to declare single-domain, single-service warfare "gone forever," the nation seems to weaken itself on the stage of international affairs, humbled by a visible reduction in its ability to exercise military power (Eisenhower 1958). David Armstrong (1995, 36) correspondingly noted that interservice agreement about the need for effective joint control of fighting forces, commonly thought of as operational jointness, is strong during times of conflict but its achievement wanes during peacetime.[14] According to the Senate Armed Services Committee report (US Senate, 1985) preceding the Goldwater-Nichols reorganization in 1986, failures in the Vietnam War, the taking of the USS *Pueblo*, the botched rescue attempt of Iranian-held US hostages, and the Grenada incursion were all failures to adequately implement unity of command.[15] The primary joint-doctrine publication asserts, "the challenges to the U.S. and its interests demand that the Armed Forces operate as a closely integrated joint team" (Joint Chiefs of Staff 2017, i), suggesting the need for both effective institutional and operational jointness.

The necessity of joint cooperation for military effectiveness is clear and it is in demand by those who solve security problems on the battlefield. However, any experienced observer of the US military would be hard-pressed to find such sentiment in debates about individual services' budgets, responsibilities, or primacy in securing the national defense.[16] In these processes, the very same services that need to cooperate on battlefields around the world compete, Janus-faced, with a veracity that seems grounded in things other than rational responses to national threats. The contrast between the need for combat cooperation and the interservice institutional battles that take place over what equipment, platforms, and systems should be developed or purchased to enable that very cooperation is perhaps the clearest example of mixed signals about jointness. The structural realities of the military-industrial complex and congressional involvement with it seem to allow and exacerbate conflict among the services despite rhetoric espousing cooperation. We do not claim to understand the full impact of the resulting dissonance; we simply note its irrefragable existence.

Ehrhard (2000, 467–477) explained that Congress demands cooperation among the services for platforms and systems that are similar in the missions they conduct. The implied thesis is that buying major defense articles—fighter jets or tanks, say—in bulk is cheaper than each service developing and buying its own model. So the services do something that they are loathe to do—agree to compromise on aspects of what they want—in order to give Congress the promise of a less-expensive military modernization program. The most prominent contemporary example of this direction and oversight is the F-35 fighter program, a jet produced in three distinct variants for the Air Force, the Navy, the Marines, and international "partner nations" but overseen for most of its existence by a Joint Program Office (JPO). Yet in the testing, fielding, and integration of these platforms and systems purchases, the services really have the final say in defining what the aircraft must do and whether the production models satisfy combat requirements. In other words, the promise of interservice cooperation, compromise, and reduced cost becomes hollow because there is no real top-level oversight. This dynamic ensures minimal efforts at institutional cooperation and incessant turf battles to acquire systems that meet a particular service's equipment preferences. These pursuits come at the expense of compromise and maximum economies of scale with an end result of an ever-growing unit-cost price tag. Indeed, the JPO is reviled by the services, Congress, and budget watchdogs alike as a protracted failure—an organization that neither controls costs nor extracts an operationally relevant level of performance from the expensive weapons platform. Over more than a decade of frustration, Congress has continually threatened that control will return to the individual services and costs *will* be tamed (Hadley 2021a).

As with military doctrine, the rhetoric of "commonality" and, later, "jointness" has filled—and continues to fill—many reports justifying large acquisition programs even as existing efforts flounder in the waste and inefficiency of institutional bickering. The TFX fighter aircraft under Secretary of Defense Robert McNamara's administration that preceded the F-35 are prominent examples (see Coulam 1977; Art 1968). However, the sources of the failure in joint acquisitions are not confined to executive military leadership and management. Harvey Sapolsky, Eugene Gholz, and Caitlin

Talmadge (2009, 128–129) took the services to task for becoming deceptive "champions of jointness" to shepherd major acquisitions projects through Congress, while burying any real ideological conflicts in bland Quadrennial Defense Review reports, papering over any serious institutional divisions and avoiding difficult recommendations that would improve the structure or performance of the armed forces.

To adapt William Martel's words about the concept of military "victory," there seems to be "no theory or precise language that permits policymakers, military officials, and the public to agree on what" (2007, 3) "jointness" means or when it has been attained.[17] To some, the term refers to effective battlefield coordination among the services (treated here as operational jointness). To others, it is a synonym for acquisition commonality among the services (a subset of institutional jointness). David Mets (1998, 81) has asked if jointness is not simply what the services pursue if they feel they are not being treated fairly, whether it be with respect to budget share, credit for victory, or favorable attention from the larger defense establishment. If one service perceives a slight in any area, it can clamor for more interservice jointness and thus bring its more prosperous teammates back in check. Understanding the disparate uses of the term is critical, especially when its use is disingenuous.

While the service bureaucracies in Washington might battle among themselves, the forces that those services provide to field commanders fighting US adversaries still need to cooperate and interact well together on the battlefield (operational jointness). The main drive of this book is to address the ambivalent, and more often than not quite dysfunctional, dynamic that the US armed services have with joint collaboration, and to ultimately alleviate the tension between how the US military as a bureaucratic organization behaves with regard to jointness (institutional jointness) and the needs of its forces in the field (operational jointness). Our study characterizes and systematizes the dynamics that promote and stifle institutional jointness between the service branches of the US military. This analytical work develops a greater understanding of the collaborative fight, as it is described here. We want to point out the complex levers that policy entrepreneurs and leaders will need to adjust when, and if, they want to succeed in joint operations and cross-service institutional collaborations. The remaining sections of this chapter outline precisely how we carry out this systematization and how our methodological choices shape the understanding we can offer.

Our Research Question and Its Application

Evidence that a desire for jointness on the battlefield exists alongside bureaucratic behaviors that do not reflect a cooperative milieu creates a quandary for those charged with leadership of the national military establishment. Just as civilian leaders may justifiably expect the military to comply with duly issued guidance, they may expect—along with the civilian population as a whole—that the military will make an effort to cooperate across branches to promote national security.[18] On the face of it, military cooperation would allow the services to maximize the amount of security they provide with the resources they have. Conversely, if the services in fact do not often cooperate or even deliberately choose to not cooperate, the nation's security is diminished. When interservice rivalry surfaces, one must grapple with the notion that the nation's defense establishment deliberately operates with inefficiency.[19] Yet there is no formal theory that scholars have applied to the military to explain why this is the case, whether it does undue harm to national security, or whether there are ways to ameliorate that harm.

Inherent in the realization that the military services do not always behave as if they have a unifying purpose in securing the United States and its interests is a knowledge gap about this phenomenon that prevents us from turning it around and using it for constructive ends.[20] Recognizing the enduring disparity between desire and reality, and with a sincere intent to support good practice, our study seeks answer the following central question: Under what conditions do US military services tend to cooperate in addressing a security problem, and under what conditions is their behavior divergent?

Our Research Approach and Design

The concept of "cooperation" in pursuit of a solution is central. Absent some form of interservice cooperation, there is no jointness. Is that a problem? Not necessarily. Cooperation is not *always* required for military success: for an operational mission of limited scope, a viable military solution can be found using only one service's toolbox. Think of a small peacekeeping effort in a land-locked country; the Army or Marines alone might do just fine. So we have to scope the problem further: it is a large-scale, national military pursuit for which joint (coordinated multiservice) effort is a precondition for success. If there is no such problem requiring that type of solution, we have to conclude that institutional jointness might be irrelevant (or potentially harmful—imagine how much more easily a *completely* united military force could effect a coup) since success in pursuit of national objectives is the only rational reason to raise and sustain a military force. A careful answer to the central research question should therefore clarify the definitional ambiguity of "jointness" and at the same time help with the pursuit of jointness' most useful side effect—military effectiveness, which might come in the form of success (defeat or deterrence of an enemy) on the battlefield or a more effective allocation of the nation's resources used to buy defense articles. Here is how we endeavor to do that.

Chapter 2 constructs a framing theory of military institutional jointness.[21] It considers theories that address jointness from conceptual and behavioral perspectives. The pertinent scholarship falls broadly into two types: (1) theories that explain organizational cooperation (or the lack thereof) in a general way that may apply to military organizations, and (2) theories that explain other military organizational behaviors that may shed some light on what is happening with joint collaboration. Several disciplines contribute theoretical insights regarding the organizational behavior of organizations that are in some ways similar to military organizations. Other research describes phenomena other than cooperation that take place within military institutions that allow us to extrapolate ideas to our framing theory about how military institutional cooperation may function.

The rich base of prior research and theoretical development in other fields and areas of study helps us identify potentially important facilitative and obstructive factors (independent variables), which we can then analyze through subordinate research questions in the case studies we conduct. From organization theory, for example, we draw questions like "[h]ow do external threats, bureaucratic politics, and political maneuvering influence jointness?" (Allison and Zelikow 1999, 155–156). Organizational cooperation theory asks whether "military services make decisions about jointness in a context of crisis" and, if that is the case, "how . . . the perceived urgency of a dilemma affect[s] decision and cooperation mechanisms" (e.g., Svedin 2009, 11). Out of theory that applies directly to the military and explains doctrinal innovation we take the question "How do the mechanisms that lead to doctrinal improvement advance or hinder operational and institutional jointness?" Research and theories on civil-military relations also inspire us to ask whether "the means of ensuring civilian control of the military in a democracy encourage or discourage jointness." While they do not necessarily help us define "jointness" outright, these areas of theory all raise questions that directly addresses the crux of joint collaboration.

Armed with the specific research questions that these bodies of theory provide, we move the examination of the development and evolution of institutional jointness into four empirical case studies. These are original case studies that we have constructed drawing on a wide range of source material and have organized to facilitate a structured focused analytical comparison (George and Bennett 2005). Our analysis of the cases is based on the variables included in the framing theory of jointness we constructed in chapter 2.

The second part of the book examines three empirical cases of successful joint (using our multiservice definition of jointness) US military cooperation. While there is a strong bureaucratic tendency for US military services not to cooperate, there are also examples of the services having worked together as institutions with visible effort to further shared national security goals.[22] Our choice to focus on success cases, arguably a form of bright spotting, follows a logic established by Rosen (1991, 5). His argument, and ours, is that the absence of cooperation (or innovation in Rosen's case) is the norm and, if we want more of something, it behooves us to study the instances when it has been successful.[23] We also examine a fourth case—the ongoing effort to develop a concept called Joint All-Domain Command and Control—one that certainly offers a potential to demonstrate joint cooperation but has not had time to develop such that we could adjudicate it a failure or success. Based on these case-study criteria and the desire for cross-case comparison, our pool of candidate cases became effectively four, so we included all four candidate cases in our study.

In the third part of the book we calibrate our framing theory of joint military cooperation, based on how the theoretically derived variables fared in our empirical case studies. The process of identifying the necessary and sufficient conditions for successful joint cooperation is iterative and does not stop with this study. Our framing theory is intended to serve as a building block from which more rigorous and extensive research can be built. This final part of the book presents a more rigorous framework for study with what we hope to be less-ambiguous and -amorphous definitions and more-consolidated mechanisms than the disparate pieces of prior research provided. Prior research informs the understanding of what facilitates and hinders successful jointness in military affairs, but the more rigorous framework serves as the basis for our conjectures about the future of the US doctrine and the material acquisition that it hopes to weave into Joint All Domain Command and Control. The framework for understanding jointness, the case studies, and our conjectures we hope will serve as a structured basis for discussion and evaluation in practice, among commanders on the ground and policy-makers in Washington.

Outline of the Book

With our core research question and its importance outlined in chapter 1, chapter 2 starts the construction of a framing theory of institutional jointness by surveying several theoretical approaches to organizational cooperation in general and interactions of the military services in particular. Our intent with this review is to establish the elements that might be relevant to organizational cooperation among military services, which, up to this point, has been poorly defined. Since the second chapter focuses on areas of academic study likely to inform our framing theory of jointness, the specific research questions that emerge are as a result multidisciplinary and establish how the questions investigated by several different disciplines create a nexus with jointness. Having surveyed the theory, our second chapter closes by highlighting how variables and dynamics established by theories examined come together to form a first cut of our framing theory of military jointness.

Chapters 3, 4, 5, and 6 each present a case study of successful (or, in the case of JADC2 in chapter 6, indeterminate) joint military action, analyzed in terms of how the variables in the framing theory (chapter 2) contribute to our understanding of what and how things happened in the case. The chapters all introduce the relevant case study and the research methodologies used to select and examine them is outlined in the appendix. In order to facilitate comparison across these cases the case studies, and the chapters that present them, are all structured the same way:

A. They each describe an area in which two or more military services exhibited institutional jointness, i.e., the services cooperated in an effort with apparent mutual surrender of institutional resources or ideals. The opening sections of each chapter explain any technical terms or details necessary to understand the cooperative area and provide some historical context of service cooperation prior to the commencement of the case in question.

B. Each chapter examines the cooperative effort in some detail, with attention to the ideas, structures, and relationships that facilitated that cooperation or inhibited it. These sections aim to follow Alexander George's *process-tracing* methodology, specifically borrowing from its *detailed-narrative* subset.

C. Each chapter also briefly summarizes how history demonstrates mechanisms for joint cooperation, intraservice conflict, or independent service efforts. These sections begin to relate the events observed to theoretical elements described in chapter 2. These chapter sections, by more closely focusing the narrative on theory, follow the *general-explanation* subset of George and Bennett's (2005) process tracing. Where applicable, these sections divide the cooperative mechanisms among the internal service level, the among services level, and the external organizational level of analysis. For example, each service has its own internal history and set of organizational dynamics (internal). Each also provides forces to Combatant Commanders—including the field commanders themselves—and works to staff, advise, and work within the recommendations of the Joint Staff (among services). Finally, the military services answer individually and as a group to the secretary of defense and the Department of Defense (DoD), which in turn must function in a security environment influenced by domestic and international politics (external).

D. The final sections of each case-study chapter attempt to tie observations about joint institutional cooperation back to disparate theoretical roots. Additional observations about jointness, unidentified in the original set of theoretical questions, emerge from the case-study observations and become material for constructing a framing theory.[24]

The second-to-last chapter, chapter 7, draws together the observations arising from the generalized explanations in the case-study chapters. The chapter begins by tying the repeated themes and behaviors from historical narrative to the established theories of chapter 2 and the behaviors observed from the historical studies. With the building blocks of theory quarried and the case studies unsheathed, this chapter offers some discussion as to how the pieces come together in a clearer framing theory.

The concluding chapter, chapter 8, offers conclusions about institutional jointness that appear repeatedly in the case studies and pretheoretical refinement in chapter 7. It offers insights shared by the close to sixty defense-establishment leaders interviewed as primary sources for our case studies. Since these individuals spent careers in the arena where the pursuit of jointness in military endeavors occurs, their views round out the historical process-tracing and theoretical facets of the study. Their observations show how defense-establishment insiders view jointness as a practical matter, and shed some insight on why the definition and pursuit of jointness have proven so elusive. In this final chapter, as a building block toward further systematic research on jointness, we summarize what we think are the most compelling considerations uncovered in this pretheoretical investigation. Based on these conclusions, we wrap up by looking forward and offering some conjectures about the future of jointness in US military affairs and the chances for success of Joint All-Domain Command and Control.

Why Read This Book?

Having laid out our intent and approach, the reader might be interested to know what benefit she will gain from investing further time in these pages. For someone like a taxpayer interested in how the resources of the US treasury are invested to ensure the common defense, this book might be of simple explanatory interest or spur someone's interest in watchdog activity. For the academic interested in the sociological phenomenon of cooperation, this study may be of interest as a specific example of how institutional cooperation can be fostered in a specific subset. Those in the academy who study military organizations in particular might find that this framing offers an additional facet of insight in how those organizations act in both a bureaucratic and a battlefield context. And finally, we hope this work is of value to practitioners, including those who try to find rare bureaucratic cooperation in the corridors of the Pentagon as well as those who organize and wield the nation's military on battlefields around the world in defense of and pursuit of US interests. It is from that audience that the initial spark of interest in this topic arose, and we sincerely hope that this returns some value to those women and men.

Finally, we wanted to offer a word about how to tailor the use of this book to each of these audiences. The academic audience may be particularly interested in reading (and critiquing) chapter 2, wishing to see how we've constructed a theoretical basis before applying it to our case studies. That application, with an academic focus, is in turn developed in the third section of each case-study chapter. Chapter 7 attempts to bring all of the theoretical development into focus. For general-audience and civil-military practitioners, skimming chapter 2 or coming back to it later might be a better approach. These readers might be most interested in the opening sections of the case-study chapters and the conclusions offered in chapter 8, which suggest how to effect joint cooperation in the real world. Meaningful, effective joint cooperation is difficult and vexing to attain, and some pointers distilled from history and theory cannot hurt. Of course, we hope that all readers find all parts of the book interesting. However, for those who need to be economical with time, we have tried to make the sections of the book stand out as a roadmap tailored for those with focused interest.

Chapter Two

Factors in Organizational Collaboration and Success Cases

Searching across Disciplines for Framework Building Blocks

This study relates interdisciplinary theories to cases of military organizations attempting to solve security problems. The well available to draw from includes two types of work. The first are *theories that explain organizations exhibiting cooperative behaviors*, especially where these organizations exhibit characteristics similar to the military such as large size or hierarchical structure. The second are *theories about military organizations that deal with behaviors other than cooperation*. Direct application of any single theory might not tell the full story, but an interdisciplinary approach might reveal consistent observations about cooperative organizational dynamics among the military services. This work presents an opportunity to refine organizational-cooperation theory as it applies to the military; it also offers a means for a better understanding of national security issues that involve multiple groups. It should reveal ties among well-studied aspects of military organizational behavior and the way they relate to cooperative behavior within the military establishment. William Martel attempted to define and explain the parameters of "victory," a term as amorphous and malleable as "jointness." Borrowing his approach, this study uses systematic observations about existing theories and their relation to interservice cooperation. In so doing, it creates a framework for a phenomenon around which ideas from many varied disciplines orbit, but for which no formal theoretical framework yet exists (Martel 2007, 6–7).

To gather bricks and mortar for our framing theory, this chapter summarizes ideas from several disciplines that might provide useful contributions to a theory of military joint cooperation. As Suzanne Joseph described, at this early theoretical stage "the emphasis is on the addition of components" (2000, 6–7). For each group of theories, a summary shows how the theory has potential application to interservice cooperation. The chapter starts by discussing theories that apply to organizations in general. Next, it deals with observations of military behavior other than cooperation per se, but whose mechanisms and explanations may contribute to our understanding of military cooperation. Many of the bodies of theory the chapter describes are vast; in-depth description of derivative ideas is impracticable. Instead, the chapter aims to draw out several broad, relevant theoretical concepts. If prevailing pairs of opposing ideas characterize some bodies of work, the discussion emphasizes the differences and considers both in parallel with respect to jointness.

General Theories Influencing Organizational Cooperation

The first source of relevant theory comes from disciplines that describe group interactions in general but have not examined military organizations in particular. Theoretical outlines from these disciplines—social science is rich with theoretical lenses focused on explaining cooperative behavior—can in turn inform a theory about the nature of interservice cooperation. (These theories have also inspired spinoff literature about military-specific behaviors discussed further into the chapter, so they make an indirect appearance again later in this chapter.) Our review of relevant theory begins, however, by hearkening to basic economic theory and its impact on group behaviors.

Groups and Public Goods Creation

Mancur Olson (1971) showed why large groups working to produce a public good that benefits group members never optimize the output of that good.[1] He further demonstrated why the desires of a rational individual would rarely drive that person to belong to an interest group that furthers those aims (64–65). According to Olson, the two factors preventing beneficial collective action in large groups are (1) the free-rider problem and (2) the problem of imperceptible contributions.[2] He proposed remedies to both problems. In small groups, social pressure can force participation and contributions where the mere shared interest of the public good does not suffice. Groups can also elect to provide conditional benefits in ways that force otherwise unmotivated participants to make a contribution or miss out on the benefits the group secures. A great deal of research has since tested the basic assertions of Olson's solution (Udéhn 1993; Sandler 2015), including the importance of group size (e.g., Oliver, Marwell, and Teixeira 1985), social pressure (e.g., Kramer and Brewer 1986; Hechter 1990), and strategic communication (e.g., Ostrom 1990; Lupia and Sen 2003) in public-goods production.

James Q. Wilson (1974) presented an early challenge to Olson's insistence that social pressure and conditional benefits alone can remedy the problems of group-collective action; he posited that appeals to purpose could overcome the inertia of rational self-interest assumed by economists (Wilson 1974; see also Coleman 1988; McAdam 2003; Williams 2004). Wilson explained the rise of pluralistic factions and the reality that while people in

America belong to ever-more interest groups that provide them a lobbyist's voice, they feel less represented within society (1974, xxii). He highlighted the increasing difficulty of cobbling together consensus among clamoring interest-group communications, many of which are infused with hyperbole designed to induce panic (see Hrebenar and Scott 1997; Pierson and Skocpol 2007; Hacker and Pierson 2010). Wilson's loss-aversion observation—that the threat of withdrawal of something a group has attained so overcomes by a large margin any promise of getting something it wants—makes the dynamic between fear of loss and potential gain a topic with obvious relevance to discussion about interservice rivalry (1974, x). We have observed that the loss-aversion dynamic seems to affect stakeholders in the larger military enterprise—Congress, community groups focused on a specific military base's economic output, and the National Guard come to mind—rather than the active-duty military itself, which generally favors rapid changes or divestment to free up capital to modernize or develop new, relevant mission capability.

Whether the arena of joint military action consists of a small group with just a few services or a sea of thousands of self-interested actors can be debated. Olson's theory, however, leads us to expect that military services may not advance the pursuit of security even though that particular public good is their chief objective. Based on Olson, we should expect some difference between the best possible outcome and the actual observed outcome of any group endeavor; joint military action should be no exception. Where cooperation toward a shared goal emerges, it may be a result of social pressure or conditional benefits, mechanisms that rely on a coercive element (Schelling 1966, 69–72).[3]

We move forward with Olson's theoretical assertions while also looking for evidence that Wilson's appeal to purpose plays into joint cooperation. However, even if the voluntary formal associations Wilson described do arise in violation of the public-goods paradox Olson highlighted, Wilson's analysis predicts a cacophony of clamoring voices in the defense establishment and frequent gridlock. The visible presence of all these elements in interservice relations would indicate the relevance of both these theories. The specific research questions derived from these competing views of public-goods are, Do the military services sometimes free-ride in the production of security as a public good? If so, do military organizations need to be spurred by peer pressure or public pressure to achieve institutional jointness, or will they seek joint cooperation if they are convinced of the purpose of that cooperation?

Organization Theory

Graham Allison and Philip Zelikow devised their widely applied model of organizational dynamics by defining three competing models of behavior.

Their "rational actor" making logical decisions based on utilitarian analysis vies with the bureaucratic inertia of an organization trying to survive for its own sake and the "pulling and hauling" of self-interested individuals trying to further their own interests in the context of a decision situation (Allison and Zelikow 1999, 4–5).

Robert Axelrod and Robert Keohane's (1985) explanation of cooperation among organizations not controlled by a superior governing body underpins the rational-actor model. They argued that the likelihood of cooperation varies with mutual interests, the need for the organizations in question to cooperate in the future, and the number of organizations involved (difficulties in cooperation increase with more participants). They also noticed a consistent tendency among the actors involved to alter the nature of the interaction "game" they were participating in (248–249). Based on these organizational approaches to understanding collaboration we formulate the following question: How do first-order threats, bureaucratic politics, and political maneuvering influence jointness?

In the realm of bureaucratic politics (Allison and Zelikow's second model) Irving Janis's offered an explanation for bad decisions that become widely adopted by an organization based on what he termed "groupthink." He also surmised that an in-group's derision of a competing out-group is a psychological outlet for any "latent jealousies and antagonisms" that surface within the in-group (1982, 257). Robert Jervis's discussion of cognitive dissonance in international politics concluded that "people must often rearrange their perceptions, evaluations and opinions" to convince themselves of "the value of the course of action they pursued when confronted with evidence of what might have otherwise happened" (1976, 406). Jeffrey Polzer further discussed the in-group and out-group roles that emerge in organizational cooperation when individuals, subgroups, and collectives have conflicting goals. Polzer found that subgroup-identification corrupted the collective's goals in cooperative efforts, both when competing subgroups are from different organizations and when a subgroup within an organization has a competitive reputation (Polzer 2004). The location of in-group and out-group boundaries, the differentiation among subgroups, and the reputation of subgroups informed Polzer's work and may be key to understanding cooperation among the military services. The question it inspires is: How do service subgroup-interests increase or diminish the likelihood of joint cooperation?

To explain cooperation among organizations, a substantial body of organizational theory references "resource dependence," arguing a mutual need for shared capabilities in resource-constrained environments (Emerson 1962, 32; Yuchtman and Seashore, 1967, 891) drives cooperation. Applied to the question of military cooperation, this line of thinking suggests that we should see cooperation among services when each has a need for the capabilities the others offer. In the context of major conflicts, this is certainly evident. The most compelling evidence available comes from large combined-arms efforts that rely on all branches performing well in their respective domains to ensure the success of others (think of World War II's D-Day landings). Where military capabilities overlap—i.e., when two or more services are able to perform similar functions or missions—or when services feel they can go it alone, the impetus for cooperating successfully seems to be missing. Arguably, this theory alone could offer a complete explanation of general patterns in joint cooperation and interservice rivalry, making it essential to include in the construction of our framing theory. The specific research question that resource dependence theory proposes for the study is: Do overlapping service capabilities advance or hinder military jointness?[4]

While the vast body of organizational theory is too rich and deep to recount here, a portion of it warrants specific examination as we establish the framing theory of joint institutional cooperation. That portion of organizational theory focuses on organizations experiencing a crisis.

Crisis Organizational Cooperation Theory

One subset of organizational cooperation theory seems to have special utility in assembling a pretheory of interservice cooperation. Lina Svedin (2009) performed statistical analysis of major observed behavioral patterns and distinct cooperation strategies found in organizations dealing with crises. The analysis of linked variables showed explicit connections among behaviors and strategies as well as how the behaviors and strategies vary among five distinct types of crisis (Svedin 2009, 93). Svedin's (2009) data set comprised a wide variety of crisis situations that affect many different types of military, nonmilitary governmental, and private organizations.

Svedin's (2009) taxonomy described five types of crises, five types of organizational behavior, and four types of cooperative strategies that organizations in crisis pursue in their interactions. The *five types of crises* are (1) unclear threat(s), (2) unclear threat(s) that persist over time, (3) threats emanating from within a crisis-management group, (4) threats seen as coming from outside a crisis-management group, and (5) uncertainty about what kind of a threat and crisis the group is facing. The *five types of observed organizational behavior* are (1) fighting, (2) agreeing, (3) talking, (4) negotiating, and (5) manipulating. The *four available cooperative strategies* are (1) bureaucratic politics, (2) concurrence seeking, (3) signaling trustworthiness, and (4) success-based helping.

In analyzing the statistical dependence of these observed characteristics from a wide sample of case studies, organizational cooperation theory offers predictive analyses of the overall cooperation climate that will emerge based on a combination of crisis type, behaviors, and strategy. Knowing one or more of these factors, a practitioner can "solve" for remaining unknowns or act to mitigate harmful conditions that hinder cooperation. For example, a crisis characterized by unclear threats over a short time horizon is likely to be typified by open fighting among organizations in decision situations, but organizations facing a persistent unclear threat over a longer time horizon are more likely to pursue a cooperative strategy of success-based helping. Organizations that perceive a threat from within the group managing the crisis are likely to resort to a strategy of bureaucratic politics but avoid open fighting.

Organizational culture and preferences are inextricable from this discussion and play important roles at two levels of analysis. The first level of analysis is what social psychologists call the "decision-situation," which describes how organizations and their representatives behave at discrete interaction opportunities. The second level of analysis is strategic; it refers to the overarching preference over the course of a crisis, a project, or the life of an institution and comprises many discrete decision-situations.

By way of example, organizations that fight as a predominant means of interacting are likely to employ "signaling trustworthiness" as their overarching cooperative strategy. Paradoxically, their open hostility over time leads to sincere communication about intent. By contrast, organizations that favor talking are more likely to use a strategy of "bureaucratic politics," suggesting that civility in communications belies intentions to obfuscate (Svedin 2009, 131–134). Using these and other identified combinations, process-tracing that reveals any pair of threat type, behavior pattern, or crisis coping strategy can reveal the "missing" element. Since service behaviors are well known or can be easily observed, this subset of theory becomes a useful tool for explaining organizational response in a crisis.

Crisis-based organizational cooperation theory offers multiple potential bases for case-study analysis. Using the taxonomy of crisis types, behaviors, and coping strategies of the kind outlined by Svedin (2009) offers a new means to analyze the mechanics of military cooperation when it occurs in a crisis context. An overarching appeal of organizational-cooperation theory as a discipline is that it poses research questions in the realm of economic theory and experimental behavioral science, including studies of group behavior with large-n samples and rigorously testable hypotheses. Expanding the aperture to examine cooperation while reexamining the conclusions drawn by students of doctrinal innovation offers additional nuance to the study of the interaction of military organizations.

Our decision to draw on a crisis-based theory warrants some further justification and discussion. Given some of the characteristics of the case studies in this volume, some of which unfold over several years, it might seem fair to argue that the time span does not suggest that these situations are crises in the sense we may know the term. However, the crisis definitions used by researchers suggest that uncertainty and threats to organizational values are as important as a short time span. A commonly used definition of "crisis" is "a serious threat to the basic structures or the fundamental values and norms of a social system, which under time pressure and highly uncertain circumstances necessitates making critical decisions" (Rosenthal, Charles and 't Hart 1989, 10). With this values-and-norms focus and the view of organizational operations as a social system, we see potential value in including crisis-focused organizational cooperation theory in the initial construction of a framing theory for jointness. Furthermore, even threatening situations that develop over a relatively long time period can be a crisis, as evidenced by, for example, the world HIV/AIDS epidemic. This disease first appeared in the United States in 1969 and was recognized by the medical community as a crisis twelve years later (Gilbert et al. 2007, 18566).

The empirical cases in this book all unfold over about a decade, but each one exhibits uncertainty and time pressure. What is more important, however, is that they unfold over similar time horizons, creating a degree of cross-case consistency that makes the conclusions they suggest more reliable. The pertinent questions about jointness that crisis organizational cooperation theory yields are, Do military services make decisions about jointness in a crisis context? How does the perceived urgency of a dilemma affect decision and cooperation mechanisms? The second question considers common organizational behaviors and cooperative strategies in context.

Theory of Professions

Alexis de Tocqueville (1835), Samuel Huntington (1957), and Morris Janowitz (1960), among others, describe the military as a profession. Borrowing Huntington's formulation of expertise, responsibility, and corporateness, the military services furthermore describe themselves as comprising a *profession of arms* (TRADOC, Combined Arms Center 2010; Dempsey 2012). Various military schools teach that the military has certain structural characteristics including entry barriers, educational requirements, ongoing education, and ethical codes that afford it the distinction of being a profession rather than a vocation. One could argue the extent to which these elements pervade each position in the military; asserting uniform military professionalism assumes a homogeneity that may not further understanding of interservice rivalry and the services' tendency to cooperate. A more fruitful discussion of professional organizations and how professionalization might influence interservice cooperation comes from Andrew Abbott (1988). Abbot suggested that "control of knowledge and its application means dominating outsiders who attack that control" (2). He put forth a definition of profession that is tied to control of the work done and a claim to the abstract knowledge necessary for the job.

Abbott's alternative definition is helpful in illuminating interservice behavior. The story of interservice conflict is often one of argument over "roles and missions," explicit definitions of the precise type of work that a military service is supposed to do—and that the other services must therefore eschew (e.g., Trest 1998, ix–xii). Sometimes it seems like the services' primary Pentagon activity is arguing over a 1948 white paper called "Functions of the Armed Forces and the Joint Chiefs of Staff" drafted by the first US secretary of defense, James Forrestal. (The paper is more popularly known as the Key West Agreements.) "It is the control of work that brings the professions into conflict with each other and makes their histories interdependent" (Abbott 1988, 19), and Abbott's observation would seem to apply directly to the military services; they have made a habit of arguing about the kinds of work they perform. Historical arguments about coastal defense, amphibious operations, and the enduring debate about the various applications of air power highlight conflict based on largely untested abstract assertions. For example, an Army-Navy agreement from 1931 reads in part, "the navy should have no part in coastal defense" (Pratt and MacArthur 1931).[5] Military missions are the essential work of the military, and the justifications put forward to defend service-specific positions are abstract knowledge in the form of doctrinal (or dogmatic) military preferences. They tend to be strictly invoked when the budget share (Topline Obligating Authority, or TOA) for one service comes at the expense of another on the line.

Abbott went further in his observation that, while technology may spur the creation of new professions, it is not the final arbiter. Addressing military air power applications, he noted that the Air Force achieved independence "only after an internal battle of several decades, and it lost a similar fight with the Navy" (Abbott 1988, 92).[6] As Jeffrey Donnithorne (2013, 227–228) related, the assumption by the Air Force that the best use of air power comes when air forces are independent from ground forces begs constant restatement and intellectual defense. Since there is no obvious reason for such a claim, it is by definition a bid for power through an abstract claim to unique knowledge or capability. The nature of the Air Force's claim on the work it does explains both its advocates' appeals to emotion and its enduring institutional insecurities. (Lobbying organizations advocating on behalf of the Air Force are unapologetic about this emotional approach. Retired Lt. Gen. David Deptula, who now runs the Air Force Association's Mitchell Institute, unironically celebrated retired Brig. Gen. William "Billy" Mitchell as "one of the fiercest . . . evangelists for airpower" in a 2021 Tweet [Deptula 2021].)

Abbott describes how professions rise and fall as "tasks are created, abolished, or reshaped by external forces" (1988, 33). Preferences for close, tactical, organic control of the air would compete with the Air Force's preference for wide, strategic, centralized direction, and predicts a strong institutional reaction if that preference is violated. The reaction would be even stronger if the Air Force perceived itself as not being held in as high esteem as other services because, as Abbott describes, the primacy of one profession over another is formed via courts and legislation, the public arena, and the workplace (62–65). The specific research question that this line of reasoning raises is identical to one raised in considering resource constraints in the context of organization theory: Do overlapping military service capabilities advance or hinder jointness?

Finally, Abbott's idea about how groups coalesce and dissolve from existing professional groups and aspirants speaks to a phenomenon noted in military services that identify closely with a specific competency. An example of this would be the tendency among military pilots to identify primarily as pilots, rather than as Air Force or Navy officers. This cross-service group identity is perhaps most salient in the pilot career fields, but it is also prevalent among other operators of specialized weapons systems (e.g., ships, tanks, submarines) and other specialized professions (attorneys, physicians, engineers). The energy required to create a new subclass within a profession invokes Stephen Rosen's (1991) description of the resources a military organization must invest to provide a legitimate means of promotion for the practitioners of a new military innovation (Abbott 1988, 174–175; see also Whitmore 2021). The question it inspires is, Do subclasses vying for recognition in their respective military services offer a mechanism for generating joint cooperation?

Prinicipal-Agent Theory

Agency theory attained gained a foothold in scholarly circles in 1976 with Michael Jensen and William Meckling's characterization of the "principal-agent relation."[7] The "principal-agent dilemma" describes the gap that exists between the intent of people who design and mandate policy (principals) and the people charged with implementation of that policy (agents). Kathleen Eisenhardt's (1989, 57) survey of agency theory at the end of the 1980s found that agency theory enjoyed widespread application across many academic disciplines and that it contributes to organizational theory with appreciable explanatory power. Agency theory makes its strongest contribution to modeling an organization's strategic choices when it accounts for both the context of decisions and the actors' (principals' and agents') preferences

(Chai 1997, 55–56; Frieden 1999, 75–76; Legro 1996, 118; Nielson, Tierney, and Weaver 2006, 130–132). Since the military establishment is hierarchical by design, but delegation of authority is necessary for its function, it is prone to frequent exhibitions of the principal-agent dilemma. Consequently, we consider it pertinent to include this theory in our framing theory describing joint institutional behavior. The principal-agent dynamic in the military establishment has several facets. Where one might think of it as limited to the boundaries of the institution itself (soldiers carrying out the orders of generals, say), it also applies to higher-echelon relationships that impact military actions. Here we are envisioning the influence of the Department of Defense, Congress, and the executive branch on the policies, actions, and strategic-resourcing decisions of the services. Furthermore, in all cases there is a delayed feedback loop between actions the services plan and begin to undertake and the relatively crude veto power of the higher-echelon authority. (Think of Congress withholding budgetary authority or demanding a report or hearing in response to a military mission that it botched or an acquisition decision that goes counter to political winds.)

The armed services have served as a subject of study for scholars wanting to test and refine agency theory and other social sciences theories. Among these efforts are Peter Feaver's (1999), Jeff Donnithorne's (2013), and Monte Cannon's (2012) studies and analyses of military jointness. Each of these authors used aspects of agency theory to explain the behavior of military organizations. Donnithorne built on Feaver's formulation of civil-military relations as a principal-agent problem; both posit that civil authority is the principal that contracts with a military agent to provide security for the state. Reflecting a fundamental premise of principal-agent theory, the interests of the military establishment and the information to which it is privy differ from those held by and available to the state acting as principal. Using an agency-theory lens, the military's behavior is a function of its interests, how much the state endeavors to monitor military actions, and the expected level of retaliation for behaviors that advance the military's interests at the expense of the state's (Donnithorne 2013, 54).

Donnithorne refined Feaver's work by appreciating that the civilian principal is not a unitary actor; the military can and does get instructions from multiple civilian branches of government. As Deborah Avant wrote, "military agents can choose the direction that best suits their interests, or they can play the two [civilian branches] off against one another to generate a policy more to the military's liking" (2007, 82). Donnithorne also incorporated Jeffrey Legro's (1996, 118) observations into his model by acknowledging the differing preferences among individual services. Finally, he expanded the temporal scope of policy development by analyzing development and implementation as separate phases of a bifurcated process (Donnithorne 2013, 62–63).

Where Donnithorne and Feaver looked at the strategic, high-level principal-agent dilemmas, Cannon (2012) applied this dilemma to a military setting, but specifically to the relationships among the military services. Specifically, Cannon looked at the relationships between the services and the combatant commanders, who use the forces provided to them by services to accomplish national missions. Cannon posited that component commanders (the domain-specific chiefs of land, air, sea, marine, space, and special operations forces) who work for a designated joint commander, such as a combatant commander, are part of a principal-agent relationship. The joint combatant commander, in the role of principal, must delegate many of her tasks to the component agents, and the act of delegation creates "the potential for divergent aims" (18) that sacrifice joint goals for "service imperatives" (150).

For the framing-theory development in this book, the most important aspects of agency theory are (1) the differentiation of service cultures and (2) the recognition that principals, be they civilian or military, are not unitary actors. Principals are constrained in what they can accomplish by the structure of the organizations they lead (or supervise) and the potential for selective obedience that those structures create. We are going to reconnect with this discussion of principal-agent relationships later on in this chapter when we discuss theoretical concepts specific to the military, such as civil-military relations, service-culture entailments, and defense-establishment hierarchy. Based on the principal-agency dynamics review earlier, however, the specific research question we raise is: Do the many principal-agent relationships evident in the defense establishment advance or inhibit jointness?

Theories and Descriptions of Military Organizational Behavior

Moving from general theories about organizations, we turn next to theory and research specifically focused on military organizations. This includes research that, while not specifically concerned with interservice cooperation, contains descriptions and explanations that have bearing on or relation to the cooperative tendencies exhibited among the military services.

Military Innovation

A large body of research examines how innovation takes place in military organizations and how those organizations may facilitate or hinder innovation. Barry Posen (1984) and Stephen Rosen (1991) are foundational researchers in this field and they describe two distinct views of military innovation.[8] In

Posen's construct, careful oversight by civilians is the only way to prevent the natural isolation of the armed services and to ensure that they pursue military capabilities appropriately reflective of technological advance while balancing "political ends with military means" (1984, 241). He viewed the sources of military innovation as almost exclusively external to the military establishment, though alliances with "maverick" individual officers may help the outside intervention succeed (Nagl 2002, 3).

By contrast, Rosen found the sources of innovation to be internal to military organizations. He also took a more nuanced view of innovation, arguing that its nature and mechanisms changed depending on whether wartime, peacetime, or purely technological aspects were the dominant environmental condition. He identified the process of peacetime innovation as slow-moving and requiring time for visionary military leaders to create new pathways for promotion along which more junior officers can carry the innovative ideas they develop (Rosen 1991, 105). He argued that wartime innovation is faster because of greater need and urgency, but outside that special case, the trajectories it follows are more difficult to describe. The changes innovation produces are fleeting, achieved at great cost, and subject to speculative gambles (Rosen 1991, 179–182). Successful innovation in war usually accompanies improved means to measure strategic effects objectively. Technological innovation is, unexpectedly, not usually a function of enemy capabilities. Instead, it reflects the struggle of inventors and military personnel to manage uncertainty. These actors deal with the unknown either by pursuing a wide range of options or by purchasing better information about anticipated needs before undertaking the expense of large-scale production (Rosen 1991, 249–250). On all three fronts, Rosen identified internal sources for innovative ideas—his theory does not rely on proactive government officials auditing the misguided paths of a static military bureaucracy. Rosen underlined the difference between his theory and Posen's by critiquing Posen's elevation of "mavericks," demonstrating that such individuals are often counterproductive to the cause of innovation.

Owen Coté (1996) outlined a theoretical middle ground between Posen and Rosen, arguing that competition between services drives innovation (Coté 1996, 338). Like Posen and Rosen, Coté's view of innovation balanced a Waltzian, system-level response by national leaders to external anarchic pressures against the political bargains and institutional inertia that bureaucratic leaders are focused on (Coté 1996, 332–334; Waltz 1979, 99).[9] Although Coté found a separate, causal source of innovation in interservice rivalry, he also affirmed the fundamental structural-realist ideas of Posen, agreeing "national leaders hold the reins of state power" (Coté 1996, 395). Coté's work highlights further nuance useful to a discussion of cooperation. He found that intraservice and interservice competition serves as a useful source of information that civilian officials would not otherwise be able to access. However, the reality that civilian intervention can stifle this information by spurring undesirable military cooperation is a "paradoxical" threat to innovation (Coté 1996, 389). (The F-35's Joint Program Office may rise as a specter in the minds of many readers at this point.) Understanding the details of military culture or innovation, for example, may illuminate understanding of cooperation, even though innovation and cooperation are not the same kind of behavior—or even related.[10] The question inspired by this theoretical area of study is: Do the sources of military innovation advance or hinder jointness?

Military-innovation theory combines with cooperation theory to yield another useful question about institutional jointness. Dean Tjosvold (1984) described a taxonomy of organizational cooperation, characterizing the context for cooperation as cooperative, competitive, or independent. Subsequent experimentation found that organizations that believed their interactions happened in a cooperative environment engaged in more helping behaviors, whereas competitive contexts led to less openness and more hostile behavior (Tjosvold 1988, 13). The distinctions Rosen (1991) drew among wartime, peacetime, and technological advances are relevant to cooperation as well. The link comes from matching the cooperation taxonomy against Rosen's three types of innovation. Wartime innovation, given a sense of urgency and shared goals for success against a common enemy, becomes cooperative (positive) interaction; peacetime innovation, marked by competing ideas and immediacy, reflects competitive (negative) interaction; technological innovation reflects both of the above, and, due to its necessarily creative aspect, adds individualistic, independent facets. The juxtaposition of Rosen and Tjosvold yields two questions: Does a given security situation constitute a crisis for the military establishment? And, if so, do the crisis conditions within which military organizations operate make jointness more or less attainable?[11]

As we continue our review of innovation theory, we consider the authors' assumptions about the nature of military cooperation. Of the three competing theories of military innovation, none assume that military services will cooperate by default. Coté (1996) explicitly identified cooperation as a detriment to innovation in the military. This is the undermotivated "going along to get along" that results in so much of the suboptimal satisficing that plagues bureaucracies. However, it is important to note that Coté's informative case studies of the development of fleet ballistic missiles for the US Navy took place during peacetime.[12] Had the opportunity for innovation occurred during wartime, Rosen's theory predicts, dynamics would have been different, albeit not necessarily conducive for superior innovation (1991, 251).[13] However, the counterfactual circumstance bears consideration, because it emphasizes the importance of context for both innovation and cooperation.

If the dynamics of innovation change in the crisis context of war, the dynamics of cooperation may change in a crisis as well.

From the study of military innovation we find another concept important to this study: pertinent levels of analysis. Military innovation scholars build their theories by (a) looking internally with a particular military service (within service level), (b) looking at the relationship among the services (among services level), or (c) looking at the relationship between the military services and their external stakeholders (external relations level). Using these three levels of analysis, we could view cooperation among the military services as the object of jointness. The services as organizations therefore are the unit of analysis (level a), or the key actor that we are trying to understand in this study while taking into account that this key actor operates in a set of situational and larger systemic contexts (levels b and c).[14]

The level of analysis that takes account of the relations between services (level b) includes those military leaders that are situated just above the services themselves and who coordinate and draw on all the services: the chairman of the Joint Chiefs of Staff, the Joint Staff, and the combatant commanders. The level of analysis that covers external stakeholders (level c) includes the larger Department of Defense, the remainder of the executive branch, Congress, and (in some senses) industry. These actors are authority-wise either above or independent of the military services (level a) and the leadership groups that coordinate them (level b).

Per Donnithorne's (2013) approach, this study does not consider the actors on levels b and c unitary actors. Though their disparate actions may influence the services in myriad ways, a commonality of purpose is not necessary at this level for jointness; only the services need to cooperate. This final aspect of military-innovation theory does not inspire an explicit question about jointness, but the levels of analysis it describes provide a consistent structure for framing the case studies in this work.

Civil-Military Relations

Because the body of literature about civil-military relations is large and diverse, this work builds on the distillations that Donnithorne (2013) and Feaver (1999) provided in analyzing the dynamics of civilian control over the military; they provide a sufficient palette for consideration in the case studies that follow (Donnithorne 2013, 37–41). In current scholarship, Huntington's advocacy for "objective control" is the seminal argument for the advancement of successful civil-military relations. He advocated a form of professional separation from society and government that would allow the military to perfect the business of armed confrontation while remaining subservient and responsive to civil authority.[15] In exchange for autonomy in internal affairs, the military of Huntington's ideal pledges apolitical obedience.

Janowitz's (1960) answer to Huntington concerned itself with a perceived unsustainable gap between military and civilian culture and norms. Rather than relying on separation to professionalize the military force, Janowitz advocated the creation of a constabulary force capable of applying pragmatic, limited force. This is the model of the National Guard, and it differs explicitly from the aims of Huntington (1957). Like-mindedness relieves the growing tension between society and the military; here the ideal military becomes incapable of undertaking missions civil society would not condone and thus societal values subjectively control the military (83).

While the various streams of civil-military relations theory diverge into dozens of rivulets, this work will concern itself only with these two foundational tributaries. Other civil-military-relations authors will make additional contributions; the conclusions of this work allude to Eliot Cohen's (2003) critique of overapplying Huntington's model, for example. For this introductory sketch, though, the competing dyad of Huntington (1957) and Janowitz (1960) will suffice. The question it yields is: Which leads the individual military services to pursue institutional jointness and effective support for one another, civilian control of the military via objective means or control via subjective means?[16]

Military Cultures, Interservice Rivalry, and Other Institutional Characterizations

Venturing into the realm of culture as an independent or intervening variable is fraught with peril. Any discussion of organizational cooperation invariably touches on the impact of organizational culture, even though the concept takes routine criticism for its lack of explanatory power (White 1975, 4n).[17] The issue of military culture, like most other aspects of this study, presents a multilevel vista. Russell Weigley (1973) described an overarching preference of the contemporary US military to fight politically unconstrained wars of annihilation with overwhelming firepower. Brian Linn (2007) and Max Boot (2002) argued that, while this might be the cultural preference, the reality was that in fact the country had fought mostly attrition-type wars with significant political restraint.

Builder (1989), Ehrhard (2000), Thomas Mahnken (2008), and Donnithorne (2013) provided well-developed descriptions of the internal military service cultures.[18] Builder's foundational study examined the cultures of the Army, Air Force, and Navy and their perceptions of one another. Ehrhard characterized the services as having "monarchic" or "feudal" power-sharing structures in his study of unmanned aerial vehicle (UAV) development through 1999. Donnithorne built on the foundations laid by Builder and Ehrhard, more fully explaining service self-perceptions, value systems, and defining cultural characteristics. In turn, he applied these institutional characteristics to explain service behaviors during policy formation and implementation.

Donnithorne, in order to avoid charges of post hoc cultural explanations, was cautious to establish narrow descriptions of service culture before applying his theory. We intend to take a similar, cautious approach, aspiring to the same rigor and for the same purposes. Therefore, the following service-culture summaries amalgamate Builder, Ehrhard, Donnithorne, and Cannon (2012), and remain intact through the case studies and analysis. If changes in service culture or anomalies emerge in the case studies, they will garner special attention and rigorous explanation.

Ehrhard characterized the Army and Navy as having feudal structures—meaning that leadership is shared among several different communities (e.g., Infantry, Armor, and Artillery as distinct combat arms with the Army) and decision-making norms respect pluralistic views. More precisely, he specified cultural "balance points," implying that the organizations can go through monarchic periods of leadership by a dominant figure but over time will gravitate back to the feudal structure. Conversely, the Marine Corps and the Air Force are monarchic; they have leaders from a single dominant community (for the Air Force, bomber pilots for an extended era after attaining service independence but then transitioning to fighter pilots in the 1980s, with a feudal disequilibrium in the 1970s) who keep bureaucratic order with strong central control and side payments to lesser constituencies in the service. This facet of service culture impacts innovation. Though monarchic structures tend not to innovate, they do provide the "centralized, top-down focus" critical to Wilson's depiction of the successful implementation of a new idea. Conversely, a feudal structure may struggle to implement new ideas, even though its egalitarian power-sharing structure allows new ideas to emerge (Wilson 1974, 398).

Mahnken's work on military culture confirms the importance of in-group and out-group interactions, differentiation among intraservice subgroups, and subgroup reputations. Mahnken noted that identification with service culture trumped subgroup identification between services. For example, although both the Navy and Air Force have significant numbers of fighter pilots, the two respective groups are more likely to divide along service culture preferences for given tactics than they are to agree as a cross-service subgroup of "fighter pilots." Mahnken's finding that the military services are effective at fostering internal group identity is in harmony with Polzer's

(2004) conclusion that cooperating subgroups can generally prevent the attainment of collective goals they may oppose.

Taken together, Polzer's (2004) general organizational theory and Mahnken's military-specific application could be discouraging to the political or high-level military leader who wishes to foster interservice cooperation. The strong group identification military services foster may in practice hinder intergroup cooperation when a collective's interests depend on it—as when the national security "collective" requires intergroup cooperation and collaboration. Mahnken and Polzer both offer an explanation to describe the deleterious effects of interservice collusion that Coté (1996) observed in his study of fleet ballistic missile systems for the Navy—we find that dynamic to be interesting. The mechanisms they offer are relevant to this study in process-tracing examples of military cooperation. Group identification and the dominant interests in each case explain how these factors influence cooperation. For this study, the overarching question is: How do services' dominant cultures and subcultures advance or hinder joint cooperation?

Defense Establishment Structure

Knowledge of the organizational hierarchies within the defense establishment is vital to explaining institutional joint cooperation. As discussed earlier with respect to agency theory, the principals directing the military within the defense establishment have many faces. While not an area of formally established academic theory, the history of organizations intended to promote jointness in the US military offers a glimpse of a concept that has proven both desirable and elusive over the life of the nation. The pattern that has emerged since 1900 is one of trying to balance the benefits of formal bodies and processes against informal consultation and the force of strong personalities. There is also a repeated trend of the military recognizing civilian dissatisfaction over obvious failures of institutional jointness and addressing these with modest internal reforms—attempts to preempt more intrusion by civilian authorities.

The turn of the twentieth century brought about major changes in the management of US military power. Polk's direct control of the military during the Mexican campaign of 1846–1848 echoed in Lincoln's authority exacted over the Union Army, and it brought to light the political liability of appointing senior military officers based on their party affiliation: returning war heroes had a chance to claim the commander-in-chief's job in the next election. From Polk's experience through the turn of the century, the pattern of formulating US military strategy changed. Instead of sending military professionals off to win narrow campaigns, civilian political leaders now voiced policy objectives and solicited policy recommendations from trusted military officers to achieve them. The rise of Mahan's (1890) ideals about the role of the Navy in protecting American shipping and the complexity of the offensive projection of naval power made the strategic calculus increasingly complex.

When serious interservice squabbles arose during the Spanish-American war, civilian leadership became concerned about the ability of military leaders to concatenate the emerging complexity of military means with national objectives into coherent strategy. Reformers within the military such as the Navy's Admiral Taylor realized this and called for the establishment of improved staffs who could address strategic issues through intelligence and deliberate planning (Godin 2004, 20).

The Joint Board's vague mandate made its role unclear and prevented it from accruing much effective power. Although the Board was influential throughout the presidency of Theodore Roosevelt, Louis Morton (1962, 132–136) described it as inconsequential for the formation of military policy after 1914. Rather than approach the group for advice, "civilian policymakers looked for military strategy recommendations outside the Joint Board, often relying upon high-ranking individual officers and the recommendations of boards within their own services" (Godin 2004, 2). The early history of joint structures is largely irrelevant to the contemporary discussion, but it illustrates that the matter over time grew to be an area of increasing executive and congressional concern as the size and complexity of the armed services grew larger.

The pertinent structures that underlie modern jointness arose with the unification of the armed services and the formal entrenchment in law of the Joint Chiefs of Staff (JCS) in 1947 (US Congress 1947). Service preferences for self-determination, along with legislative requirements to openly report disagreements within the Joint Chiefs to the secretary of defense and president, created a tendency for the Joint Chiefs to come to internally compromised agreements rather than subject themselves to executive or congressional scrutiny (Jones 1982; Donnithorne 2013, 295). Huntington thus argued that the JCS is a committee that must subject its decisions to standard legislative compromise procedures, making it a weakly united body (Donnithorne 2013, 60). If the Joint Chiefs of Staff resort to legislative tactics such as logrolling and making use of ambiguous language to elude making controversial decisions, this tendency is important to a pretheoretical understanding of jointness.

External to the services and JCS is the Department of Defense, an executive organization that has gained power and increased budgetary authority—at the expense of the military services—since its creation in 1947.

The secretary of defense has increased in influence by subsuming the cabinet-level authority the services held before reorganization. The secretary also has proximity to and influence over the chairman of the Joint Chiefs of Staff, who is himself not in the military chain of command, but who represents the other service chiefs to the president. Reforms during the Eisenhower era worked to further diminish the independent authority of the service chiefs, subverting them to the influence of the chairman, while further strengthening the authority of the secretary of defense and the office of the comptroller (Donnithorne 2013, 290). The additional mid-1980s reforms of the Goldwater-Nichols Act continued most of these trends, strengthening both the Joint Staff and the authorities of the unified and specified combatant commanders. Donnithorne's case study of the legislative background is an admirable review; this work borrows its descriptions of the effects of the act as a jumping-off point for understanding the internal service (endogenous) influences, the among services coordinating (meso-organizational) influences, and the external organizational (exogenous) influences on joint cooperation. The questions raised by this line of inquiry are complex: Do the among services coordinating structures, i.e., the Joint Chiefs of Staff, the Joint Staff, and specified and unified combatant commanders, further or hinder joint cooperation? Do the external powers of the secretary of defense further or hinder joint cooperation? (Donnithorne 2013, 290).

Knowledge of service cultures, the formal JCS and Joint Staff organizations, and the Department of Defense is but a partial picture of the hydra that is the defense establishment. The final area of theory this work considers is congressional influence, which impacts the military's ability to function in a joint manner on many fronts.

Congress has legislative authority to declare war, but generally conducts its oversight in more nuanced ways like hearings, investigations, and reports about defense activities— all centered on its role as the steward of defense spending and, perhaps just as important, military promotions.[19] Again, to limit the scope of consideration to a tractable level for a pretheoretical endeavor, this work considers congressional influence on the single area of defense acquisitions, discussed next.

Other External Factors

The last lens of analysis in the military-specific set is an all-encompassing category that looks for influence on jointness from external influences beyond the immediate influence of the Defense Department. For example, discussion of joint cooperation would be incomplete without acknowledging the direct and indirect effects that implementation of the Goldwater-Nichols

Act has had on defense procurement. Military materiel is part and parcel of interoperability, one of the explicitly codified requirements for jointness.

Units equipped with incompatible radios cannot communicate; fast-moving, high-flying aircraft may not be able to differentiate ground targets from friendly forces with sufficient fidelity. Because the act has created two de facto procurement systems—one to support service priorities and one to support the requirements of combatant commanders—contemporary battlefield needs, provided they are articulated in persuasive language, can subvert any attempt by services to control their own strategic procurement plans through the budgeting process. The dynamics of these processes affect all of the case studies in this book.

Since more than 150,000 military and civilian personnel work in the acquisition system, spending money at a rate exceeding twenty million dollars per hour for the US Department of Defense, the scope of the endeavor is immense and its effects impactful for the issue of jointness (Charette 2008; Pincus 2012). The context of military acquisitions is the planning, programming, and budgeting system (PPBS) developed during the McNamara defense administration in 1961. As "the primary mechanism for determining fiscal needs and funding programs," its outputs influence and inform presidential budgetary recommendations as well as congressional appropriations and outlays.[20]

This work stays deliberately clear of most academic theory pertinent to military acquisitions or the PPBES; while it is a rich and deep area of study, the level of detail is distracting from the focus on institutional cooperation the way that a deep-dive into joint doctrine (Buede and Bresnick 2007; Boyle, Higgins, and Rhee 1997; Pursell 1972; Mills 1956) would be. The present review does, however, incorporate a simple appreciation of the military acquisitions process, which is that the large amounts of money spent in the process create dedicated interest-group factions in the form of industry and local constituencies. From major weapons systems to footwear, these interest groups have some opportunity to interface with military leaders to convince them of the military value of the goods they produce. They have a greater opportunity to interact with congressional representatives, who seek the dual economic boon of jobs and district spending as a prize for which to fight (Lexington 2014a, 31; Sapolsky, Gholz, and Talmage 2009, 70; Pursell 1972, 38). The nexus of interests in military acquisitions therefore usually occurs in the context of congressional oversight of the defense budget. This study relies on congressional hearings, reports, and budget documents to follow the money in understanding how interest groups coalesce around defense acquisitions projects. The primary question it raises is: Do defense acquisition processes advance or hinder joint cooperation?

A second phenomenon that this work relies on to shape an understanding of jointness in the acquisitions context is a pejorative mnemonic applied to the services by groups like defense department offices and congressional staffers. Phrased succinctly as "Dumb, Devious, and Defiant," the unofficial summation takes aim at the Army, Air Force, and Navy, respectively. Jones and McCaffery have a well-developed explanation of the derivation and meaning of the expression as defense budget officials use it. The stereotypes break down as follows. The Army fails to submit complete budget information. Ironically, the apparent administrative failure often appears to be a way to gain congressional negotiation leverage by making budgetary elements questionable or deniable. The Navy may not comply with comptroller requests, instead providing answers in a format suited for its purposes, justified with explanations that the service analyzes its budget with a "more thorough review process" than other services or DoD could. The Air Force "puts on the best show" for the department, attempting to dazzle reviewers with technology and elaborate presentation. The intent is to gain budget share and gain additional flexibility from Congress, but the approach has led to mistrust when, with a seemingly always-changing story, the service clearly masked past program overruns or failures (Jones and McCaffery 2008, 382–383).[21] Washington insiders have applied the cliché to everything from recalcitrance before base closure panels to weapons testing (*Charleston News and Courier* 1988; Umansky 2001). Nonetheless, the issue is a part of service-culture description that jointness pretheory must consider in answering questions broached by the consideration of acquisition processes and other aspects of military business touched upon by the legislative branch.[22] The question is: Do external organizations' perceptions and stereotypes of the services advance or hinder joint cooperation? Specific case studies yielded other insight about other external influences on jointness—these are summarized under this category as well.

Adapted from Donnithorne and combining elements described earlier, table 2.1 offers a summary of relevant cultural factors.

Having described some ten areas of theoretical study that have likely bearing on joint cooperation, we close with a summary in table 2.2 listing the central ideas within each theory, and the questions about joint cooperation the theories raise. These areas are by no means a comprehensive list of potential influences on joint cooperation, but seem to be of outsized utility based on a reading of contemporary social theories and defense topics.

Summary

So far, we have established a dichotomy of understanding about the concept of military jointness, arguing that a determination of why military services

Table 2.1: Relevant Military Cultural Variables

	Navy	*Marine Corps*	*Army*	*Air Force*
ENDS	Armed American embassy: any-where, anytime	Warriors from the sea; anywhere, for anything	Apolitical servants of the nation	Air controlled by airmen
WAYS	America and her Navy prosper together	Survive to serve	A land force of last resort	Ubiquity of influence at minimum risk
	Enlisted order, commissioned judgment	Elite warrior identity	The Army way of battle	Decisive strategic potential
	Independent glory of command at sea	Faithful stewards of the national trust	Synchronizing the fragments	Command the air, first and always
				Centrally controlled flexibility
MEANS	Professional and permanent Army	Every Marine a rifleman	Fielding an Army: regulars and the militia	Technology, airplanes, and beyond
	Size matters: bigger-as-better		Soldiers, units, and leaders	Flyers and technicians
DOMINANT STRUCTURE	Feudal	Monarchic	Feudal	Monarchic
Pejorative acquisitions/ congressional stereotype	"DEFIANT": ignores oversight, does not comply, faith in internal review processes		"DUMB": provides incomplete information, attempts to "out-cooperate" others	"DEVIOUS": attempts to dazzle, lies, elaborate claims about technology

Note: That control of the air by airmen is an "end" in Air Force culture speaks volumes about a unique organizational conceit. To an unbiased observer, the command-and-control method used to employ air power in military action would seem to be a means rather than an end.

as institutions do and do not cooperate is worthy of rigorous academic study. Next, we outlined the scheme of this book in which we hope to advance that study. Because several academic disciplines promise explanatory power with respect to jointness, we have taken a multidisciplinary approach, using Martel's approach to "victory" to refine a definition and propose a pretheoretical framework for "jointness." Borrowing from general theories about organizations and specific theories about the military, our survey of the literature

Table 2.2: Summary of Applicable Theories

	Theory	Relevant idea(s)	Question(s) for jointness
General theories of cooperation	Public Goods	Free riders & suboptimization	Do services act as free riders in producing security?
		Success with appeal to purpose	Which better encourage jointness, social pressure or appeals to purpose?
	Organization	Type I/II/III interactions	How do threats, bureaucratic politics, and political maneuvering influence jointness?
		In-group/out-group competition	Who creates stable cooperative structures?
		Resource competition	How do service subgroup interests advance or inhibit joint cooperation?
			Do overlapping capabilities advance or threaten jointness?
	Crisis Cooperation	Crisis conditions influence org. behavior/strategy	Do military services make decisions about jointness in a crisis context?
			How does the perceived urgency of a dilemma affect decision and cooperation mechanisms?
	Professions	"Traditional" professionalism vs. control of basic competencies	Do overlapping service capabilities advance or hinder jointness?
			Do subclasses vying for recognition in their respective military services offer a mechanism for joint cooperation?
	Agency	Principal-agent dilemma	Do the many principal-agent relationships evident in the defense establishment advance or inhibit jointness?
		Implementation slack	
Specific theories about military organizations	Military Innovation	Within service vs. among services vs. external sources of innovation	Do the sources of military innovation advance or hinder jointness?
			Does a given security situation constitute a crisis for the military establishment? If yes, do the crisis conditions make jointness more or less attainable?

continued

Table 2.2 *continued*

Theory	Relevant idea(s)	Question(s) for jointness
Civil–Military Relations	Objective vs. subjective control of military	Which leads to better joint cooperation, civilian control of the military via objective means or control via subjective means?
Service Cultures	4 distinct service identities Feudal vs. monarchic	How do dominant service cultures advance or hinder joint cooperation?
Defense Department & Joint Staff	Among services/ external orgs. growing stronger than services themselves	Does the structure of the Joint Chiefs of Staff, the Joint Staff, and specified and unified combatant commanders further or hinder joint cooperation? Does the secretary of defense further or hinder joint cooperation?
Other Exogenous Factors	Primary means of congressional defense oversight External orgs.' service stereotypes	Do defense acquisition processes advance or hinder joint cooperation? Do external organizations' perceptions and stereotypes of the services advance or hinder joint cooperation?

endeavored to show how several types of existing theories might raise pertinent questions about jointness. Having surveyed and selected several theoretical tools with potential applicability, the next three chapters contain case studies that provide practical examples of observed joint institutional cooperation; a fourth introduces an important new arena of multiservice interest where we *might* hope to see effective institutional cooperation.

Chapter Three

The Army and Air Force Collaborate on AirLand Battle, 1973–1991

> *The Departments of the Army and the Air Force concur that the opportunities are right, the level of joint interest is high, and that valid military requirements exist to initiate an agreement of inter-service cooperation in joint tactical training and field exercises based on the AirLand Battle doctrine as promulgated in Army FM 100-5, Operations, 20 August 1982.*
> —General Edward C. Meyer and General Charles A. Gabriel, US Army & US Air Force Chiefs of Staff, April 1983

Introduction and Background

Overview of AirLand Battle

Our first example of interservice interaction is an effort undertaken by the Army and Air Force. According to the definition we established in chapter 1 for jointness, the effort is an example of interservice cooperation. Both participating services hailed the effort as an example of joint cooperation.[1] Both services worked on the effort for many years before seeing its fruition, witnessing plenty of resistance from naysayers along the way. Not everyone has hailed it as a triumph, either. This case provides a first key lesson: real joint cooperation is difficult to attain because it sits at odds with views of zealots and even the influential public leaders who repeatedly broadcast them as the de facto service positions.

Regardless of how it has been viewed after the fact, AirLand Battle (ALB) started as unilateral conceptual work by the Army in 1973. By 1977, it had morphed into two-service conceptual work that would mature into ALB as a warfighting model (Cardwell 1992, 70). This mutual effort reached its apex with more than a dozen significant agreements between 1983 and 1986 (Davis 1987, ix–xi). ALB also did more than burnish the resumes of garrison-bound generals: it answered the pressing military problem of its day and imbued a significant part of two services' thinking with unified structure and vision. The external defense and national security stakeholders interacted with the concepts as they developed, in turn influencing and being shaped by service actions. ALB renewed emphasis and debate on the concept of "operational art," the systemic linking of actions on the battlefield to meet grand strategic goals (Joint Chiefs of Staff 2013, I-7, I-8; Piatt 1999, 1; Dolman 2005; Joint Chiefs of Staff 1984a). Significantly, it influenced both services' requests and justifications for budgetary share; it was not a mere paper tiger (Wickham and Gabriel 1984). ALB also influenced the Navy and Marines to couch their war plans and funding justifications in terms of the concept, eventually unifying the entire defense establishment, if only in lexicon and only for a few years (Johnson 2014; Skinner 1988, 33–34).

ALB is a complex concept. To a cynic, the term is mere military newspeak, squelching necessary debate via linguistic manipulation and insinuating jointness by shoving two combat domains together (Chomsky and Barsamian 2013, 130–141; Hayden 2014). Contemporary and latter critics, observing the swollen US defense budgets of the 1980s, see ALB as the unrepeatable product of a unique era. Others see a tricky and elaborate swindling—of either the Army or the Air Force, depending on point of view. Our interviews revealed, though, that ALB's progenitors were sincere and sought, of their own volition, to improve military capability and interoperability, and shared their effort with outside audiences so that soldiers and airmen might imitate them. ALB was a sweeping model for how the Army and the Air Force might fight a future war together (Winton 1996, 116). Its stated intent was to combine service capabilities effectively and efficiently to allow the United States to prevail against a numerically superior adversary in a conflict. Though ALB often emphasized its applicability to *any* military conflict, the vision that inspired it was a Soviet invasion of Central Europe, with the United States fighting alongside its North Atlantic Treaty Organization (NATO) allies to repel and defeat Warsaw Pact troops that outnumbered NATO's (US Army 1986, 1; United Nations 1955; Yost 1998, 27–28).

Themes of This Chapter

The narrative arc of ALB and the cooperative initiative it inspired took shape from a well-publicized international security threat. The rapid growth of the Soviet Union's military power and that of its Warsaw Pact satellite states worried the West because it constituted the apparent capability to quickly mass and invade the free states of Central Europe. ALB, after years of fine-tuning, became the framework for a military response to that threat by the United States and its Western European partners in the NATO alliance.

ALB's meaning varied among US services and NATO allies, but common concepts led to a common tactical vocabulary that was relevant for significant interservice cooperation issues. This was useful for both the anticipated battlefield and the halls of the Pentagon. The advocacy and leadership that made ALB a dominant concept for more than a decade were important ingredients in a complex reaction. If Soviet aggression planted the seed, decisive leadership and initiatives from senior military figures served as its protective greenhouse until it matured. A succession of generals, who at first were working to repair the Army's image and ideological foundations after the ignominy of the Vietnam War, encouraged service-wide debate about competing doctrinal concepts for war against the Soviet Union. After an abortive attempt to dictate top-down doctrinal concepts, the service incorporated the ensuing criticism into a widely accepted concept for warfare, inculcating it in its forces through training and maneuver exercises.

The successful shepherding of ALB required more than just one service's participation to lend credibility though, particularly if it was to become a concept that influenced the entire defense establishment. While the Army's debate sparked a US strategy for an anticipated war in Central Europe, it reached out to its recently emancipated younger sibling, the Air Force. ALB's intellectual and doctrinal concepts were a plausible response to the first-order Soviet threat, but also happened to satisfy several organizational requirements, within the services (level a), among the services particularly among the Joint Chiefs of Staff (level b), and in the relations to external defense establishment stakeholders (level c). An early coalescence of interests made ALB the cognitive framework for battle against a conventional Warsaw Pact threat, but then it took on another incarnation as a service defense mechanism in a bureaucratic struggle over defense reform. Antireform advocates used ALB as proof that the Pentagon could innovate and reform from within and did not require legislation as a forcing function. ALB allowed the insular defense establishment to close ranks as the defense reform caucus closed in, demanding that military failures and the excesses of defense acquisition be addressed by departmental reorganization.

ALB, as one of several Pentagon resistance schemes, did not prevent reorganization, and major defense reform came to fruition in 1986 in the form of the Goldwater-Nichols Act. This outcome is not surprising in its historical context (Scroggs 2000, 215; Odierno, Amos, and McRaven 2013, 3–5). However, ALB's influence continued after defense reform was legislated, and remained a driving force for defense organization, acquisition, and interservice battlefield cooperation into the 1990s. Though combat on the European plains that had inspired its main ideas did not test ALB's mettle, its central ideas and military hardware did shape the way the United States fought its only major conventional conflict of that decade, the Persian Gulf War of 1991, which ejected an Iraqi invasion force from Kuwait. This conflict accompanied changing US national security challenges and diminished Air Force support for ALB, effectively ending its run as a leitmotif for the organization, training, and equipping of US military forces.

ALB's Meaning: Distinct among the Services

AirLand Battle Eludes a General Definition

Although we would like to provide a formal definition for ALB at this point in the book, we have to let it develop along with the process-tracing history of the next section. The central idea of ALB was and is truly in the eye of the beholder. Both service's respective histories are enthusiastic in their embrace of the cooperative interservice atmosphere that led to ALB, but they develop different pictures of its nature. Richard Davis's 1987 summary for the Air Force History Office described it "as an example of bi-service harmony," "a case study of innovation," and a valuable template for "future Air Force leaders concerned about change within the service and about the background of bi-service relationships" (Davis 1987, 1). Meyer's signature alongside Gabriel's on the interservice memorandum formally introducing ALB further develops the picture of jointness (Meyer and Gabriel 1983, 1). However, the Air Force never gave ALB the same weight as did the Army. While ALB became *the* doctrinal and organizational construct of the Army for an era, the Air Force never adopted it to the same degree. It never even achieved the status of received doctrine within the tactical-air organizations that it affected the most. Evidence suggests that the two services did not view ALB as even the same *type* of doctrinal thinking, which makes the breadth of cooperation it drove even more remarkable (Chapman et al. 1998, 1; Kem 2014). To describe one of the most substantial interservice barriers ALB overcame, a few words about military doctrine and what it means to each service are in order.

We begin by peering through the service-culture lens of military doctrine. ALB influenced Army and Air Force doctrine as soon as the two services began working together. However, the impact differed between the two because of how they view and develop their essential organizing concepts. The question of doctrine and what it means to an individual service requires analysis through a service-culture lens.

For the Army, doctrine is malleable, but new, novel doctrines become guiding principles relatively quickly. If the meaningful measure is an ongoing, iterative, prolific process, no other service approaches the enthusiasm the Army has for doctrinal creation. Regarding its dominance within the joint force, David Johnson claimed, "The Army captured the doctrine-writing process in the late 1980s" (2014) our investigation found no disagreement about this assertion. About its enthusiasm for doctrine's utility and sway, William Lind added, "The Army has over a thousand doctrinal manuals, and the theory is—for any situation—if you just find the right book and the right page it'll tell you what to do" (2014). Immersion in doctrine happens for soldiers faster than the members of any other service, and the Army's tradition of doctrinal adherence runs deepest. Robert Futrell described how the Army hands down "age-old principles of war" derived from Napoleon, Clausewitz, and Jomini, while revising its doctrine with remarkable passion and frequency (1989a, 6).

An Army officer will often say, "That's not doctrine." This is a serious charge, and she will expect those involved to fix the deviation. Army doctrine addresses a wide range of topics, sometimes blurring the line between what other services might think of as "doctrine" and mere "tactics, techniques, and procedures" (TTPs). A caveat in an Army field manual gives insight into the Army's view of its doctrinal literature, which it defines with a low, wide bar. A wide range of topics, from relationships among component commanders and a combatant commander in the context of a major theater war to the proper method for heating a field ration, can be defined in doctrinal literature. Army doctrine consists mostly of tactical reference manuals, but it also includes esoteric discussions about the nature of warfare and its relationship to strategy (Ancker 2013, 50–52; Demarest 2010, 19; Henke 2008, 99; Conway 2007; Dunlap 2007).

The close relationship implied in Army manuals between doctrinal principles and specific tactical standards against which forces train and evaluate themselves is by design, and has its roots in Gen. William DePuy's campaign for reform in the 1970s. As David Johnson put it, "Army doctrine really starts with Bill DePuy as the TRADOC Commander" (2014). The importance the Army assigns to its four-star Training and Doctrine Command (TRADOC), as well as the sheer volume of professional articles published about the subject, testifies to the importance of doctrine DePuy wrought in the Army's professional culture. The Army puts its doctrine to regular, practical use. An infantry officer, for example, will be familiar with the different field manuals associated with his profession and will have exercised the troops he leads according to their principles. Because they touch on a wide range of topics frequently affected by technological progress, Army doctrinal publications see frequent updating and are maintained by melding headquarters guidance with "lessons learned" gleaned via robust feedback processes. The Army's approach to doctrine makes the mechanisms of a continuously learning organization a requirement.

Military writing overall, though, tends to define doctrine as immutable principles of the nature of war that do not shift as quickly as technology or other factors that set the character of war. The Army, compared to other services, treats doctrine as a more malleable artifact. There are certain dearheld beliefs—among them the idea that "no major conflict has ever been won without boots on the ground"—and the Army eschews the idea (popular with both the Navy and Air Force) that a single strike from a distance can be decisive, even going to far as to label other services "interlopers" in the domains where they fight (US Army 2012, 1–1, 1–4). The Air Force's and Navy's approaches to doctrine devote less routine attention but also induce less variability. This is a post–World War II shift. Huntington described the Navy's well-established doctrinal roots predating World War II as stronger than those of the contemporary Army (Huntington 1961, 399). The Air Force's postwar push for self-determination on a doctrinal basis temporarily jostled the Navy's Mahanian base, but it rapidly righted itself with its theory of carrier airpower as the new means of naval force projection. Thomas Cardwell noted that "the naval services do not have, like the U.S. Army and U.S. Air Force, basic or capstone doctrine" (Cardwell 1992, 52) but allowed that they did provide a strong naval perspective informing joint-warfare concepts. Although the Navy and the Marine Corps have since written capstone doctrinal publications, Cardwell's observation reflects the spectrum of services' doctrinal reliance (Cichowski 1993, 39; US Navy 1994; Hall 2014).

In stark contrast lies the Air Force. Somewhat surprisingly for a service borne nominally from innovation, Air Force doctrine is nearly immutable, but operational concepts can evolve and form a basis for cooperation. An outsider might guess that the Air Force has relatively less interest in doctrine than does the Army since it leans on the rapid evolution of advanced technologies. After gaining its independence from the Army, the Air Force's doctrine manuals confirmed this (US Air Force 1953, ii). Carl Builder (2002, 23) argued that the Air Force slipped away from the original doctrinal and strategic arguments that drove its quest for independence. The Air Force places less emphasis on creating doctrine than does the Army. There is no Air Force equivalent of the TRADOC four-star general; typically, colonels with independent charters write doctrinal revisions that do not elicit much service-wide debate.

Relatively little doctrine activity at more senior levels of Air Force leadership implies neither lack of interest nor insecurity, however. Although routine review of doctrine quickly faded, several analyses advance the idea that the Air Force places great value in its doctrine, believing, in fact, that it has created an "infallible doctrine" (Wacker 1967, 49–50; Futrell 1989b, 711). The service tends to make few substantive changes to the foundational ideas in its capstone doctrinal publications and has not altered its priorities for air missions since the 1930s. When Air Force doctrinal pronouncements appear, they seem focused more on educating others about received truth rather than on grappling again with fundamental questions of existence in the manner of the Army. In 1979, critics of the Air Force's latest doctrinal publication critiqued its slick presentation as evidence that it was more of a public relations campaign than a serious attempt at an intellectual engagement of doctrine

(Williamson 1983, 89; Futrell 1989b, 735). In short, an insider's perspective might suggest that the service has not drifted away from its doctrinal roots, but rather the roots have grown so deep that they anchor the service whether it discusses them openly or not.

What is the Air Force's central doctrinal idea? It might appear that the Air Force has evolved from ideas of strategic bombing (the 1930s and World War II), through nuclear deterrence (the Cold War), to a view that precision weapons delivered by air platforms can end wars quickly (the Gulf War, Balkans). A complete treatment of this topic would need to include the service's eternal hope that technology would eventually deliver on the promises of history's air-power theorists. It would also need to discuss the controversial but enduring Effects-Based Operations (EBO) concept, an attempt (made despite James Mattis's efforts to quash it)[2] to differentiate military outcomes (i.e., ordnance impacts, explosions, electronic jamming, and so on) from the systems that do it. However, these topics are mere sideshows; air power advocates, theorists, and zealots alike hold to a few core concepts:

Air power is a unique form of military power, the effects of which transcend the usual constraints of war, often enabling strategic goals without other types of military power.

A single air commander's centralized control of unified *theater* air power is inviolable and any effort to parse authority over limited air assets among ground commanders is wasteful; air power must be commanded centrally by "Airmen," who are alone intellectually capable of understanding it.

Air superiority is of first importance in any campaign, for with it the actions to attain ultimate victory are possible (US Air Force 2011, 14–20).

David Mets's (1999) summaries of air power theorists show that these core doctrinal statements reflect an extension of similar ideas with historical roots. Guilio Douhet argued that strategic bombing would lead to "humane victory." Hugh Trenchard emphasized breaking the enemy's will. Billy Mitchell emphasized the superiority of the Air Force over other military services. John Warden's "Ring Theory" retained the central theme that air power could "function independently to achieve decisive effects." Throughout its history, air superiority—the friendly force's unfettered use of the air domain and its attendant denial to the enemy force—is an unshakeable core of air power doctrine. Despite "revolutionary" periodic revisions of fundamental Air Force doctrine, its essential doctrinal beliefs have held constant since before the Second World War and are enumerated in the preceding list (Cichowski 1993, 19).

Air power theorists' ideas are remarkable for their exclusivity and audacity. The first real doctrinal contribution these ideas made, as reflected in the

1943 version of Field Manual (FM) 100–20, *Command and Employment of Air Power*, asserted claims about air power's unique utility and argued for centralized control under a specially trained leader (War Department 1943, 16). Cichowski noted that Air Force doctrine had matured since the days of proponents who believed "air power alone . . . was sufficient to win wars" (1993, 49). We would simply offer that air-power theorists have never willfully accepted a subordinate role for air power or the notion that it cannot sometimes be a war-winning tool independent of other services.

Deep-seated confidence aside, the Air Force frequently finds itself on the losing end of the national debate about what its doctrine *should* be. When facing Congress or other critics, who may not appreciate the cunning of enduring air power tenets, the service reliably defends its contributions to national defense and war efforts. Andrew Abbott (1988, 60) described the importance of public opinion in maintaining professional credibility. For the Air Force and all military services, the upshot is that internal arguments alone will never suffice; public engagement about the utility of services provided is an enduring reality. In chapter 5, we unpack a somewhat humiliating example of this for the Air Force. When engaged or challenged thus, the typical Air Force approach lists numbers of missions flown, bombs dropped, and enemy personnel wounded, emphasizing danger, boots-on-the-ground, and an in-the-mud combat focus. This reflects the insecurity the Air Force feels when the land services get attention for risky missions that lead to individual heroism. Yet it is also politically astute; the US defense establishment gives singular attention to messier forms of combat, irrespective of their contribution to national goals (Lexington 2014b, 33). Frequent calls for the Air Force's abolition come from outside the service. Perhaps the Air Force's lack of apologists elicits reverse psychology in the Army, or maybe it is annoyance over its younger sibling's brashness. Either way, the "indispensably needed" continental service asserts its own reasons for existence from time to time, even if no one asks (Scroggs 2000, 123).

Because the Air Force, its own existential confidence aside, must at times engage in public debate about its doctrine to stay politically relevant, criticism of its shortcomings in this arena periodically reveals doctrinal gaps for which it must seek remedy. These criticisms commonly arise over the questions of air support to land forces and give the service a reason to concentrate on support functions that would not otherwise make a short list of essential doctrinal concerns. For the Air Force, ALB grew out of the Tactical Air Command's (TAC) significant involvement with TRADOC under DePuy. For the younger service, however, ALB was an operational concept but never an essential doctrine.

For as much utility as ALB had for the Air Force—it did have great use as an organizing, training, and equipping concept—there is some adamancy in self-appointed keepers of service hindsight that ALB was never an Air Force "doctrine," even as broadly applied as that term is at times (Deptula 2013). The Air Force throughout its short history has simply been too confident, too immutable in the doctrinal concepts to which it holds to seek a new definition of the way it fights, even in the face of extreme changes in the Clausewitzian "grammar" of war (Clausewitz 1989, 75, 87, 605). Technology seems to drive the service asymptotically closer to the vision of precision and instantaneous strategic effect about which the Air Corps Tactical School (ACTS) planners dreamed at Maxwell Field in the 1930s. Former Air Force Chief of Staff Gen. Mark Welsh gave an Air Force–stereotypical view of ALB when he compared it to the Air-Sea Battle concept, an emerging idea during his tenure:

> For those of you who remember AirLand Battle . . . it was just a kind of a conscious approach to, "How do you make the Army and the Air Force work better together?" When I was flying A-10s back in those days, we couldn't talk on a radio to the Army tactical operations center for the unit on the ground. . . . We had an FM radio, but it didn't work very well with theirs. We couldn't speak in a secure means at all. And so, that was one of the objectives of AirLand Battle: to talk to each other—you know, create the technology, the equipment, the tactics so you could communicate. (Welch 2014)

This characterization is revisionist, greatly downplaying the significance of the construct the Army believed it had created in making ALB its official doctrine. It also indirectly illustrates the Air Force's dismissiveness toward new doctrines. However, the scope of cooperation that took place in the name of ALB suggests that both services assigned it substantial gravitas in its era.

We distinguish nuances with which the Army, the Air Force, and, when applicable, the Navy and the Marine Corps viewed the concept. The initial interservice agreement kept points of friction and division about the concept muted. Differences came to light, though, when prominent Air Force voices repudiated ALB after the Gulf War, marking the beginning of the model's demise. Despite the services later distancing themselves from ALB, there is little doubt that it and the efforts predating it built trust between the Army and the Air Force and provided a common vision for how air and land power would fight together in a conflict between superpowers.

A History of Army-Air Force Cooperation on AirLand Battle

Tracing the history of AirLand Battle's (ALB) development into a cooperative joint effort requires going back one step before ALB in the Army's doctrinal evolution. The Army's reinvention and focus on a new security problem had at its roots the discouragement and ignominy of the Vietnam conflict.

CHANGING SECURITY CLIMATE

A sea change in the strategic perspective of the United States followed Vietnam. In the 1950s, the projection of strategic power was an organizing tenet for US foreign policy and military doctrine. National security in the 1950s and 1960s centered on checking the Soviet Union, offering an ideological response to Communism, and acknowledging the possible use of nuclear arms (Poole 2013, xi). Eisenhower's New Look, massive response, and "balance of terror" gave way to Kennedy's Flexible Response and "assured destruction" (Weigley 1973, 442–446; May, Steinbruner, and Wolfe 1981, vii). These concepts and the policies they inspired raised the stock of the respective nuclear power–projection branches of each service. By the 1970s, the ideal of containment as a response to Soviet expansionism was waning, giving way to Nixon's realist détente in the wake of Vietnam. Carter's foreign policy promoted human rights as an ideological counter to Soviet aggression until the invasion of Afghanistan in 1979 caused a temporary reversion to containment (Kissinger 1976, 170; 2008, 45).

As security elites shifted, so did the military. Henry Kissinger (1957, 20) noted that all of the services' mid-1950s congressional budget hearings emphasized strategic, long-range power-projection capabilities, even at the expense of more traditional military capabilities. He highlighted resulting interservice rivalry: "The more the other services have extended the range and power of their weapons, the more closely they have approached what the Air Force considers its primary mission, thus opening the way to endless jurisdictional disputes" (20).[3] He also critiqued the uneasy truces under which the services operated and the chilling effect this had on developing a national strategic doctrine, which he hoped would issue from the Joint Chiefs of Staff and the National Security Council (237–238).

The two decades after World War II witnessed burgeoning attention on strategic weapons, with the Air Force's budget share rising as high as 48 percent in 1957 at the height of New Look—reflecting its custodianship of two of the three legs of the nation's nuclear deterrent triad (Lewis 1990, 15). The Key West agreement established the Strategic Air Command (SAC) as a specified command under the Joint Chiefs and made the Air Force chief of staff its executive agent (Futrell 1989a, 200). Long-range bombers shrank the globe and emphasis on deterrence via a survivable first-strike capability established SAC as a dominant defense organization during the era (Futrell 1989a, 8–10). Intercontinental ballistic missiles (ICBMs) and medium-range nuclear missiles deployed to Europe solidified this dominance for several years. The Army was absorbed by self-pity in this strategic environment

(Brownlee and Mullen 1979, 112). The Navy's struggle to develop fleet ballistic missiles, wresting away one leg of the nuclear triad from the Air Force, served to drive the defense budgets back toward the relative interservice parity that has been observed since the 1970s (Lewis 1990, 13–15).

Nuclear-arms focus waned each time actual security problems deviated from the blatant Soviet aggression strategists had envisioned. By the mid-1970s, the immediacy of unlimited nuclear war began to fade. The emerging security outlook also emphasized containment-inspired "brushfire wars" epitomized by Korea and Vietnam to a lesser degree. The possibility of significant conventional engagement in Europe with the Soviet Union was waxing in the stead of these previous ideas. If the Soviet Union seemed to accept the strategic futility of a nuclear exchange, its ambitions for influence and territory in its near abroad still filled NATO's front window. These international geopolitical trends influenced the military services, who studied intelligence reports and the remarks of the international-relations intelligentsia as a stockbroker might examine a firm's quarterly results: carefully, repeatedly, and with an aim to forecast such that they might position themselves for enduring strategic advantage in Washington's ceaseless budget battles.

1973–1976: Growing a Sense of Urgency

The Army's official TRADOC history attributed the post-Vietnam innovations in the doctrine of land warfare that the Army pursued to General DePuy (Romjue 1984b, 4–11). DePuy's analysis of both the US Army's experience in Southeast Asia and the Israeli experience in the 1973 Mideast War caused him to make significant changes in the way his institution thought about and prepared for war (DePuy 1994c, 69–74; Leonhard 1991, 130). John Romjue speculated that significant changes in Army doctrine were "bound to come" after Vietnam, but DePuy's opinion that the existing body of training literature the Army used had "ceased to be valid" (1984b, 3–4) in light of the lethality of contemporary weapons systems seems to have accelerated change. DePuy gets credit for laying a foundation for what would become ALB and setting in place its trusses of interservice cooperation (Davis 1987, 33).

Active Defense became that new capstone doctrinal concept. Under DePuy's leadership of TRADOC, the post–Vietnam Army's soul-searching had evolved into the Active Defense concept by 1976. In the Army historical narrative of ALB, DePuy used two phases to accomplish this (Chapman et al. 1989, xv). First, he seized the growing threat in Europe as a mechanism to motivate the Army (Lock-Pullan 2003, 488; DePuy 1994h, 194). He believed that prescriptive guidelines, promulgated as "how-to-fight manuals," would reinstill discipline and confidence in an institution that had seen both qualities shattered during its involvement in Southeast Asia (Herbert 1988, 7; Brownlee and Mullen 1979, 187; Lind 2014).

DePuy did not invent the idea of Soviet menace to advance his agenda

—government executives, intelligence analysts, and other security experts inside and outside the military establishment had validated it as real. Studies acknowledged the difficulty of assessing NATO against the Warsaw Pact but nevertheless identified risk in an offensive assault against NATO forces along the Central Front that separated West Germany from Eastern Europe (Congressional Budget Office 1977, xii). Herbert London defined a stark imbalance: 93 NATO divisions against 176 Warsaw Pact Divisions; 14,400 tanks to face 42,600; and a mere 11,500 artillery tubes to square off against 35,000 (London 1984, 29). The Chinese strategist Sun-Tzu wrote a well-known maxim about the utility of desperation; if a confident force might not respond to a leader's exhortation, one in "a position from which there is no escape" will be loyal to the death to the general commanding them (Sun-Tzu 2005, 214–215).

Bleak assessments of NATO's position on the Central Plains of Europe show that DePuy and his flag-officer cohort used this leadership-by-exaggeration technique flawlessly. Former Cold War Army personnel still describe the expected life span of their units at the "Gap" in minutes, yet simultaneously speak proudly of their commitment to such battle plans. Histrionics aside, the apparent directness of the strategic situation seems to have had a dramatic effect on freeing up a moribund doctrinal-development process. DePuy borrowed from the clear signals handed him by the national security dialogue, made the Soviet Union the hardened objective of that focus, and sharpened Army doctrine on the whetstone it afforded him. The 1976 Field Manual (FM) 100-5, over which DePuy presided (and which defined the Army's fundamental doctrine), became known as "Active Defense," and it directly addressed "the prime strategic problem the Army faced: a U.S. force quantitatively inferior in troops and equipment on an armor-dominated European battlefield" (Romjue 1984b, 5). The manual, which DePuy admitted was "oversimplified" to reach his congressional audience, professed to teach the "real" lessons of modern warfare, the capabilities of a Warsaw Pact enemy, the terrain of Central Europe, and the weapons in the inventories of the potential combatants at the time (DePuy 1994i, 143–150). In emphasizing the danger of unpreparedness against threats others had defined, DePuy channeled external factors into a sense of urgency in his organization, a prerequisite for "gaining needed cooperation" (Kotter 1996, 36) within and among large organizations.

DePuy's clear vision and the intent he conveyed to his subordinate doctrine writers also drove a centralized intrainstitutional debate about Army doctrine, a vehicle for change that has remained in place long after his tenure. He built the foundation of his movement by writing to the schools and training centers under TRADOC's sway, using a "pot of soup" metaphor to encourage them all to contribute a wide collection of ideas about the changing nature of warfare and the doctrine that should accompany it (DePuy 1994a, 121). He infused the pot with his sense that the 1973 Arab-Israeli War demonstrated a new lethality in military weapons that US doctrine, training, and tactics were ill equipped to handle, particularly against better-equipped, better-trained Warsaw Pact forces (DePuy 1994b, 161–163; US State Department 2014). DePuy viewed doctrine development as inseparable from the Army's acquisition efforts, deliberately altering doctrine for which his command had responsibility to ensure that the service would prevail in its quest for its "Big Five" weapons systems—an infantry combat vehicle, a main battle tank, an attack helicopter, an assault helicopter, and a short-range missile defense system (Herbert 1988, 77–78).

DePuy used collaboration to good effect. He started discussions in 1973 with Tactical Air Command (TAC) that became seeds of ALB's later interservice cooperative bent. Within his own service, DePuy involved senior leaders of important Army organizations. Forces Command (FORSCOM) is the Army's largest command and the one responsible for deploying combat units to meet the needs of joint force commanders: DePuy convened a FORSCOM seminar at Fort Knox in October 1974 to discuss combat techniques and tactics for company- and battery-sized units (Romjue 1984b, 4). Learning of a hunger for new doctrine within the Army at all echelons, DePuy held the first of several conferences with subordinate commanders in December 1974 and delegated to them writing his "how-to-fight" manuals based on the weapons of the day, the expected Soviet threat, and the most recent example of modern combat, namely, the Arab-Israeli War (DePuy 1994b, 193–195; 1994d, 427–435).

DePuy elevated the importance of the new FM 100-5 by moving responsibility for it from Fort Leavenworth, home to TRADOC's subordinate Combined Arms Center, to his headquarters at Fort Monroe. Conferences throughout 1975 incorporated the perspectives of US Army commanders in Europe and reexamined the anticipated role of air mobility in that theater (Romjue 1984b, 5; DePuy 1994b, 179–183). The July 1976 FM 100-5 was a product of the institutional Army. It used clear language, incorporated specific examples of how to fight against new enemy weapons systems, and established an imperative to win the "first battle." The first-battle concept held emotional significance for the Army; this new doctrine dismissed the idea that the United States would have the luxury of a long time to mobilize and bolster its force in Western Europe if the Soviet Union exhibited an aggressive push in that direction. In so doing, it addressed an incipient but growing concern that the service had a dismal record of failure in the opening battles of wars in which it became involved, and required the mobilization of ever-greater force before eventually regrouping and prevailing overall—or, in the case of Vietnam, never attaining victory (Heller and Stofft 1986). In spite of all these positive aspects and a clear attempt to address pressing Army issues, DePuy's product did not receive acclaim throughout the Army. It took a successor's modified approach to achieve that.

After DePuy's doctrinal concepts were solid, the interpersonal foundations of interservice trust were laid. While Nixon pursued a bit of Cold War warmth via détente with the Soviet Union, the Army and the Air Force enjoyed their own version of the concept. The Air Force chief of staff, Gen. George Brown, sought to improve interservice relations with his Army counterpart, Gen. Creighton Abrams (Futrell 1989a, 530). TAC-TRADOC dialogue carried the overture to two of the services' most influential four-star commands. While discussing the new lethality of war with the wider Army, DePuy's dialogue with TAC Commander Gen. Robert Dixon began to imbue his thinking with "the vital role to be played by tactical air in the air-land battle" (Romjue 1984b, 5; DePuy 1994e, 1–2; DePuy 1994f, 1–2). DePuy emphasized an institutional relationship with the Air Force throughout his command of TRADOC (Nielsen 2010, 43). After April 1974, DePuy had a free hand in this endeavor, compared to what a non-service-chief Army general would otherwise enjoy, as General Abrams's fell terminally ill and thereby yielded his characteristically close control. DePuy and Dixon started the Air Land Forces Application (ALFA) Agency in 1975 to coordinate cooperation between their respective commands, an organization that has endured until now, albeit under today's all-inclusive joint umbrella called Air Land Sea Applications Center (ALSA). The transformation from ALFA to ALSA is proof of a certain corollary of joint cooperation—once two or more services go to the trouble of forming an interservice construct, labels proclaiming its jointness are inevitable regardless of how effective it might be.

Motivation for TAC's eager cooperation with TRADOC is as nuanced as ALB itself. Davis speculated that impending conventional-force reductions, manpower uncertainties induced by the end of the draft, and the two organizations' relationship developed during Vietnam pushed them together again (Davis 1987, 24). Thomas Ehrhard noted that the Air Force had since the mid-1960s been going through a change in its "monarchic" leadership structure (Ehrhard 2000, 102). Between 1965 to 1982, a "feudal" interlude marked transition from domination by SAC's bomber pilots to TAC's fighter pilots (Worden 1998, x). SAC's slipping influence within the Air Staff was evident in Maj. Gen. Arthur Agan's 1965 conference of fighter aces—one needed fifteen or more kills to get an invitation—to discuss the need for a new, dedicated air-superiority fighter. Efforts like this pushed institutional thinking in the 1970s to focus on attaining air superiority through fighters (Futrell 1989b, 471). Moreover, the changing weapons and international dynamics of strategic deterrence diminished the appeal and prestige of the formerly dominant SAC (Worden 1998, 217–220).

Defense Department leadership stood in opposition to TAC's fighter-focused gambit, instead acquiescing to Army pressure for more fixed-wing close air support (CAS) capability. Secretary of Defense Robert McNamara and Harold Brown—then the department's director of research and engineering and later the secretary of the Air Force—"demanded that the Air Force procure some less expensive aircraft specifically for ground attack support missions," which led to the decision to purchase a redesigned version of the Navy's A-7 *Corsair II* for CAS missions in Vietnam (Futrell 1989b, 471). The Air Force chief of staff, Gen. John McConnell, described the decision as means to sate demands of "certain elements in the military and certain elements in Congress" (US Senate 1969, 123). Without clarifying the exact source of congressional or DoD pressure, he elaborated that the A-7 decision came about "to demonstrate that we did want to give the Army every possible means of close support" and because the Air Force believed that it could "do it better than they can," a reference to the Army's bid for the AH-56 attack helicopter (US Senate 1969, 124).

Despite efforts of senior officers in both services to compromise, the interservice dispute over CAS and the Army's requirement for additional attack helicopters simmered from the mid-1960s through the end of Vietnam (Futrell 1989b, 516–531). The services reached an agreement in mid-1975 that the attack helicopter, while a complement to CAS, not impinge on the Air Force's role of providing CAS (US Senate 1976, 5639). Gens. Dixon and DePuy personally helped broker the agreements, commenting that they saw clearly the first-order threat in Europe: they had "grown up"; they had gotten past earlier fights about roles and missions; they saw a clear need to cooperate given "the overwhelming size of what we have to do" in prosecuting "the air-land battle" (Dixon and DePuy 1975). Both services were looking for redemption in the aftermath of a demoralizing and unsatisfying Vietnam experience (London 1984, 2; Nielsen 2010, 36–37). The Army, following the example of other land forces after a defeat or substantial setback, turned to doctrine for this redemption (London 1984, 1). Containment and exclusive focus on nuclear superiority were dead, and the Army had to style itself to fight a numerically superior Warsaw Pact in Central Europe.

In sum, although the first-order threat drove the cooperative spirit, individual personalities and external pressures had marked effects. The prominent senator Barry Goldwater characterized Senate hearings on CAS in 1971 as an effort "to answer the question of whether we need one, two, three, or four tactical air forces. . . . My concern is duplication, a very costly duplication" (US Senate 1971, 105). Given this scrutiny, the Army and Air Force's mutual—but relatively sudden—decision to settle previous close-support arguments indicates that first-order threat, organizational self-interest, and bureaucratic politics were all in play. Both TAC and the Army proved adept at reading congressional tea leaves. In 1975, 75 percent of the defense budget went toward general (as opposed to strategic nuclear) forces, a dramatic shift from New Look. Fully 25 percent went toward buying tactical fighter aircraft (Johnson and Schneider 1991, 217). As much as "appeal to purpose" appeared in both services' histories dedicated to ALB's evolution, the concept seems also to have provided an immediate solution for a raft of self-interested problems and opportunities that the Army and the Air Force—more specifically, TAC—faced in the years after Vietnam.

Service leaders started their attempt to effect change via a top-down approach. The first stepping-stone era prior to full-fledged development of AirLand Battle came to a close with the release of the 1976 edition of FM 100-5 and the Army's response to the document. This response grew to become quite negative. Part of the reason may have been DePuy's approach to doctrinal change. Our most compelling evidence for this is the response that his TRADOC successor, Gen. Donn Starry, received in developing ALB. Starry, rather than focusing on top-down change via institutional elites, "took pains to include the Army at large" (Romjue 1984a, 7). The difficulties of top-down leadership, especially in a highly bureaucratic, change-resistant organization like the Army, are legion and well documented (Easterly 2008, 99; Berman 1978, 12–13). The feedback Active Defense received did not doom it to failure; rather, the impetus for doctrinal change it ushered in cleared the way for ALB. DePuy provided Starry enough momentum to continue a large, interorganizational project that reinvented Army doctrine and gained Air Force buy-in, an impressive foundation that eventually witnessed Active Defense's transformation into ALB. The evolution continued with the Army's churning reaction to the 1976 field manual.

1976–1981: REACTIONS TO ACTIVE DEFENSE

The 1976 FM 100-5 elicited responses that spanned a spectrum from enthusiasm to pessimistic skepticism, but pessimism came to dominate discussion. The TRADOC command historian was understated in describing "pointed and lively doctrinal debate" (Romjue 1984b, 6). In general, positive responses cited the unequivocal and authoritative voice with which the document spoke. In accord with DePuy's vision and forceful style, there was no doubt that FM 100-5 was the Army's new doctrine. It was accessible, widely distributed, designed to be "current and readable" (Romjue 1984b, 4–6), and formed a foundation upon which more tactical publications would build. While pleasing to those who wanted the Army to return to fighting expertise, it gave pause to those who saw gaps in its battle plan in Central Europe. Desire for institutional redemption after Vietnam remained a steadfast backdrop through this era. Instead of apathy or a subdued response that might accompany a favorable state of affairs, Active Defense was hot or cold, eliciting either enthusiasm or disdain. Whether they loved or hated its specific ideas, students of doctrine in the Army were looking for change, and Active Defense propelled them down that road.

The manual did earn some quick positive-leaning reviews, but with caveats. Luminary strategist Colin Gray called it an "excellent new master operations manual," but his context was a broader article about the challenges of force planning with the constraints of American democracy and the reference was a passing one (Gray 1978, 13). Archer Jones admired Active Defense's clarity and return to Clausewitzian basics, but gave it the backhanded label of "the new defensive doctrine" (Jones 1978, 36), questioning its analysis of NATO defenses, optimistic military-intelligence assumptions, and characterization of European terrain. Dan Loomis (1977, 67–69) commended the manual's overarching assumption of a nonnuclear land war in Central Europe, and then critiqued its defensive stance against the Soviet bloc forces and the great costs it would entail. Loomis also foresaw the controversy the release of this FM 100-5 would bring.

The focus the 1976 FM 100-5 placed on defensive battle in Central Europe attracted its strongest criticism. Critics decried its singular focus on a defense against a "central front." The concept that the Warsaw Pact would continue to feed its forces predictably and linearly into an easily identified portion of the European plains while NATO forces maneuvered to mow them down with concentrated firepower seemed fanciful. William Lind (1977, 54), in particular, savaged the manual for its emphasis on fighting defensive battles at a numerical disadvantage and on the importance of the first battle, for the implication of attrition warfare, and for tactical prescriptiveness.

Lind's list of deficiencies, which he outlined in the Army's flagship doctrinal publication, assaulted the 1976 manual like waves of Warsaw Pact armor maneuvering against outnumbered NATO defenses. He flogged it for neglecting friendly electronic-attack and precision-guided munitions advantages, but also pointed out that NATO lagged its potential foe in ability to deploy forces. He argued that Europe was not prepared to use or face the latest antitank weapons used to great effect in the latest Middle Eastern war, and speculated that allied forces were perhaps not as mobile as hoped. In short, he questioned the assumed superiority of the defense (1977, 55).[4]

Lind (1977, 56) even turned on the famous "can-do" attitude of the Army, asking whether careerist commanders were too timid to speak out against foolhardy strategy. John Sloan (1977, 111) questioned the mechanisms by which defeat would come, citing the enemy's superior artillery ratio and the defenders' risk of encirclement. Lind asked whether the Army's first-battle focus might put it in a tough position to win follow-on battles, given Soviet doctrine of continuous echeloned attack. He decried Active Defense as a return to elevating attrition over maneuver while assuming away the enemy's own ability to maneuver, in the worst case forcing NATO into a linear defense against a maneuvering Soviet threat. Archer Jones (1978, 35–36) pointed out that combat in Central Europe would be far from clean. With history punctuating his attack, Lind dealt the manual a decisive intellectual blow, accusing the Army of a "Maginot mentality" and "preparing to lose a war honorably" (1977, 56–64).

Lind's critique would not merit recounting here in such detail but for the breadth of his influence: the Army made great efforts to answer all of his criticism in its next effort. Phillip Karber (1976, 27) attempted a defense, and TRADOC weighed in on its own behalf about the maneuver-attrition charge, but negative reactions quickly overwhelmed its viability. Richard Lock-Pullan argued that the evolution from Active Defense to ALB, given that it happened during a period of "strategic stability," debunked "the common perception that failure leads to innovation" (2005, 680). While the point remains intact that doctrinal innovation occurred absent a tangible failure of a fielded army, the criticism of Active Defense indicated its failure on paper to many observers. If it was not an outright failure, the specter of future failure haunted its reception, thus fanning the flame of urgency still burning among post-Vietnam US military organizations. To quench that flame, the Army needed to find answers for the charges leveled against Active Defense. In that search, the service would be pushed into ever-closer alliance with the Air Force's TAC.

General Starry took command of TRADOC in July 1977, in the midst of heated FM 100-5 discussions. A former corps commander in Europe and a chief contributor to the current version of the manual, Starry knew the existing material well and had developed an informed perspective about the theater of war toward which it looked. He had performed a detailed analysis of the force structures and possible battles that might take place. A study Starry had overseen as the V Corps commander purported to simulate how battle against the Warsaw Pact forces might unfold, mathematically calculating the interaction of weapons systems, forces, weather, terrain, and other battle factors (Swain 1994, xi). He concluded that defense was stronger than previously believed but that delaying or destroying second-echelon forces was important and that CAS would be essential to any winning effort (Romjue 1984b, 23–24). He saw three roles for himself at TRADOC: (1) conduct a continuation of the "battle calculus" he had started at V Corps; (2) put Army doctrine on secure footing for the long term; and (3) solve the problem of second-echelon, follow-on forces that the Soviet machine could generate against NATO (Romjue 1984b, 24; Meyer 1983).

Starry produced an analytical framework called the Battlefield Development Plan (Romjue 1984b, 25). In 1979, TRADOC developed an "extended battlefield" (Romjue 1984b, 26) concept to envision warfare against

the Warsaw Pact. It demanded that brigade, division, and corps commanders "see" fifteen, seventy, or 150 kilometers, respectively, beyond their fur-
thest-forward troops into enemy territory. The concept also emphasized
the time dimension of fighting, further asking commanders to look forward
twelve, twenty-four, or seventy-two hours into the future, anticipating and
countering the movement of forces the enemy might bring to bear (Romjue
1984a, 7).

The new battlefield was bigger, in both size and time, than any previous
Army doctrine had attempted to envision. Since their potential Warsaw Pact
opponents would initially outnumber the forces available to fight in Central Europe, the doctrine created new requirements for special-operations
forces, long-range reconnaissance, and long-range interdiction to sense
and attack the deepest echelons and forthcoming sallies of the enemy force.
While developing the novel operational concepts, Starry was also absorbed
with formalizing a doctrinal-development process for the Army that made
these concepts the starting point for doctrine and all Army publications. He
returned responsibility for doctrine writing back to the instructors at various
combat schools, rather than segregating it within a pool of dedicated doctrine writers (Romjue 1984b, 30).

Starry recognized the difficulty of some of the abstract concepts his development process created. He wrote that the "integrated battlefield" (an
operational concept for interdiction with tactical nuclear options) and the
"extended battlefield" (combining the operational concepts of geographic
depth and time) threatened to become Army "in-words" that were bandied
about without significant meaning or distinction (Romjue 1984b, 37–38). To
get the essence of his comprehensive idea across in a term that would resonate with wide audiences, Starry elected to use—and went to great effort to
promulgate—the "AirLand Battle" terminology (Starry 1981). While it was
helpful to define and use a common term, the problems of "seeing deep" and
conducting deep operations into enemy territory exercised the intellectual
capacity of TRADOC. The answer emerged as an increasing reliance on an
interservice partnership. A "twenty-star" meeting at Fort Monroe in October 1979 brought together the service chiefs, the Army's vice chief, the TAC
commander, and General Starry. The discussions stressed "attack of the
second echelon" and broached the sortie requirements for air-interdiction
operations, establishing a nexus of Army and Air Force interests (Romjue
1984b, 32).

The United States' overall national sense of urgency increased in 1979 as
well. The Carter administration, marked by hope that international agreements and multilateral frameworks would provide global stability, "awoke"
after the Soviet Union invaded Afghanistan in December ("Welcome to the
19th Century" 2014). The simultaneous Iranian hostage crisis spotlighted

the dark side of international relations and a dearth of US military-response options. Following Reagan's 1980 election, the urgency to respond to the Soviet Union grew. Candidate Reagan had chided Carter for being "too soft" on the Soviet Union and lackadaisical in developing intelligence assets to thwart its base intentions; President Reagan would not fail the same way (Brinkley 1998, 14). The new director of Central Intelligence, William Casey, "usually devoted some part of his day to the Soviets" (Woodward 1987, 171). Strategic thinking about how to defeat what would become the "Evil Empire" was in vogue in Washington, DC, and throughout the US security establishment (Goodnight 1986, 390). The military would follow suit.

The Army's concept of interdiction became more refined and specific, advocating an approach that would not just interrupt enemy capacity, but would "set the terms of battle" (Dinges and Sinnreich 1980, 14–17). The Army and Air Force would spend a great deal of time refining interdiction to have specific, focused effects on the battlefield. The collaboration led to a hybrid air-power application that came to be known as battlefield air interdiction (BAI), bridging the gap between deep interdiction and CAS. In March 1980, the revision of FM 100-5 began again, and the guiding principles gripped up in the name "AirLand Battle"—one that synchronized "deep attack" with the fighting at the forward line of troops (FLOT)—provided an intellectual framework that demanded joint cooperation between the Army and the Air Force during the next few years.

1981–1986: Army-Air Force Cooperation Reaches a Peak

General Starry formally published the ALB operational concept in March 1981 alongside an Army reorganization study called *Corps '86*. This release marked a campaign throughout 1981–1982 to inform military and general audiences about ALB (Romjue 1984b, 44–45). The groundwork and previous interservice coordination had gelled relationships between the Army and Air Force, and April 1981 saw publication of a collaborative TAC-TRADOC pamphlet on joint suppression of enemy air defenses (JSEAD). In keeping with the distribution of responsibilities across the deep battlefield, the JSEAD agreement gave both services a voice in target nomination and put the air component commander in charge of the overall effort (Davis 1987, 31–32).

Conspicuous cooperation continued with an offensive air support (OAS) agreement in May 1981. The concept was called Joint Attack of the Second Echelon (JSAK), and its significance was to wed both services to a tactical concept for air-land battle that NATO had published (Davis 1987, 30). BAI supported ground commanders by attacking enemy formations that would threaten friendly forces, but unlike CAS, it remained completely under the control of the air-component commander. The agreement was a remarkable compromise on a topic otherwise marked by mutual dissatisfaction throughout the history of air power. Both Army and Air Force headquarters staffs endorsed the agreement, with the Air Staff subsequently designating it "authoritative Air Force doctrine" destined to appear in appropriate "doctrinal manuals" (though the service neither adopted the idea nor put it in official doctrine as much as the Army might have wished) (Romjue 1984b, 744; US Air Force 1984, vii; Kent and Ochmanek 1998, 2, 9). The May 1981 agreement also set in place the method for apportionment and allocation of air missions under a joint force commander, a system of nomination and execution that still holds sway in joint doctrine (Davis 1987, 30–31).

The agreements signed in 1981 took place while the Army revised FM 100-5 for a new release in August 1982, an effort led by then Lt. Col. Huba Wass de Czege, aided by Lt. Col. L. D. Holder and Lt. Col. Richmond Henriques (Romjue 1984b, 43). This version enjoyed greater acceptance within the Army than had its predecessor, and it also gained more cachet at the among services and external stakeholder levels of analysis. Warm reception for the 1982 FM 100-5 arose in part from disillusionment with the 1976 version. However, Starry and his deputy also took positive steps to give the new manual a better chance for wide acceptance. By removing any language that might reduce warfighting to formulae and by returning classical military theorists into the manual, they sought to show they were returning the Army to timeless, immutable, incalculable truths. Four themes emerged: initiative, depth, synchronization, and agility (Romjue 1984b, 52–57).

Rather than use DePuy's top-down approach, Starry sought early-draft feedback from the field. He directed his team to brief and staff the forthcoming 100-5 throughout the Army starting in February 1981 and incorporate comments elicited. He reached out to previous critics, including Edward Luttwak and Lind. This effort led to the inclusion of the "mission orders" concept, the idea that subordinate units informed of their commander's intent could execute a campaign successfully when the fog of war prevented centralized control (Romjue 1984b, 58). It paved the way for inclusion of the center-of-gravity concept—a point of main effort, attack, interest, or vulnerability (Romjue 1984b, 59). Efforts to get Army-wide buy-in fleshed out the ideas at the core of the document, stressed the depth of the battlefield (including the importance of defending rear areas), and emphasized flexibility in approaching the Central Battle. The Clausewitzian imagery of a "shield of blows" emerged, bringing better clarity to an idea originally in Active Defense and reducing concern about a solely defensive posture (Romjue 1984b, 59). Most significantly, the ALB doctrine of 1982 required significant involvement from the Air Force to engage as it described.

The Army and the Air Force advanced ALB cooperation again when Generals Meyer and Gabriel signed a joint service-chief memorandum pledging "enhancement of joint employment of" (Meyer and Gabriel 1983) ALB. Progress accelerated under a new Army chief of staff when Gen. John Wickham reaffirmed his commitment via another biservice memo with General Gabriel. This established terms of reference for a "joint . . . force development process" capable of creating "the most combat effective, affordable joint forces necessary for airland combat operations" (Wickham and Gabriel 1984). These terms of reference and the development process they encompassed led directly to the "31 Initiatives," the clearest and most tangible evidence of joint cooperation between the two services observed throughout the ALB era.

It is worth noting here the commitment leaders in both services devoted to the effort. From the outset of ALB development, the principals of both organizations committed to a long-term relationship and negotiations. General DePuy remained involved in Army doctrine after his retirement. He, along with the former commander of the US Army in Europe, continued to foster interservice cooperation centered on ALB. Despite their shared doubt "that the effort would result in significant change" (Davis 1987, 36), they both reviewed the 1983 terms of reference that the Army and Air Force drafted. Regular TAC-TRADOC discussions started in 1973 continued unabated in this era, establishing a pattern of cooperation and commensurate trust that witnessed the release of the ALB operational concept in 1981 and continued through its incorporation in the 1982 version of FM 100–5.

Using Starry's grassroots approach, Air Force and Army leaders' effort leading to the 31 Initiatives centered on small, empowered teams who investigated the key problems impeding successful ALB implementation. After signing the joint memorandum of April 1983, they commissioned a biservice team of a few midlevel officers with extensive joint experience to undertake the intellectual heavy lifting for the promised cooperative effort (Davis 1987, 35). The Joint Force Development Group (JFDG) formed on the basis of the 1983 agreements had twelve members, six from each service, and consisted of mid-level officers chaired by two colonels (Davis 1987, 40). Lacking the constellation of high-ranking commanders DePuy assembled to write doctrine and "how-to-fight" manuals, it retained the endorsement of both service chiefs and so remained sufficiently empowered to be effective.

Though this progress suggests spontaneous, cooperative interservice momentum building, other factors were at play that dilute its validity as evidence of organically grown jointness. Calls for defense reorganization, long echoing in Washington, were gaining volume in the early 1980s. The military establishment was quite sensitive to the matter. Not only did a commission appointed by the chairman of the Joint Chiefs, Gen. David Jones, recommend several initiatives to "increase jointness" and "improve joint activities," it also acknowledged that any actions taken would "have maximum impact

only if the civilian leaders—the President, the Secretary of Defense . . . —
actively support the JCS organization and solicit and use its products," im-
pugning the Pentagon's cold shoulder for any external reforms (US House
of Representatives 1982, 702–709).

An idea that well-publicized displays of jointness could stem the building tide of reform infected the Pentagon. Even the Navy, despite a well-
documented stance of defiance toward reform efforts, decided it could make
a show of jointness for its own sake. General Gabriel and Adm. James Wat-
kins, the chief of naval operations, signed a memorandum of agreement in
September 1982 addressing interoperability between the Air Force's airborne radar platform (the AWACS) and the Navy's aviators ("Air Force and
Navy Agree to Closer Ties" 1983, 30). The repetition of the word "joint" in
the title of a memorandum of agreement between the two service secretaries, "Joint USN/USAF Efforts for Enhancement of the Joint Cooperation,"
might belie some awareness of, or even cynicism toward, the signatories' external audience (Orr and Lehman 1982). Even between the Army and the
Air Force, the effectiveness of their substantial paper agreements must be
adjudicated in light of the increased military capability they brought about
before we may draw a valid conclusion about jointness.

There were some very specific entailments to the "31 Initiatives." The
Air Force and the Army announced the entire scope of the 31 Initiatives at
a joint press conference in May 1984 (Davis 1987, 35). The body of work
had proceeded rather quickly. Following the signature of the Meyer-Gabriel
memorandum in April 1983, General Wickham (now the Army chief of staff)
and General Gabriel signed another agreement in July 1983 that pledged to
"submit a single joint package for AirLand programs needed for the attack
of enemy follow-on forces" (Davis 1987, 36). This, in one move, tied ALB,
a formal US Army doctrine, to the accepted NATO war plan for Central
Europe and pledged the Air Force and the Army to work together in the
budget arena to gain funding for the initiative. It was a significant mutual
commitment. Work on the 31 Initiatives commenced under the services' respective deputy chiefs for operations, Lt. Gens. Fred Mahaffey (Army) and
John Chain (Air Force), who in turn appointed Cols. Raoul Alcala (Army)
and Howell Estes (Air Force) to draft the aforementioned terms of reference
(TOR) for the interservice discussions.

The TOR were a significant example of joint compromise and understanding, extending a vision of a tripartite battlefield comprising "close"
(formations engaging in close contact), "rear" (behind friendly forces), and
"deep" (behind enemy forces) operations areas, with the close and deep battle areas divided into three zones based on distance. Each conceptual subdi-
vision of the battlefield envisioned a specific type of battle and an appropriate
level of command and control, with the greatest need for "integration and

synchronization of friendly air and ground elements" (Wickham and Gabriel 1984, 3) occurring in the close battle area. The TOR set the highest
priority for force development to address a possible enemy breakthrough in
the closer zones, and reemphasized a commitment to economic efficiency in
the acquisition of systems that would be able to defeat projected Warsaw Pact
capabilities in the 1990–1995 timeframe (Davis 1987, 36–39).

The TOR established the aforementioned JFDG, which was critical for
building continued trust between the services and imbuing the joint effort
with credibility—the officers composing it were both familiar with the tactics
of their respective services and experienced in working interservice matters
in the Pentagon or at the field commands from which they were selected. The
JFDG divided command, control, and force employment into eleven mission areas that it tackled in two- or three-person teams (Davis 1987, 40–41).
The protection given the JFDG by the service chiefs and supervising senior
flag officers, along with its low profile while work was underway, freed it from
the high-level scrutiny that normally accompanied the tightly controlled,
choreographed process of military and interservice coordination. Absent the
politics and parochialism of sensitive Army-Air Force matters, the working
group could make assessments based on effectiveness and cost alone (Davis
1987, 43). The 31 Initiatives adopted the same scheme of grassroots empowerment protected by senior-officer custodianship that led to ALB's original
success.

The JFDG presented its final report to the service chiefs in March 1984
and offered thirty-two recommendations. All but one of these, fusing the tactical battlefield intelligence of both services, received approval, and the "31
Initiatives" moniker was born. Wickham and Gabriel emphasized that the
report was not a terminus, but "an initial step . . . of a long-term, dynamic
process" (Davis 1987, 45), directing its presentation to the Joint Chiefs of
Staff, the secretary of defense, combatant commanders, and selected members of Congress. Having cooperated on an interservice effort at the internal
service level over many years, the service chiefs made short work of presenting their efforts to the coordinating and external relations levels of the
defense establishment. Their move reflects the belief that their approach was
appropriate for both the existing military threat and the domestic defense
politics of the era.

Somewhat predictably, the 31 Initiatives had a mixed reception. Joint force
commanders were generally in favor of the collection of ideas, but special interests within the services had objections. The Military Airlift Command did
not like the idea of giving Air Force special operations rotary-lift capability to
the Army; TAC thought that an interdiction initiative gave too much say to
Army commanders; Army missile commanders feared losing systems to the
Air Force, among others (Davis 1987, 45). Skepticism from within did not

stop the service chiefs from announcing the initiatives in triumphant tones in May. Some media reports highlighted the distrust engendered among these special interests, and critiqued the agreement for not addressing redundant capacity in the Air Force's fixed-wing CAS fleet and Army attack helicopters (Hiatt 1984, A3).

The 31 Initiatives included broad areas of cooperation and singular, standout efforts. The series of initiatives on air defense accounted for both the Air Force's expertise in airspace control and added the Air Force's voice in the design of surface-to-air missile systems, previously an exclusive domain of the Army. The proposal mandated the study of air defenses in general, allowing for the possibility of transferring the Army's missiles to the Air Force. General Wickham's approval of this proposal was surprising to many observers, but it met the Army's concern for guarding specific sectors according to their priority for the ground campaign. Conversely, the Air Force benefited from Army expertise in installation security via the initiatives centered on defense of rear areas, and was able to rely on Army assistance and training in an area for which it did not show great affinity. The first twelve initiatives reflected increasing interdependence between the services in their battle to defend Central Europe as well as compromise over issues spanning a spectrum from minor to quite contentious (Davis 1987, 47–54).

Efforts toward suppression of enemy air defenses (SEAD) reflected both a nod toward the Air Force's technical expertise in electronic warfare and the Army's desire to gain better targeting information about significant nearby threats. Though they removed the Army from the jamming business, these proposals were intended to tie the two services closely together in forming a unified targeting system populated by both tactical aircraft and ground-based weapons systems. The initiatives related to special-operations forces (SOF) and search and rescue (SAR) took a comprehensive look at the interrelated capabilities of both services' special operations units. One of the most controversial of the initiatives was the proposal to rebalance SOF capabilities related to SAR and transfer SOF rotary-wing lift responsibility to the Army. Munitions initiatives expanded the range of existing Army artillery systems, but pledged to limit the acquisition of redundant systems that could range the same targets (Davis 1987, 54–58).

Among the initiatives that would prove to have combat utility later were those devoted to specific combined-arms battlefield tactics and systems. Initiative Number Twenty established a standard for night combat proficiency that has shaped tactical-air power to the present day (Toppe 1953, foreword). Number Twenty-One solidified the concept of BAI and the new concepts of apportionment and allocation of air assets within a joint force. The procedures for BAI also promised to strengthen Army-Air Force cooperation at the command-and-control nodes critical for effective CAS and BAI employment, an area of chronic, historic weakness at the outset of conflict. Along with separate initiatives on CAS, tactical air control party (TACP) procedures, and close-support aircraft acquisition, the services took an ambitious step toward cooperation on their most consistently volatile source of interservice dispute—though criticism from within the Air Force never really faded (Rasmussen 1978, 11–13). The influence of TAC-TRADOC dialogue, ALB, and the 31 Initiatives is demonstrated with some clarity by BAI, because it is a concept that runs counter to Air Force doctrinal preferences and received pointed criticism from senior Air Force leaders from its first introduction as a joint term of art (Momyer 1981). Further initiatives about air interdiction, ground-surveillance aircraft, high-altitude, intelligence-gathering aircraft updates, and airborne reconnaissance platforms completed the weapons-system proposals. Initiative Number Thirty mandated closer cooperation on defining intratheater airlift needs, second only to CAS in earning the Air Force accusations of deliberate underperformance from the Army.

In sum, the thirty battlefield-centric initiatives of the original 31 Initiatives (Number Thirty-One was administrative in nature, but no less remarkable—a joint budgetary process for ALB programs) mandated greater interdependence of combat forces and required closer links among command-and-control organizations. The proposals were sweeping, upsetting several parochial interests within individual services, but, for an era, ending previously common charges that the services engaged in treacherous dealings with external defense organizations to scuttle one another's weapons-system programs (DePuy 1994g, 159–160). This demonstrates the effectiveness of the JFDG and the willingness of the service chiefs to tolerate turbulence in their organizations to advance AirLand Battle. But how were the results? Analysis continues by examining implementation of the 31 Initiatives and the outcomes they attained.

Our observations of the 31 Initiative's implementation and ability to effect constructive joint behaviors are positive on balance, though some of the more complex and contentious issues exhibited slower progress than the initial tranche of quickly executed initiatives. Along with the external defense-reform subtext pushing interservice cooperation, a robust process of implementation and continued attention from senior officers drove the initiatives the last mile to common practice, normally the step that proves to be most difficult in the administration of public policy (Sabatier and Mazmanian 1980, 538; Sabatier 1986, 21–22; Berman 1978, 12–13).

Evidence of cooperation visible in the public sphere did not diminish at all. Yet another joint-service agreement arrived in June 1984. It established an exchange-officer program to place Air Force and Army officers in each other's headquarters staff, demonstrating ongoing commitment to ALB concepts and the "free exchange of ideas and concepts between" staffs

(Mahaffey and Chain 1984, 1). A Joint Assessment Office followed to aid in implementing and monitoring the 31 Initiatives (Davis 1987, 66). This office created an implementation-tracking scheme and process that gained acceptance in the bureaucratic structure of the Pentagon, successfully consulting with joint combatant commanders and assimilating additional initiatives (beyond the original thirty-one) from them. The initial phase of the effort witnessed twelve initiatives reaching "closed" or "implemented" status in thirteen months (Davis 1987, 78–87).

The low-hanging fruit of initiatives that required little effort, engendered few disputes, and thus could be easily implemented ran out in September 1985, but progress did not stop. Between then and June 1986, another fourteen of the original initiatives reached closure, along with three additional initiatives added later (Davis 1987, 78–87). Reflecting the origins of ALB and the 31 Initiatives, work on initiatives related to the integration of air support into the land battle continued between TAC and TRADOC. In addition to top-down leadership and tracking of progress through a central Pentagon office, implementation also involved a collaborative element, as responsible service commands received a say about how efforts within their respective area of responsibility should proceed. While this gave the bureaucracy a chance to lobby for the status quo in some areas identified for change, it also prevented the outright rebellion against changes mandated from above.

Significantly, some planned service programs were canceled or modified as a result of the joint process, allowing more than a billion dollars to flow toward other priority requirements. A May 1986 agreement on manned-aircraft systems replaced the 1966 McConnell-Johnson agreement, representing a new equilibrium on the perennially contentious matter of duplicative air fleets. This agreement marked the first time the two services had signed a joint statement during a period of cooperation rather than in the midst of a bout of visible interservice rivalry. The most notable areas of cooperation were in new weapons systems, combat interdependence, and budgetary cooperation (Davis 1987, 80–83).

A few of the initiatives came to involve Navy and Marine Corps participation, ironically under the rubric of the internal coordination process. One example was cooperation on munitions development (Number Nineteen), but the most promising all-service initiative from the period was Number Thirty-One, which had originally specified Army-Air Force budgetary cooperation. In December 1984, the Navy and Marine Corps joined in the signing of a four-service agreement of mutual participation in service-budget activity. The comprehensive agreement was slightly less ambitious in scope than the original biservice agreement, but pledged collaboration for systems deemed essential to combined-arms fighting. At its peak, the interservice process developed to oversee implementation of the initiatives had triservice support (Air Force, Army, and Navy) for inclusion in the curriculum of their senior professional military-education schools. Participation from the organizational level that coordinates the services themselves (analysis level b) gained momentum later in the implementation phase as well, with unified commanders providing another forty-four initiatives to the assessment office, further validating its utility as a clearinghouse for joint concepts despite its status as an internal service agreement (i.e., not formally joint under the rubric of the Joint Staff) organization (Davis 1987, 83–87).

A relatively small handful of issues reflected substantial deviation from the original intent of the initiative or outright failure. The Precision Location Strike System (PLSS) fell victim to cost overruns, precluding the intent of initiative Number Fourteen to share its data with Army units, though this was achieved via different means through ground-surveillance aircraft (Futrell 1989b, 546; US House of Representatives 1981, 32–33, 345; Davis 1987, 74; Johns et al. 1987, 149–150). Rotary-wing lift for SOF proved too politically contentious, becoming part of a larger conversation about SOF in general but negating the intent of Number Seventeen. The addition of another initiative by the Army for a comprehensive CAS review brings up the question of whether the three initiatives pertaining to CAS and BAI inspired sufficient confidence, and we evaluate the question of whether the 31 Initiatives had a material effect on CAS capability later in this chapter. Special interests, operational challenges, and complexity conspired to blunt the intent of some of the 31 Initiatives, but ALB's biservice support at the highest levels of the service establishment gave it effective clout in that realm, and its implementation procedures also became accepted by the most influential figures among the services—the joint combatant commanders.

Recall that internal service and among service organizational dynamics were not the only factors in play, though. Earlier we mentioned some of the service-culture barriers to adopting sweeping change based on a new doctrine, particularly within the Air Force. Though ALB as the Army's overarching doctrine received more support than had its predecessor, criticism of its underlying ideologies continued. Attacking from the other side of the attrition-maneuver spectrum (though with less ardor than Lind), Herbert London questioned whether the deep attacks ALB envisioned were viable, given the NATO politics and a deescalatory bent (1984, 29). London's critique of ALB was gentle overall, however, and may have reflected a broader agenda. The question of an external pressure source on the interservice cooperative effort merits more attention, including a closer look at the motivation for books like London's.

Congressional pressure played a role in the ALB dynamic. Harold Winton asserted that the ALB partnership "was not foisted on the services by outside pressure for greater joint cooperation" (1996, 113), and noted that the relationship had its roots in the mid-1970s, more than a dozen years prior to the passage of the Goldwater-Nichols Act. This analysis is correct inasmuch that Congress did not order TRADOC and TAC to cooperate, nor did it explicitly guide the post-Vietnam Army-Air Force relationship. In this instance, Huntington's description of Congress existing "off to the side, an ever present threat to the symmetry and order of the military hierarchy" (1957, 259) held true. In the absence of direct, legislated influence, though, the influence of congressional and other external influences factored throughout ALB's life span. The interplay between ALB and the congressional reform caucus, which traces its roots well prior to passage of Goldwater-Nichols, is our prime example. The examination of periodic defense spending bills, both the content of the bills and the opinions revealed in related hearings, also illustrates congressional involvement.

We believe Winton's assessment that Congress had negligible impact on ALB based upon the date of passage of the Goldwater-Nichols Act does not capture the full scope of congressional influence on military affairs. James Lindsay and Randall Ripley argued that the study of defense legislation passed by Congress is an ineffective means of analyzing its influence on military matters, as it would lead one to conclude, "Congress does not appear to matter much" (1994, 8). They found that more accurate measures of congressional influence exist in the way it creates situations that drive executive response, the changes it induces in executive decision-making processes, and the way it frames defense policy issues (8–9). Here we find relevant congressional influence on ALB. Richard Davis wrote that part of the impetus for the Army and Air Force pursuing their thirty-one initiatives based on ALB was that the "public, *the Congress*, and the DoD had consistently pressured the armed services to cooperate fully and to avoid wasteful duplication" (1987, 36). The Army briefed G. William Whitehurst and Newt Gingrich, both members of the congressional defense-reform caucus, on ALB with intent to inform them about its doctrinal developments, the transition to *Army '86*, and new weaponry coming into production and deployment" (Romjue 1984b, 66). While this was no doubt a solid political strategy for the advancement of the Army's interests, a partial impetus for the display of interservice jointness might have been to convince the caucus that it did not need to take significant further action. A *New York Times* feature on ALB drew an explicit connection between the new doctrine and the caucus, asserting that "many of these changes have been urged by the military reform movement" (Shapley 1982, 262–263).

We found that internal service components (analysis level a) and coordinating groups among the services (analysis level b) were aware of a perceived jointness problem on Capitol Hill and that they wanted to shape the debate about potential reform. A 1984 memorandum circulated among the

Joint Staff calling for better, faster advice to the national command authority and deemphasizing service-specific interests in that counsel (Joint Chiefs of Staff 1984b; Donnithorne 2013, 343). Donnithorne cited an uncanny proximity between the date DoD provided Congress its own draft legislation for defense reform and the memorandum of understanding that led to the Army-Air Force cooperation on initiatives under the rubric of ALB—both emerged in April 1983 (2013, 344–345). Col. Raoul Alcala (2013) confirmed that the services wanted to gain maneuvering space with respect to defense reform and put their own stamp on impending changes.

James Locher gave an even clearer picture of General Wickham's motivation. Despite a genteel reputation, Locher accounted an uncharacteristic response from the general, characterized by "huffing and puffing," in which he flung vitriol upon the draft legislation at a meeting with Senators Barry Goldwater and Sam Nunn in February 1984. Wickham delivered a "lengthy harangue" that emboldened some of the other service chiefs present to display behavior bordering on disrespect, in turn eliciting a stern response from the legislators and solidifying their convictions that DoD required major reforms (Locher 2002, 4–8). While we cannot say congressional pressure alone drove Wickham and Gabriel's continuation of cooperative efforts underway when they became service chiefs, Wickham's clear opposition to reform suggests it was a factor.

Another example of pre-Goldwater-Nichols congressional influence on ALB arrived in 1984. Senator John Tower, generally considered to oppose efforts toward defense reform, then chaired the Senate Armed Services Committee. His introduction to London's book on ALB seems innocuous; he promoted a holistic consideration of defense issues, focusing on training and doctrine instead of only the major weapons systems that dominated the defense conversation of the day (London 1984, ix–xi). In the subtext of Tower's opposition to defense reform, however, ALB served as proxy to prove his belief in the defense establishment's capability for internal reform. Frank Barnett picked up Tower's theme and gave it a finer point, citing the paucity of attention that would-be reformers had "given to the doctrinal and training innovations which have been instituted by the military themselves" (London 1984, vi–viii). Tower's embrace of ALB suggested he would support the status quo and not push for further reforms.

An absence of debate can reveal as much about what legislators deem important as those topics that dominate the record. Hearings for defense-spending bills through the mid-1980s made frequent, uncritical reference to ALB as an organizing concept (see US House of Representatives 1984, 516–768). Later in the decade, however, lawmakers put explicit language in legislation demanding that ALB and other related concepts for the European battle be reconsidered in light of Soviet and Warsaw-Pact force reductions (US Congress 1989).

After appearing regularly starting in 1984 in the *Congressional Record*, the last reference to ALB happened in 1992, and that was a mere reference in tribute to a retiring Army service chief (Nunn 1991, 11998). Despite a residual use of the "Airland" descriptor for congressional committees concerned with weapons-system acquisition, the last material statement about ALB *doctrine* appeared in 1989, the aforementioned direction to study how it must change in light of the enemy's unilateral stand-down (US House of Representatives 1997, 11840). By 1992, congressional language reflects concern with specific weapons systems rather than the organizational structure of US military forces or doctrine. After Desert Storm, with post–Cold War defense-budget cuts impending, representatives made mention of the Apache attack helicopter, the A-10 attack (CAS) aircraft, and JSTARS surveillance aircraft (DeConcini 1992, S5596). This marks a return to familiar territory, with "marquee items . . . taking up most of the debate about defense acquisition" (Sapolsky, Gholz, and Talmadge 2009, 65).

ALB's congressional subtext shows that there was some bureaucratic resistance behind the scenes, but that the net outcome was to push two services toward more jointness than they otherwise might have pursued. Perhaps two services became *more* joint in trying to *resist* the congressional concept of jointness that might otherwise be forced upon them. And what of the Navy in this era? As Donnithorne summarized it, "the Navy's overall posture . . . remained steadfastly opposed to reform" (2013, 346). Navy Secretary John Lehman elected to use the DoD bureaucracy as a scapegoat for military inefficiency, and entertained no suggestion of reforms (Locher 2002, 172–176; "Navy Secretary Strafes Bureaucrats" 1983). The overall response served as a great example of the Navy having earned its "Defiant" moniker on Capitol Hill and shows how a single congressional input can have disparate effects on the services. For the Navy, it hardened a suspicion of centralized control into yet another minor open rebellion, but without meaningful consequences for the Navy.

While there is little doubt the Army and Air Force sought to project the appearance of jointness in the lead-in to Goldwater-Nichols for Congress's benefit, spotty implementation of their internal program does not quite match the shiny public-relations effort. Two explanations are apparent. On one hand, the services could have been behaving cynically, flaunting cooperation to thwart the reform caucus. On the other, initial reforms may have been sincere, but normal bureaucratic politics choked out leadership to become the dominant dynamic. The former explanation makes ALB an elaborate example of deliberate preemptive shirking by a subordinate agent to fight Congress in its control of the US military—an agency-theory horror story. We find it unlikely that false motivation sustained an effort for over a decade, and more probable that the ALB initiatives were both productive efforts toward jointness as well as remarkable examples of cooperation in their own right. One pass of Occam's razor leaves us to conclude that the bureaucratic politics explanation is more plausible. Fortunately, that view is more palatable from a US viewpoint of civil-military relations as well.

Academic commentary and its impact on ALB merit mention because it was in view throughout the doctrinal development process; academics who rotate between an executive administration's advisory positions and Washington think tanks remain a prominent fixture in any discussion of military jointness. John Mearsheimer criticized an ALB-like concept he labeled "Mobile Defense" (1982, 107–109) and defended Active Defense in the early 1980s, arguing that Active Defense was a reasonable strategy to defeat the Warsaw Pact, but more so that the former doctrine was too abstract to be successfully explained to the Allies or executed in the event of Soviet aggression. London argued that the whole need for a new strategy was because Soviet nuclear parity after the 1970s exploded the myth of an "ambiguous balance" and revealed "clear Soviet superiority" (1984, 27).

Thus the role of academic critique shows itself to be of mixed utility; it has a central role in describing, if not shaping, questions about joint cooperation. Lind's critique of Active Defense did much to usher in its replacement. London's description of ALB, if indecisive as a single salvo in a broader attempt to stop defense reform, further entrenched the concept in the ongoing external conversation about defense policy. Mearsheimer's critique and others like it forced ALB to adopt some of NATO's defense strategy, muting some of the offensive flavors that enthusiasts embraced. ALB walked a fine line between its own ideals of "offense as defense" that satisfied American complaints about Active Defense and a static, prepositioned resistance dictated by European politics. It managed to blend questionable ideas with accepted norms to a degree that allowed it to satisfice the demands of both audiences. Would-be practitioners of jointness may take a lesson about academia's role in tearing down or building up ideas about defense around which effective coalescence can occur.

We have to acknowledge that the defense budget, at least for a portion of ALB's tenure, was an outlier on more than one front. One of the most impressive facets of ALB is the fact that it succeeded where we have observed that joint cooperation is most difficult to nurture—in the bureaucratic and budgetary realm. Rarely do the services' *figurative* guns come out more quickly than when budget share is on the line. Then again, the defense environment during ALB's period of influence was characterized by a unique lack of budgetary pressure. From 1980 to 1987, the overall defense budget grew by approximately 40 percent (Schneider and Merle 2004, E1). During the same period, the number of humans in the US military population grew by fewer than 5 percent, leaving the bulk of spending for additional weapons system acquisition, modernization, and training (Rotsker 2013). Though subsequent defense secretaries attempted to increase military populations, various political strictures thwarted their efforts (McFadden 2014, A18; Gelb 1975; Kaplan 2008). The effect was to keep the size of the military population at post-Vietnam levels.

Strategic security questions dominated the 1970s defense debate after Vietnam. Missile deployments, bomber capability, and the nuclear triad received more attention than conventional forces. The Carter administration focused on cutting the defense budget through its first two years before making a slight reversal of the trend in its waning months. Against an oversized Soviet force, these constraints required *creativity of doctrine and warfighting technique* rather than reverting to the American tendency to match force through numbers and brute firepower. Though a reasonable person could have expected the size of the military to grow during the Reagan-Weinberger years, a tacit restriction on the number of military personnel remained largely in place; growth was only modest. Accelerated spending on missiles, aircraft, and the Strategic Defense Initiative substituted for more personnel, reflecting a conscious political effort to stimulate the defense industry. Weinberger's restrictive standard for committing US forces also tended to limit the demand for more troops (Yoshitani 2011, xi–xiv).

Though ALB was received favorably in most parts of the defense establishment, it did not silence all naysayers. The 1982 version of FM 100–5, with ALB central, did not put to rest all criticism of Army doctrine. Robert Leonhard charged the Army with enduring devotion to attrition warfare, giving air power a failure-prone "exaggerated role" in the doctrine, and relying too much on flawed simulations of Soviet tactics in wargaming computer models (Leonhard 1994, 4, 61, 135–155). A new edition of FM 100–5 duly appeared in 1986, but it was still unmistakably ALB doctrine throughout. The funeral dirge for ALB would not come primarily from internal service-doctrine critics, but rather a host of other changing outside circumstances. The next section discusses the practical tests ALB underwent while it was the dominant idea in US defense policy. We exhibit its successes and failures to expose conditions and mechanisms that permitted joint cooperation.

Testing AirLand Battle

Evidence of Success

Regardless of how one assesses the severity of disconnects between the rhetoric of ALB's operational concept and the interservice cooperative agreements put in place by the 31 Initiatives, the effect on jointness did not stop in the mid-1980s. Beyond the services' efforts to implement a discrete plan, other lasting joint advances occurred.

A Return to Operational Art

Shimon Naveh (2004, xvi, 1–3, 287–292) argued that operational art as a discipline had been lacking in the United States since World War II and that its return via ALB marked a true revolution in military thinking. If imitation is the sincerest form of flattery, the Soviet Union paid ALB a compliment soon after it was unveiled by mirroring the doctrine in its own "Operational Maneuver Groups" designed for deep thrusts against NATO's rear support and supply areas (Middleton 1983, 12). Consideration of operational art continues to dominate joint and service doctrine discussions today, making it a welcome intellectual contribution of the ALB era.

ALFA/ALSA as an Enduring Joint Organization

The TAC-TRADOC dialogue that General DePuy initiated in 1973 received organizational manifestation as the Air-Land Force Application (ALFA) Center in July 1975. ALFA continued to host many of the important interservice discussions about ALB during its era and did not disappear after ALB faded from view. As previously mentioned, ALFA became ALSA in August 1992, incorporating the Navy and Marine Corps to make it an all-service joint cooperative organization. ALSA remains a source of joint doctrine as well as a clearinghouse for multiservice techniques, tactics, and procedures manuals for the employment of air power functions. Given the minor importance services put on joint doctrine relative to service doctrine, and given the wide acceptance of the joint tactics that ALSA releases within the communities it influences, the latter contribution is arguably more important. Either way, this artifact of ALB collaboration has endured and remains a viable contributor to jointness.

In this context, geographic proximity's effect on jointness merits mention. TRADOC, at Fort Eustis, Virginia, is just miles away from TAC headquarters (now Air Combat Command) at Langley Air Force Base. If not inevitable, a TAC-TRADOC partnership was more easily built. By contrast, the nominal home for Air Force–wide doctrine is located at Maxwell Air Force Base, near Montgomery, Alabama, home to the service's educational facilities for active-duty personnel. The Air Corps Tactical School, responsible for developing the strategic bombing doctrine that held so much sway during World War II, was located at Maxwell; legacies like this (and the golden handcuffs of congressional pork) remain the impetus for retaining the service's doctrine center there today. The geographic separation seems to be reflected in an intellectual separation from Washington politics and the larger Air Force, putting a barrier between the ideas Maxwell germinates and the audiences for whom those ideas are intended. There is little surprise that the Army found it easier to work with a subgroup of the Air Force located nearby its more central doctrinal center than with the Air Force as a whole, and it was not solely because Air Force doctrine beggars change.

APPORTIONMENT AND ALLOCATION: A COMMON VOCABULARY FOR AIR OPERATIONS

One of the most useful practical successes from ALB was the arrival by the services at a common definition of the terms "apportionment" and "allocation," which as terms of art still describe the distribution of limited aviation assets. Apportionment is the "determination and assignment of the *total* expected effort expressed in a percentage or priority" (Joint Chiefs of Staff 2014b, 9), a decision that falls to a joint force commander in current US doctrine.[5] Allocation is turning the apportionment decision into a tangible number of sorties for execution, which is a role handled by the air component commander (Joint Chiefs of Staff 2014b, 15). Thus, terminology originally established over a topic that is controversial every time the services go to war together has endured for well over a quarter-century without significant dispute.

The central joint tenet to arise from ALB was recognition that, to minimize duplication and maximize efficiency, both the Army and the Air Force needed to participate in target nomination, prioritization, and execution in a way that used both services' inputs and resources. This led to a common vocabulary and concept for making these decisions in a structured combat environment led by a joint force commander. It also led to in-depth discussions about the joint command-and-control organizations and systems required to affect the outcomes these terms represented. As later discussion will reveal, the progress here was a mixed result, one in which the participating services exhibited a dearth of confidence.

OPERATION DESERT STORM DEEP BATTLE

Since there was no Soviet invasion of Central Europe in the 1980s (Soviet incursions into the sovereign affairs of the Warsaw Pact aside), ALB never received a test against the first-order threat that inspired it. It does, however, receive broad credit for establishing the US military on firm footing for its quick, lopsided victory over Saddam Hussein's forces in the first Gulf War. The descriptions of the strategy for executing the first war against Iraq from Gen. Colin Powell and Secretary of Defense Dick Cheney referred to ALB's rapid, armored thrusts (Gordon 1991, 1). Conrad Crane offered Operation Desert Storm as "evidence of its success" (Langer 2011, B8). Rick Atkinson's widely cited history paints ALB as "best suited to armored warfare in the open desert," as no other type of battlefield terrain "allowed a commander to look deeper, move quicker, or seize initiative faster" (1993, 253). It is clear that the scheme of operations for Desert Storm borrowed from a cognitive framework informed by thinking about ALB. Gen. Norman Schwarzkopf's battle plan involved deep maneuver and the use of air power for deep strike well prior to the beginning of the ground war. He avoided a battle of attrition in favor of one that moved rapidly to outflank and cut off retreat options for the Iraqi Army entrenched in Kuwait. The battle plan was full of arrangements to "delay, disrupt, or destroy the enemy's potential before it [could] be used against friendly forces" (Cardwell 1992, 101), including the Scud hunting in the western desert and efforts to ensure that reinforcements could not flow south from their positions around Baghdad.

Benjamin Lambeth summarized Desert Storm's January 1991 Battle of Khafji as an air-power attack against a second column of Iraqi armor that decimated the column before it could reach its objective or make contact with allied forces (Lambeth 2014, 61–62). JSTARS aircraft allowed the United States and its allies to see the column movement deep into enemy territory, and precision munitions allowed for lethal engagement, with the air-component forces diverting from their preplanned missions to destroy over six hundred Iraqi vehicles and artillery pieces. This anecdote offers a clear illustration of ALB's more salient points. A deep picture of the battlefield, including one that took into account enemy follow-on force movements over time, allowed the flexible targeting of enemy forces as they flowed toward the front, but before they could mass to become an overwhelming force. It was a textbook application of ALB, albeit against a much smaller and less sophisticated enemy than the doctrine originally envisioned.

The circumstances of the Iraqi-US engagement prevent complete validation of all the praise that has been heaped upon ALB's role in bringing about a decisive win in 1991. Robert Scales's account claimed that the Army since 1973 had become a new organization, one that emphasized defeating an enemy first by intellectual capability rather than simply accumulating more firepower (Scales 1997, 36n). Though the Gulf War campaign employed operational-art and maneuver-warfare concepts aided by technological and doctrinal advances, the huge military buildup that occurred in the Persian Gulf before the start of military operations and the ease of movement in open desert terrain serve to temper any claims that the Gulf War marked a complete abandonment of the American predilection for advancing on an enemy with the firepower to simply overwhelm the opposing force (Park 2012, ii).

Evidence of Failure

INTERSERVICE RIVALRY IN WARTIME

The US military intervention following the Iraqi invasion of Kuwait in late 1990 offered a chance to test almost every aspect of ALB's operational concepts, but it first offered a chance to see if the idealism reflected in optimistic joint proclamations could endure in actual combat. The Air Force had an opportunity, via a preinvasion air campaign to soften Iraqi resistance, to lead off with a show of deep interdiction capability. In some ways, the campaign before the start of ground hostilities signaled a departure by the Air Force from its commitment to joint action. Conceived by Air Force officers who believed in an independent-action theory of air warfare, the Desert Storm air plan would have, had it been executed as offered, attempted to destabilize the Iraqi government, defeat its fielded forces, and destroy any remaining resistance to coalition demands, almost completely by itself. The high expectations of air-power capabilities from its proponents—and interservice resentment of the lofty hopes inspired—appeared even before the war started. Secretary of Defense Richard Cheney removed Gen. Michael Dugan from his post as Air Force chief of staff after a series of comments to media traveling with the general in Saudi Arabia (Schmitt 1990).

Dugan's insistence that air power was "the only answer available to our country in the circumstance" recollected the brashness of nonjoint independent-action theories about air power put forward in the 1930s that so rankled Army and Navy leadership (Broder 1990). While most media sources attribute the firing to the inappropriateness of Dugan's speculation on classified war plans, the deleterious effects his remarks had on joint military harmony and their implied rejection of combined-arms interdependence seemed to be a factor as well. Secretary Cheney remarked, "Statements . . . to the effect that the Army and the Marines would provide for diversionary activities while basically the Air Force carried the ball were inappropriate" (Schmitt 1990). Cheney's comment represented the value the external defense establishment places on mutual respect among the services.

CAS EXECUTION IN DESERT STORM

Despite Dugan's display of air-power chauvinism, the Air Force had a good showing in Desert Storm. The lesson offered about CAS is by negation rather than observation, though. The combat air power employed in Desert Storm was primarily interdiction and BAI, not CAS. The Air Force, by an agreement worked out with Gen. Norman Schwarzkopf, performed only "emergency CAS" within five kilometers of Army troops, meaning it was a rare practice. Army helicopters assumed responsibility for CAS inside this range. Reported CAS sorties constituted 32 percent of ground-attack sorties, but since most of these occurred without visual coordination by a ground or forward air controller, they did not meet the doctrinal definition of CAS other than in the effects they provided (Joint Chiefs of Staff 1984a, 35). The inability and unwillingness to attempt a serious CAS effort revealed a major execution gap for a doctrine claimed to enable rapid maneuver of forces against a numerically superior enemy by relying on a well-oiled CAS delivery system. A "reaffirmation" of the importance of CAS was Number Twenty-Four of the 31 Initiatives, designed to address "the traditional distrust the two services felt toward one another on the issue" (Davis 1987, 60). If the Army and Air Force did not have enough confidence to execute this type of tactics against an inferior and smaller Iraqi force, though, one can imagine that it portended unfavorable outcomes should its use have been required against the Warsaw Pact.

Besides the Army-Air Force coordination shortfalls, differences in the Air Force's centralized command-and-control system vis-à-vis the Marine Corps's organic system came into sharp relief again in Desert Storm. Marine message traffic referred to General Horner as merely the "joint force air coordinator," snubbing his given title of Joint Forces Air Component Commander (JFACC). Paralleling this passive-aggressive shirking of joint component authority, the Marines' contribution to the overall air war through the Tactical Air Control System (TACS) dwindled from 50 percent of their total capacity on the first day to "running an independent air war" at the end of the conflict (Horwood 2006, 185). Horner's system of "push CAS" he adopted for the air assets under his control reflected a Marine Corps preference to allocate aircraft at regular intervals to areas where ground forces anticipated enemy contact (Clancy and Horner 1999, 244). Very few CAS sorties flew in support of the ground forces to which they were first allocated; a rapid ground advance after a heavy bombardment of the enemy forces created little need for "pure" CAS (Costello 1997, 31). Unwillingness by Schwarzkopf to force the Army and Air Force to coordinate in any meaningful detail concerning the fire support coordination line (FSCL) eliminated the chances that true CAS would occur. It also served as a reminder that as joint as ALB may have been, it was by no means compelling across *all* of the services.

Desert Storm revealed that ALB left significant operational challenges in place despite its ability to facilitate Army-Air Force cooperation in the 1970s and 1980s. Debates over the placement of the FSCL, in light of long-range artillery and attack helicopters pursuing the deep attack methodologies of ALB, made the overlap between Army and Air Force areas of responsibility even larger. While giving commanders more insight over a wider area of terrain as intended, ALB also seemed to have given them more to fight over with their joint counterparts. With no commensurate agreement about how commanders would orchestrate these operations, the likelihood of friendly fire and other combined-arms failures was high (Horwood 2006, 185). In Desert Storm, deliberate operational decisions kept those risky areas from being tested. In effect, Schwarzkopf agreed to plans that eliminated potential Army-Air Force contentions about CAS by ensuring that the services would not have to work together in that arena.

Theoretical Explanations of Jointness

This section hearkens back to the questions raised in chapter 2 by individual areas of theory that might apply to jointness. This section begins to tie the

Table 3.1: Observed Joint Accomplishments of AirLand Battle

	Summary of AirLand Battle Joint Accomplishments	
Area	*Tangible Improvements*	*Enabling (Inhibiting) Actions*
Institutional Jointness	Biservice helping behaviors over more than a decade	Credible Warsaw Pact threat; fear of failure of Active Defense
	Doctrine & organizational concept w/ biservice impact	Congressional demands for jointness; expanding early-1980s defense budget; Army, AF "moving on" post–Vietnam
	Adopted as de facto DoD, US national strategy for NATO	DoD and Joint Staff required credible internal plan for Europe
	Joint budgetary cooperation (AF-Army Initiative #31)	Trust & habitual cooperation over time: geographic proximity of TRADOC & TAC; leaders who established and reinforced routine joint working relationships; well-publicized, senior leader-endorsed terms of reference & mutual agree-ments; top-down & collaborative approaches; empowered mid–level officers
	ALFA/ALSA became enduring multiservice tactics organization	
	Elicited response (not adoption) by Navy and Marine Corps	
		Air Force's immutable doctrine; Navy's resistance to defense reform
		NATO unease over offensive strategies
Operational Jointness	Viable Cold War doctrine & war plan	Met needs of endogenous, meso-org., and exogenous defense orgs.
	Increased geographic and time depth of operations	Creative/visionary senior leadership
		Evidence from contemporary conflicts
	Battle-tested org. scheme for Gulf War	Joint training at national training centers (incomplete preparation)
	Revived "operational art"	Army's doctrinal culture & interplay with academic community
	"31 Initiatives" accomplishments: Air defenses (#1–5, 12); Rear area defenses (#6–11); SEAD[1] (#13–15); Special Operations (#16, 17); Munitions (#18, 19); Night operations (#20); Battlefield Air Interdiction (#21); Joint Target Assessment (#22); Threat Interdiction (#23); CAS/TACP/CAS aircraft (#24–26); JSTARS (#27); U2/TR-1 & Tac Recce (#28, 29); Intra-theater airlift (#30); Intelligence sharing (#32)	Trust & interdependence: AF airspace expertise; Army ground defense expertise; allocation/apportionment compromise
		Lack of confidence in Desert Storm
		Compromise/incomplete OT&E
		Technology
		Complexity, perceived difficulty
		Hubris, insecurity
	Joint ideals not reflected in combat	

1. SEAD stands for "suppression of enemy air defenses" (Joint Chiefs of Staff 1995).

Table 3.2: Theoretical Observations on AirLand Battle Joint Cooperation

	AirLand Battle Observations
Theory	Observed Outcomes for Institutional and Operational Jointness
Public Goods	No deliberate free-riding by any service; an appeal to purpose sufficed to create cooperation and a sense of trust
	Perceived clear threat and strategy led all services to adopt ALB's operational concept
	Operational jointness pursued in good faith for duration of threat
Organizations	Threat response drove collaboration and systems development; overcame bureaucratic politics
	Social psychology enabled by senior military leaders overcame common bureaucratic barriers to trust
	TAC grew influence within the AF; coalescing interests with Army and exogenous defense establishment provided credible deterrent
	TAC/TRADOC/Pentagon proximity facilitated extensive cooperation
Crisis Cooperation	Enduring perception of urgency from Warsaw Pact conventional military threat provided impetus for cooperation
	"Unclear/long-term" crisis provided trusting, cooperative atmosphere; many helping behaviors observed
Professions	Operations envisioned for ALB minimized conflicting interests in AF and Army
	Interdependence in operational concept encouraged cooperation, but size of perceived threat did not create overlapping capabilities
	TAC's standing enhanced within AF and in larger DoD by its role in ALB
Agency	Little evidence of implementation slack for ALB
	Good-faith response to threat established implicitly and explicitly by exogenous defense establishment and other security analysts
	Primary sources for this investigation are unlikely to uncover evidence relevant to agency theory
Military Innovation	Military establishment demonstrated endogenous innovation in creating tactics & technology to enable the ALB operational concept
	Top-down & consensus-building methods both used in the development of ALB
	Cold War was a hybrid of wartime & peacetime conditions
	Lack of true test for C2 and conceptual integration prevented complete realization of some of the doctrine's most critical components

continued

AirLand Battle Observations

Theory	Observed Outcomes for Institutional and Operational Jointness
Civil-Military Relations	Congressional dissatisfaction with military's ability to conduct joint operations served as a subtext for ALB
	Ongoing defense reform caucus a possible motivation for services to exhibit enthusiasm for jointly cooperative endeavors likeALB
	To the degree ALB is a demonstration to Congress, it reflects a traditional desire for objective civilian control, but the process was subjective in nature
Service Cultures	ALB provided opportunities for both services to operate within their preferred strategic culture
	ALB offered a vision of an "unconstrained" battlefield that cleared service palates after Vietnam, though a shooting war did not test the reality of that perception
Defense Department & Joint Staff Structures	ALB served the interests and met the needs of the meso-organizational and lower exogenous defense establishment
	Joint Staff and DoD adopted the language & premises of ALB; influence as unifying concept spread throughout defense establishment
	DoD adopted and amplified Army's call for more CAS aircraft and capability
Other Exogenous Factors	Congressional interest in defense reform played a unique role in giving ALB multilevel appeal
	Congressional influence on cooperation is evident on CAS dispute and for doctrinal clarity

historical process-tracing narrative back to the foundational theories, offering a summary of ideas that are more fully developed in chapter 7. The overall structure of the section follows the pattern established in chapter 2, beginning with general organizational theory and proceeding to military-specific theory. When applicable, it identifies mechanisms as corresponding to the three identifiable levels of hierarchy in the defense establishment in which the services exist. Table 3.2 lists a summary of relevant observations.

General Theories of Organizational Interaction

Public Goods

Do the services act as free riders in the production of national security? Do military organizations need to be goaded through public pressure to reach jointness or will they seek it with sufficient appeal to purpose? Which better encourages jointness, social pressure or an appeal to purpose?

Neither free-riding nor a particular need for small-group pressure in co-ercing one branch of the military to contribute to solving a national security problem seems operative here. Responding to a first-order threat of Soviet conventional power in Central Europe, a concept of operations for meeting that perceived danger arose. If the Army as an entire organization did the earliest, most thorough job of defining that problem, it was probably because it had no other significant role—such as the Air Force's concentration on maintaining a sizable arsenal of strategic nuclear weapons on alert or the Navy's consistent pursuit of full-time power projection—to distract from moving on to a new definition of its role in a postcontainment, post-Vietnam security environment. The Army's effort to define its force structure in terms of the Warsaw Pact did not meet with charges of guile, though. On the contrary, the Army followed the lead of the external defense establishment and the community of Western security analysts. The Air Force, particularly TAC, was happy to follow the Army's lead and accept a role that appealed to its desire to be of central importance in the strategy for the victory of NATO against Moscow and its satellites.

ORGANIZATIONS

How do threats, bureaucratic politics, and political maneuvering influence jointness? How do service subgroup interests advance or inhibit joint cooperation? Do overlapping capabilities advance or threaten jointness?

The three distinct types of organizational behavior are all in view in the ALB case. The first-order threat of the Soviet Union played a substantial role in the dynamic. Also notable were the abilities of senior military leaders to triumph against bureaucratic inertia by advancing bold visions based on those threats. While theories of bureaucratic politics generally describe any resistance to outside influence and are more appropriate for describing *resistance* to an external policy change like Goldwater-Nichols, the service behaviors of the Air Force and Army also reflect the politics of subgroup coalescence. As Donnithorne noted, baseline organizational theory did not provide satisfactory explanatory power with respect to Goldwater-Nichols, as it would predict unified service opposition to the law, when in the event the actual responses among the services were nuanced (Donnithorne 2013, 331). Later analysis touches on the congressional influence on ALB.

CRISIS COOPERATION

Do military services make decisions about jointness in a context of crisis? How does the perceived urgency of a dilemma affect the decision and cooperation mechanisms?

The Cold War, though it unfolded over decades, meets all facets of the definition of "crisis" put forward by academia. The unpredictable behavior of leaders of the Soviet Union, along with its undisputed military might, made the Warsaw Pact and the specter of European aggression an existential threat to all of NATO. As mentioned, defense leaders exploited this aspect of the crisis to advance their ideas about ALB. At the same time, there was a certain comfort that arose from the constancy of having the Soviet Union as the sole opposing superpower. Aside from the scare of a nuclear exchange that arose sporadically through the period, the staring contest made for stability, however tense it may have been. As a long-term crisis with conditions that did not change rapidly and came to be well understood, the conditions that spawned ALB constituted a promising environment for interorganizational cooperation. Empirical research has demonstrated that organizations are more likely to adopt crisis-coping strategies that signal trustworthiness to other organizations when there is less uncertainty about the definition of and preferred response to a given crisis (Svedin 2009, 120).

Professions

Do overlapping service capabilities advance or hinder jointness? Do subclasses vying for recognition in their respective military services offer a mechanism for joint cooperation?

Per Abbott's discussion of professional competition, the need to compete is strongest when there is uncertainty about what different groups do, where their competence lies, and what distinct contribution they would make. ALB diminished the need to engage in this type of competitive behavior because the warfare it envisioned assigned distinct roles to both the Air Force and the Army. The "deep battlefield," with NATO forces arrayed against a numerically superior and aggressive enemy, offered both services plenty of opportunities to engage the unconstrained, all-out warfare for which both yearned in the wake of Vietnam. ALB made Central Europe "a place where the Air Force could do, doctrinally, absolutely what it wanted, as could the Army" (Johnson 2014). The Army knew it would require deep airstrikes to disable the follow-on forces that the Warsaw Pact could generate. But it did not fear for its relevance to the overall battle, because the clash of armor and infantry at the front would be significant as well. With disputes over air mobility and CAS solved (albeit on paper) neither service felt a need to defend its position in the defense pantheon. If ALB inspired insecurity, it may have been on the part of the Navy and Marine Corps, which had to scramble to reposition their contributions in the context of ALB; this dynamic offers a partial explanation for their participation in the 31 Initiatives (Johnson 2014).

TAC, a rising subgroup within the Air Force that was in this era taking over the monarchic leadership role previously held by the strategic-bomber community, was looking for another interest group in the defense establishment with which it could combine resources. It found it in TRADOC and the larger Army, whose attention was primarily focused on the type of war to which TAC could make the most significant contribution. With ALB, the

Air Force and the Army were making an argument to their political masters in the external defense establishment about the quality of their expert
labor. They were, borrowing Andrew Abbott's formulation of professional
dialogue, arguing that their recognition of and answer for the challenge of
defeating the Soviet threat merited validation in the form of extra resources
(Abbott 1988, 20, 69–77).

AGENCY

Does the large number of principal-agent relationships evident in the defense establishment advance or threaten jointness?

We did not find apparent implementation slack, shirking of principal-
assigned duties, or other maladies suggested by agency theory that were relevant to the development of ALB. To be sure, there are agency dynamics at
play anytime the DoD is going through organizational changes. Donnithorne
exposed the extent of those in applying agency theory to his model of service
response to the Goldwater-Nichols Act, which was contemporary to ALB
(Donnithorne 2013, 323–326). However, since ALB was an internal service
effort to innovate and solve a pressing security problem, it does not exhibit
the same top-down policy implementation characteristics that make agency
considerations relevant. Though the service coordinating level (level b) and
external relations level of analysis and organization (level c) came to adopt
the cognitive framework ALB provided and even used them for their own
purposes, the initiative itself remained a development of the services (level
a) and was executed according to the vision and leadership of its participants,
not Congress or some other external body.

Military-Specific Theories and Discussion

MILITARY INNOVATION

Do the sources of military innovation advance or hinder jointness? Do crisis
conditions advance or hinder jointness?

On its face, ALB is an internal service innovation. To the degree that the
doctrine was a hybrid of peacetime (with no shooting war taking place), wartime (the specter of the Cold War and uncertainty about Soviet plans loom-
ing), and technological change (sensing across the deep battlefield required
advanced technology), it exhibits elements of each of the traits outlined by
Stephen Rosen in his baseline theory: namely, the Army and the Air Force
evaluated "the future character of war" (peacetime); they developed "new
measures of strategic effectiveness, effective intelligence collection, and an
organization able to implement the innovation within the relatively short
time of the war's duration" (wartime); and they developed "strategies for
managing uncertainty" (1991, 52) (i.e., technological change). The official

Air Force history of the 31 Initiatives frames that part of ALB explicitly as a case of internal military innovation and leading organizational change (Davis 1987, 1). A plausible subtext of wartime innovation arose from the sense of urgency spawned by the unfavorable conventional force ratios NATO perceived itself facing in Central Europe.

Also relevant to the success of ALB innovation is the leadership atmosphere under which it occurred. "Involved senior leadership" best describes the approach the Army took toward developing the doctrine and gaining a sister service's participation and support. Though ALB was largely the intellectual output of an effort overseen by General Starry, General DePuy's early effort to reinvent the Army's main battle doctrine created an intellectual environment conducive to change, and his earnest pursuit of relations with TAC created a long-term, trusting relationship that was able to persevere over several leadership changes. Finally, the empowerment of small groups of mid-level officers represented by the JFDG epitomized in microcosm Rosen's recognition that senior leaders need to provide top cover for innovative ideas (Rosen 1991, 251). Though the implementation group did not attain a specialized promotion system per se, the demonstration that their ideas would hold real weight in the development of interservice operational concepts overcame initial skepticism on the part of its members and provided an equivalent mechanism (Davis 1987, 42).

CIVIL-MILITARY RELATIONS

Which leads to better joint cooperation, civilian control of the military via objective means or control via subjective means?

In advancing ALB, the Army and the Air Force strove to show trust in each other, to demonstrate an ability to cooperate on combined-arms efforts, and to prove to an external audience that they merited objective responsibility for national defense. Ironically, in responding to the concerns of security analysts outside the military who perceived the Warsaw Pact countries as the next big security issue, the overall effort reflected some degree of *subjective* control, with the military reading and responding to external security preferences. This characterization of the services' actions could be chilling to those who think Huntington's formulation of civil-military affairs is superior to others. However, concern diminishes with the realization that military-intelligence estimates in large part informed the civilian opinion, and a significant military mismatch did not exist elsewhere in the world. In short, the same conclusion would have accrued whether pursued by objective or subjective control.

Where the effectiveness of subjective control does seem to make an important contribution is the subtext of congressional pressure for defense reform that characterized the early 1980s. If ALB was a good idea because it made the services fight together with greater efficiency against a potent enemy, it was an even better idea in the political context of the day. To use this type of subjective control to bring about jointness can have a dual nature, however, as exhibited by the Navy and Marine Corps responses during Goldwater-Nichols deliberations. Where two services saw a means to position themselves for better negotiating positions by exhibiting cooperation and jointness, the remaining two adopted a position of defiance, hence remaining largely separated from ALB development.

SERVICE CULTURES

How do services' dominant cultures advance or hinder joint cooperation?

Though Donnithorne predicted and found support inside the Army for Goldwater-Nichols reforms based on the service's culture of selfless compliance, he also showed that the service attempted to improve the quality of proposed reform legislation by influencing the details. A cooperative institution, demonstrating its competence against the nation's preeminent threat through an effective ALB doctrine, was a predictable image for the Army to adopt. It fit its own perception of its values while providing a means under which it could lobby DoD and Congress without violating its institutional ethic. By contrast, the Navy viscerally opposed Goldwater-Nichols, with its most severe reactions reserved for policy changes that created a single military advisor to the president, thus limiting the direct access of the Navy (Donnithorne 2013, 317–318). As testimony before Congress demonstrated, the Navy and Marines adopted defiant opposition to the measure, which explains in part why they made only token gestures of additional jointness during the era before the act's passage in 1986 (Hayward 1982, 99, 101, 201, 246, 252–253; Kyle and Schemmer 1982, 61–67; Donnithorne 2013, 323–326).

Among the most culturally illuminating aspect of this case study were the distinct views of doctrine the services took. As a unifying concept, ALB had to be presented to the Air Force as something other than a doctrine for it to take hold. The Army rather readily adopts new and novel doctrinal concepts that fit its current problem set, but the Air Force, wed as it is to immovable and oft-unspoken doctrinal principles, requires less rigid operational concepts that will not pose a challenge to these ideals if it is to become a meaningful partner in a joint initiative.

DEFENSE DEPARTMENT AND JOINT STAFF STRUCTURES

Does the structure of the Joint Chiefs of Staff, the Joint Staff, and specified and unified combatant commanders further or hinder joint cooperation? Do the powers of the secretary of defense further or hinder joint cooperation?

Inside early 1980s DoD, any effort that made reorganization seem less necessary was likely to garner at least tacit support. Secretary of Defense

Caspar Weinberger believed that defense reform and reorganization were unsuitable substitutes for defense budget increases, noting that "you can't buy airplanes and bullets and rifles and submarines and things like that with reorganization plans" (Donnithorne 2013, 329; Weinberger 1998). He neither advanced nor thwarted ALB. In contrast, service coordinating and utilizing groups, including the Joint Chiefs and combatant commanders, adopted ALB's language and bureaucratic procedures, finding they helped to advance desired programs. While this probably gave the internal service innovation more credibility within the larger defense establishment, the structure itself neither helped nor harmed the initiatives. The more significant observation is that the service coordinating level and the external relations level of the defense structures can serve as building blocks in coalitions that give joint initiatives momentum, credibility, and strength.

OTHER EXTERNAL FACTORS

Do defense acquisition processes advance or hinder joint cooperation? Do external organizations' perceptions and stereotypes of the services advance or hinder joint cooperation?

ALB came about during a unique period of congressional involvement in questions of defense organization. These seem to arise only every few decades in American politics, so ALB's context is exceptional. It is, however, something for which a practitioner seeking to advance a joint cause should account. The dynamics of defense reform, when present, provide a context for any effort—even if only in the spirit of Rahm Emanuel's "never let a good crisis go to waste" mind-set—and will be available for exploitation by both those who would favor and those who would oppose service sovereignty vis-à-vis additional movement toward cooperation.

The aspect of American defense politics that remains relatively constant is a high degree of congressional interest and involvement in any military procurement decision. ALB and its attendant agreements led to new weapons systems. As it turned out, the permissive defense budget allowed DoD to buy several major new systems during the 1980s, and some of the notable ones, including the A-10 and surface-surveillance JSTARS aircraft, were directly attributable to ALB.[6] That the procurement requirements arrived prestamped with biservice support from within the defense establishment gave members of Congress seeking to advance them additional political credibility. As a bonus, they provided an answer to the pressing question of a perceived Soviet menace. Thus ALB was an effective mechanism for bringing into quick coalescence interests that resided within the individual services, the joint staff and commanders coordinating among services, and the external executive and legislative components of the defense establishment.

Given the internal structure of the services and the fact that the leaders that coordinate and draw on the services (analysis level b) come from those services, a dynamic of interaction among service and congressional interests will endure. The services need not become lickspittles over congressional wishes, as they will usually retain the most credibility when it comes to stating equipment requirements, but ALB and the 31 Initiatives showed that engaging, entertaining, and showing progress on legislative concerns seem to generally have more positive outcomes than does a strategy of outright defiance. The relative constancy of the Navy's budget and clout in Washington demonstrates that there is more than one path to attaining service interests, though.

Conclusion: A Mixed Result of Cooperation

The context of the ALB cooperative effort has its roots in the moribund post-Vietnam military. The path to cooperation between the services seemed anything but inevitable. Army-Air Force relations had frayed alongside the morale of individual services as the Vietnam conflict wore on. After the final removal of US troops, the Army faced the task of rebuilding itself from a wartime nadir.[7] If the vision that inspired ALB was first dressed in the green fatigue uniforms and black leather boots suitable for combat in the deciduous forests of Central Europe, the concept saw its only real combat action wearing the light brown fatigues and beige boots chosen for the deserts of the Middle East. The force assembled in 1991 to eject the invading Iraqi Army from Kuwait is the only substantive military test of ALB's efficacy. Since spoils and the right to slant history go to the victorious nation—and its military services, legislatures, and defense contractors—it is not surprising that ALB appears in many descriptions as a key to US success.

A more objective view is that the Iraqi Army was a threat of less significance than first perceived and that so many incomplete tests of doctrinal concepts occurred in a short time that none could have been definitive. The truth lies somewhere in the swampy middle ground of service-specific historical narratives. Whether or not the ALB doctrine *itself* ensured victory, the operational concepts it outlined led to substantial segments of the force structure and military capability with which the United States fought Operation Desert Storm. The era ALB defined ended with some sharp diversions in (or reversions to old) ideology of the Army and Air Force after the Gulf War, but even the apparent fatigue the sister services showed in pursuing joint cooperation serves as evidence that a genuine effort had existed. In the final assessment, ALB appears to have corralled other existing forces, driving meaningful biservice cooperation in a genuine search for a means to counter a military threat that required a response by the US military establishment.

Case Epilogue: Repudiation by the Air Force, Abandonment by the Army

ALB was never perfect. And while we assert it drove meaningful cooperative efforts for over a decade, we concede that its unraveling had a half-life comparable to that of a spit-shine meeting a muddy battlefield. Even at its peak, an insider related a senior-officers off-the-record view of ALB: " 'when we say we agree with the AirLand Battle concept . . . we agree that the concept is a good concept for the Army' " (Gordon 1983, 1274). Just as the Soviet threat to national security peaked in the mid-1980s, Goldwater-Nichols reforms came into effect, diminishing both first-order and bureaucratic-politics inspirations for ALB. Amid *glasnost* and *perestroika*, Congress and academia came to hold the eastern bloc with less of the awe that had swelled defense budgets earlier in the decade (Proxmire and Almquist 1987). In addition to questioning the military might of NATO's rival, they noted the trepidation ALB caused in Moscow (Almquist 1987, 17). While validating ALB's effectiveness in helping to end the Cold War, the end of the threat also diminished its relevance to all stakeholders. In the early 1990s, the West came to view the once-great Russian military-industrial network as defunct (Erlanger 1993, A1). Congress focused on making sure the same thing did not happen in the United States, advocating for weapons systems based on Gulf War effectiveness rather than ALB principles (DeConcini 1992, S9956). Without the red menace, military services struggled to redefine their respective niches against the backdrop of shrinking defense budgets. They quickly took their battles into the public arena, with the Air Force's "Global Reach—Global Power" facing off against the Army's "Strategic Force—Strategic Vision."

Whether the Air Force would have ever completely embraced the role the Army envisioned for it in formulating its Cold War–era doctrine is doubtful —the Air Force's doctrinal rudder seems locked. The loss of the Soviet threat made movement away from ALB in the late 1980s a certainty, though, and some continue to argue that ALB never reached the status of "official" service doctrine (Cardwell 1992, 551). Retired Lt. Gen. David Deptula, now firmly established as one of the most ardent air-superiority apologists, drove this point home in a personal interview (Deptula 2013). Then Lt. Col. Deptula's epigraph opening this chapter distills to the same point, though adding an implication that the Army had through its partnership with TAC nearly deceived the large Air Force into accepting the concept as its organizing doctrine (Rice cited in Deptula 1991). It seems more likely that the Soviet threat and TAC's need for an alliance in the defense establishment made it a consenting partner; nevertheless, both of these conditions have faded from view.

In 1993, the Army's new version of FM 100–5, acknowledging congressional prodding that both the Cold War and an era of interservice doctrinal cooperation were complete, mentioned ALB five times, but allowed that its new doctrine "causes AirLand Battle to evolve into a variety of choices for a battlefield framework and a wider interservice arena, allows for the increasing incidence of combined operations, [and] recognizes that Army forces operate across the range of military operations" (US Army 1986, ii).

Practical means of bureaucratic cooperation ended along with the shift away from ALB, including the fiscal coordination put in place by Initiative Number Thirty-One. Consistent conversations in the Pentagon and personal observation allow us to state with confidence that very little coordination on budget matters occurs between the Air Force and the Army at the action-officer level (Babcock 2014). High-level coordination does occur when there is an OSD policy decision that the services collectively find odious; senior flag officers still get together to put together a unified front of opposition. The services are collectively nervous, however, about proactive cooperation to advance programs of mutual interest of the type demonstrated during the ALB years, and have not sustained long-term, standing working groups of the kind that distinguished the 31 Initiatives. Reduced defense budgets tend to bring cooperation to an end. The difference in the interservice fiscal atmosphere compared to the ALB era serves to underline the remarkable effect defense budgets can have on joint initiatives.

Another significant explanation for ALB's reduced ability to drive joint cooperation was that Congress's appetite for organizational change had been sated, at least for a while. The passage of Goldwater-Nichols gave the reform caucus a big victory and temporary pause. The Gulf War, even if accompanied by increased interservice bureaucratic fighting, was a widely hailed victory. The fact that the two very public faces of the military effort—Central Command's Gen. Norman Schwarzkopf and Chairman of the Joint Chiefs Gen. Colin Powell—held positions that the legislation had strengthened lent credence to the idea that the reorganization of the defense establishment had helped it emerge from the debacles of Vietnam, Grenada, Desert One, and Beirut. The popular perception held that after successful emergency surgery by Congress, the US military was again well equipped to handle national-security contingencies. If ALB's final examination was the military's performance in the eyes of military hagiographers and congressional staffers eager to prove their contributions to national defense, the entire effort would be graded a success.

This is not the only reasonable basis for adjudication, though. If Desert Storm indeed tested ALB, an objective operational assessment of that campaign yields mixed results about the effectiveness in combat of procedures set forth on paper. The lack of confidence in CAS procedures is an indictment, though the widespread practice and effectiveness of BAI, a mission set identified and prepared for under ALB's auspices despite the Air Force's deep-seated resistance, are a bright spot. The fact that weapons systems initiated in the name of ALB went on to become key components of joint warfare over the next forty years demonstrated that the doctrine did have some of the flexibility it claimed, a narrow focus on Europe notwithstanding. Far from questioning any doctrine, any military methods, any command-and-control structures, or the suitability of joint military leadership, most historical analyses of the Gulf War stress the dominance of US forces and the outstanding military equipment with which the campaign was waged. As to operational art and the operational level of war, we hold that the test was not severe enough to settle the issue.

Yet Desert Storm need not serve any role in our assessment of ALB. If, finally, it is judged against a standard that calls for ongoing, meaningful interservice communication and cooperation, ALB succeeded in leveraging the circumstances of a particular era—including an existential threat, congressional pressure, growing defense budgets, increasing lethality of arms, and burgeoning information technologies—to promote significant joint cooperation for a while. ALB remained effective in this context until a host of changing circumstances rendered it irrelevant. We will see in our next case studies that this is a common outcome across other instances of military cooperation.

Chapter Four

Cooperation in Peacetime

The Joint Primary Aviation Training System,

1988–Present

> *I am encouraged by the cooperation and progress we have made*
> *in bringing jointness to flight training and hope that it serves as a*
> *model in other areas.*
> —John M. Deutch, Deputy Secretary of Defense,
> October 24, 1994

> *We like the airplane . . . but flight training at Pensacola is not*
> *joint at all. . . . The Navy and the Air Force run two*
> *completely separate enterprises.*
> —Captain James Vandiver; Commander, Naval Aviation
> Schools Command, April 2014

Introduction and Background

Prelude to Joint Aviation Training: The T-46 Debacle

At about the same time as AirLand Battle (ALB) cooperation reached its peak, the Air Force was bringing one of its acquisition programs to an unceremonious end. Despite powerful advocacy from the New York and Arizona congressional delegations, service anger killed the T-46 *Eaglet*, which it had come to see as an overweight, overbudget "turkey" (US House of Representatives 1986). After competing to become the Air Force's replacement for its 1950s-vintage T-37 basic flight training aircraft in 1981, T-46 cost overruns had forced it out of the service's plans for training new pilots. Fairchild Republic's design and manufacturing gaffes—including an unflyable demonstration aircraft rolled out of the factory in February 1985 with wooden and cardboard parts—likewise pushed it out of the aircraft industry (Kross 2014). Though the aircraft eventually flew in October 1985 (Pike 2011), the Air Force's September 1985 request to remove all T-46 procurement funds from its budget reflected deep-seated displeasure. In addition to the growing expenses, the sense of entitlement displayed by Fairchild's CEO offended the secretary of the Air Force so much that even the protection of Senators D'Amato (for the airframe) and Goldwater (for the engine manufacturer) could not save it (Sladek 2014). The formal death of the replacement-trainer program was a fact by 1987 (Emmons 1991).

Almost fifteen years after the hearing that killed the T-46, the T-6 *Texan II* began to appear on primary-flight training ramps of both the Air Force and the Navy, replacing their T-37B *Tweet* and T-34C *Turbo Mentor* trainers. The Joint Primary Aviation Training System (JPATS) aircraft (the T-6's acquisition-process name) supported full-scale training of military pilots by October 2002, quickly spreading across both services' pilot- and aviator-training bases (Colarusso 2002; Orton 2006). The Air Force's T-6 transition was complete before 2014; the Navy was flying all T-6s for primary training in 2017 (Groendyk 2007). Blood spilled under the T-46 gallows fertilized the ground for a new program that came to exemplify jointness (DeGarmo 2014). Despite an unanticipated extension of the T-37's service life by a quarter-century, the T-46's demise also rang in a unique, previously improbable example of interservice cooperation—the partnered acquisition of a trainer aircraft by the Air Force and the Navy. Even before the T-6A had supplanted the T-37 at all of the Air Force's training bases, the first Navy T-6Bs had taken their place at Pensacola Naval Air Station's Training Wing in 2005, and appeared at the Navy's Whiting Field for pilot training starting in September 2009 (US Navy 2009).

As a peacetime acquisition project undertaken between the Air Force and the Navy, this second case study differs in its scope and essence from the ALB case. The program's success in hindsight seems a foregone conclusion, but at no point was it guaranteed. That the Air Force and the Navy would undertake any joint project is itself remarkable. This is not unlikely simply because of the differences inherent in land-based and naval aviation. The two services have a historical record of institutional conflicts, deep-seated distrust, and, most significantly, examples of failed joint acquisition efforts. The question of partnership between the two services—which are most alike in their strategic preferences and doctrinal philosophies, yet most often prone to self-destructive disagreement when they conflict—is worthy of attention in our search for the conditions that allow interservice cooperation.

Our Air Force-Navy case study highlights an area of core interest to both services: the production of new aviators via military flight training. This chapter's process-tracing narrative tells the story of the acquisition JPATS—a label that encapsulates the T-6 aircraft and attendant ground-training assets. Overcoming habitual institutional distrust, the Navy and the Air Force collaborated to procure a shared aircraft and even—for a time—cooperated to form a common training pipeline in which both Air Force and Naval Service aviators could matriculate. JPATS had more external significance attached to it than the mere training of prospective military pilots. It initially had the attention of the wider defense establishment as an issue of jointness and civil-military relations, and received periodic attention as gender issues, cost increases, and corporate protests over the military's contract-award process arose over JPATS lifespan.

The architects of the program, no doubt taking measure of the then recently passed Goldwater-Nichols Act, shrewdly packaged their efforts in voguish jointness. The novel program set its mainsail in line with the winds of civil-military change. It retained cachet in both the services and the larger defense establishment in part because of this political savvy, allowing the joint procurement effort to develop a momentum resistant to dissociative forces that individual services applied to it.

Themes of This Chapter

The acquisition of the JPATS trainer and, to a lesser extent, the establishment of joint primary aviation training programs were successful endeavors for about a decade. This joint cooperation shines even though the Air Force and the Navy are perhaps the two services least apt at cooperating, especially on an acquisition program. While at least one Navy source claims that "examples of USN-USAF cooperation have been legion," this interservice relationship is probably better known for a rivalry that "has at times been particularly intense, even legendary" (Swartz and Duggan 2011). The two services often find themselves working in close proximity in the same theaters, which has driven a frequent need for civil, if not brotherly, coexistence. However, when role and mission disputes do arise, they reflect unbridled hostility. The focus of this chapter is therefore on a most-difficult or least-expected example of joint cooperation. Phillip Meilinger wrote that Brig. Gen. Billy Mitchell's "inordinate and near-neurotic hatred of the Navy . . . left a legacy of animosity between the two Services that has never fully healed" (1997, xv). We begin with some historical examples to underline this tendency and offer a brief discussion of the importance the military services place on basic and core-specialized training programs. In addition to imparting specific technical knowledge, these programs impart a unique service identity on each new recruit, and are therefore unlikely areas for close interservice cooperation.

A second theme of this chapter is the constant input of energy above and beyond the standard bureaucratic churn required to keep this objective alive. Several issues at different times in the life of the JPATS program could have derailed both the joint acquisition project and the multiservice aviation training associated with it. Individuals who had detailed knowledge of their technical roles in the program, a subtle appreciation for the interest-group politics that affected it, and an ability to exploit the spirit of jointness that marked the JPATS era appeared repeatedly in the story and acted to keep its underlying vision alive. There is no one person who can claim to have spawned, incubated, and hatched the JPATS program, but several can claim to have overcome legitimate obstacles capable of preventing the jointness we observed. A major characteristic that emerges in surveying the cast of characters responsible for bring the JPATS to fruition is singular, explicit focus on jointness.

While expectations of joint JPATS development were reasonable in the Goldwater-Nichols environment, claiming it to be inevitable would be a stretch; we emphasize the sheer willpower that military officers exhibited in keeping it alive. For mid-level "action officers"—the nonexecutive staffers who get work done in the bureaucracy—the effort was most extreme, and often at its most frustrating, when combating their own services' parochial views (Walters 2014). Interservice friction was on display too, however. Raymond "Doc" O'Keefe (2014), a Naval officer who was at the focal point of the project during its most critical years, said the Navy was "really not interested in participating," but was "eventually told to participate and to find a way to pay for the program." Despite internal obstacles and this less-than-enthusiastic joint foundation, the Navy found itself embraced by an Air Force eager to make an interservice program work, and found enough benefit in the accommodations offered to go along with the project as a joint undertaking.

The third theme of this case study is the outsized role of external forces in creating jointness. In its role of defense oversight, Congress frequently acted as an instigator that spurred the Air Force and the Navy to action through frequent public attention and its spending power. DoD appears as an external driver of jointness later on in the process, but usually seems to follow congressional direction. In one notable instance, the While House focused attention on JPATS. Despite opportunities for jointness afforded by Congress's attention and positive reinforcement given to efforts in that direction, initiative for creating a functioning cooperative program nevertheless remained with the services. Their fears and interests interacted over many years with legislative priorities and direction; both the internal service level and external relations level of analysis played substantial, symbiotic roles in forming JPATS.

The fourth theme is the strength of a congressional mandate relative to one issued by the Pentagon or the service bureaucracies. An interesting entailment of JPATS, joint flight training, was an initiative of DoD. Secretary of Defense Les Aspin gave an official directive to the services to "consolidate initial fixed-wing aircraft training for all Services and transition to a common primary training aircraft" in April 1993, Deputy Secretary of Defense John Deutch approved the services' plans for implementation in October 1994, and the services wasted no time in visibly complying with that explicit, outside actor order (Aspin 1993b, 1–3). Responding to Congress's direction to "paint everything purple" in the wake of Goldwater-Nichols, DoD forced the Air Force and Navy to restructure aviation training programs in a way that would permit joint primary flight training in the new aircraft. In the joint spirit of the day, DoD brought the services together—for a time—in a way they would not have pursued on their own, if only to remind Capitol Hill that they were clear on legislators' intent. Recognizing the nested outside interests at play, the services complied with this direction for more than a decade before preference and convenience led them to shirk the (long-departed) defense secretary at a glacial pace.

John Law (1989, 114–115) described the "heterogeneous engineering"—coalescing technology advances and human initiative—that allowed Vasco de Gama to sail around Africa into the Indian Ocean and bring the transoceanic spice trade under Portuguese domination for the next century. The specific events that made de Gama's journey possible are not relevant to the case of the JPATS development, but Law's treatment of the attendant technological advance and stabilization, incorporating as it does theories of constructivism and systems engineering, is a helpful framework for understanding this case study.

What makes the T-6 and joint aviation training almost as remarkable as conquering an impassable trade route is the number of complex adjustments to military systems and ways of thinking that had to happen to allow the innovation. Technological advances made during the development of the airplane and associated training constructs may not rival the compass, the advent of celestial navigation, advances in shipbuilding, and the discovery of ocean currents, which occurred over the centuries leading to de Gama's journey. The weaving together of systems by a number of disparate actors, though, united only by a beckoning promise of joint aviation training, is in some ways as remarkable as the African passage. Political forces aligned against joint cooperation can be as formidable as the capes, shoals, and coastal currents that made sailing from Europe to India a fool's errand for centuries. A team of heterogeneous engineers, united and focused in their aim to make joint primary aviation training possible, exploited associative forces and overcame dissociative elements to stabilize a system that achieved their vision. Using this analogy, we turn to some of the obstacles that had to be overcome in pursuing the JPATS.

Factors Influencing Joint Acquisition and Training

The Navy-Air Force History of Interservice Rivalry

If the Army and the Air Force have bickered like siblings from the same family, Air Force-Navy feuds have a darker, more strident tone. Whereas the tension between early incarnations of the air service and the Army existed over the question of the newer service's independence and mission priorities, the two services at least tacitly always acknowledged that the modern battlefield would require the services of each. In contrast, the Air Force and the Navy early on in the JPATS story seemed to have disagreed in ways that suggested they viewed each other as existential threats.

The Air Force and the Navy had a strained institutional relationship prior to Air Force independence. The early years of the Air Service and Air Corps witnessed the Army's aviation branch battling the Navy on two main issues: long-range shore-based patrols and antisubmarine warfare. The air service opposed the Navy conducting shore patrols that flew from land, and the Navy objected to land-based aircraft hunting subs (Swartz and Duggan 2011, 11). Brig. Gen. Billy Mitchell's flamboyant air power advocacy got under the skin of several senior Navy figures. Rear Admiral William Moffett, the first leader of the Navy's Bureau of Aeronautics, derided Mitchell as "of unsound mind and suffering delusions of grandeur" (Emme 1979, 69). Mitchell raised the larger Navy's ire by making it the butt of one of his most notable aviation experiments. In September 1921, while attempting to "prove" bombers' superiority over battleships, Mitchell oversaw the sinking of the captured German battleship *Ostfriesland* and other warships anchored for the test (Hurley 1975, 66–68).

Beyond showcasing aerial bombardment's effectiveness against a maritime military target, Mitchell claimed that the Air Force could take responsibility for coastal defense and implied the Navy's impending irrelevance. Despite his bluster, Mitchell did not achieve Air Force independence or halt construction of battleships, but his methods "enhanced Mitchell's public standing . . . [while they] poisoned already strained relations with his superiors and the Navy Department" (Trest 1998, 36–38). An ironic outcome of the demonstration was that it spurred the Navy's pursuit of aviation for its own ends (Pavelec 2010, 305). Mitchell followed up his first demonstration with vitriolic criticism of the Navy and War Department following the crash of the *Shenandoah* dirigible in 1925, which led to his court martial and retirement (Swartz and Duggan 2011, 10).

Even after Mitchell's exit, the Army and the Navy struggled to find agreement over the roles and missions of their air assets. The Navy resented its treatment under Army aviators, enduring the ignominy of Army supervision for its land-based aviation missions, which flew as far as three hundred miles over the ocean. While pursuing aviation programs to project fleet power, the Navy could not seem to escape coastal defense for the Army, which itself exhibited quite a bit of apathy toward the mission. Salting the wound, the Army made several efforts to stop the Navy from procuring more land-based aircraft for experimentation or other uses for which it might determine later need, thereby intensifying the rivalry. The 1931 Pratt-MacArthur Agreement established naval air power as that which would be "based on the fleet and move with it" to complete naval missions that did not include coastal defense. Even that brief truce began to spin apart following Admiral Pratt's retirement as chief of naval operations, and the Navy again competed with the air service for supremacy in maritime scouting and coastal defense (Trest 1998, 55).

World War II allowed both services to highlight contributions to national power without bickering. Cooperation even arose, including Army Air Force bombers launching from USS *Hornet* to strike mainland Japan under Lt. Col. James Doolittle's command (Doolittle and Glines 1991, 2–8). After World War II ended, however, the services tackled the apportionment of roles and missions among the US military services, and the brand-new Air Force and the Navy again became wrapped up in a dispute with zero-sum overtones. The Navy had an epiphany during the war that its new power-projection strategy was based on carrier-launched strike aircraft, and further that nuclear strike would be prominent in the US defense strategy. During the Key West Conference to settle the issue of service roles and missions, the Navy fought to retain aviation capabilities. Key West hardly put the matter to rest, though, with each service emerging with "its own interpretations of what had been agreed to, which could neither be proven nor falsified by reference to a written transcript of the proceedings" (Barlow 1995, 123). When the Air Force's long-range bombers began to usurp aircraft-carrier funding, Navy leadership took drastic action. The "Revolt of the Admirals," a campaign to discredit the Air Force's B-36 bomber, demonstrated that the Navy was willing to undertake even subterfuge to maintain its significant aviation fleet and place of prominence in the strategic arsenal (Trest 1998, 128).

The "New Look" and "Balanced Response" eras fueled more competition, mainly with respect to the nation's nuclear force. Wresting control of both bomber- and ICBM-delivered warheads, the Air Force's share of the overall defense budget surged to nearly 50 percent (Lewis 1990, 15; Coté 1996, 234). The Navy was determined to stay in the business of strategic force projection, and invented its leg of the nuclear triad by developing submarine-launched ballistic missiles (SLBMs). Owen Coté (1996, 2) credited the interservice rivalry as powering the Navy's development of the Polaris missile and the Navy's most innovative contribution to US nuclear-weapons doctrine. Peter Swartz acknowledged that a " 'culture of rivalry' greatly overshadowed [the] record of cooperation" (Swartz and Duggan 2011, iv, 6) between the services.

While frequent tiffs colored the aura of interservice relations when planning for the JPATS began, previous outside involvement in joint procurement may have been even more damning to a shared acquisition effort than previous rivalry. Both services retained recent institutional memory about the TFX program, intended as a fighter with deep-strike, fleet-defense, and nuclear-delivery capabilities. Robert McNamara's Office of the Secretary of Defense (OSD) directed "commonality" in procurement for the jack-of-all-trades system, and then chose a different defense contractor's design over the one the Navy and the Air Force had determined to be the best.

The decisions to undercut the services' equipment-procurement prerogatives infuriated military leadership, but Secretary McNamara held firm, keeping the civilian service secretaries in his camp and deflecting congressional and public fury over his decisions. Despite the unprecedented display of civilian control of the services—especially remarkable for a major procurement effort—the effort fell far short of McNamara's vision. Only the Air Force purchased the TFX, which became the F-111 *Aardvark*. As we noted in the first chapter, one contemporary meaning of "jointness" is a euphemism to substitute for and overcome the political baggage "commonality" earned during the TFX debacle. While not causing friction among the services per se, TFX and other programs like it—the F-35 has become another (Lorell et al. 2013, 9–17)—inculcated distrust of joint platform acquisition, and strengthened service proclivities to purchase their own equipment whenever they could (Lorell et al. 2013, 9–17).

Shift to Specialized Pilot Training and Interservice Aviation Training Differences

Neither a history of interservice bickering nor botched previous acquisition programs were the most immediate obstacle to be overcome for aviation-training cooperation. A critical shift in philosophy for the Air Force with regard to its training of pilots involved the movement from a universal approach to undergraduate pilot training (UPT) to a specialized, or "tracked," system. An underlying premise for JPATS was the idea that both services would train their fledgling aircrew in a specialized undergraduate pilot training (SUPT) program rather than a general scheme. As the name suggests, UPT students flew the same syllabus, which included a phase in T-37s followed by training in the supersonic T-38. Though the Air Force had previously had separate pipelines before 1959, in the short-term institutional memory of the military it seemed a tectonic shift (Emmons 1991, xxi).

Fast-jet training was less applicable—even demotivating—to students who did not end up flying fighters, raising questions about UPT's ability to produce a healthy crop of enthusiastic pilots (Roberts 1982). Resistance to change remained. A 1976 study identified cost savings and improved quality of training inherent in specialized pipelines; individual major commands and a reluctance to undertake a major trainer replacement program during a period of national fiscal austerity overwhelmed it (Emmons 1991, 13–14). A perception reinforced by four-star generals held that a pilot who did not fly the T-38 was a "second-class citizen" (Chain cited in Emmons 1991) of Air Force aviation. The SAC commander demanded that his bomber pilots fly T-38s despite a blunt-nose B-52 flying very much unlike the pointy trainer. Cycling all Air Force students through the same two planes subjected the T-38 to lots of fatigue stress; using the T-1 business jet for tanker and transport pilots extended the T-38's useful life-span. With budgetary breathing room afforded by the Reagan administration's cash infusion, a T-38 fleet being used up by generalized UPT, and dissatisfaction with the now-elderly T-37, "specialized" undergraduate pilot training gained economic momentum and won the day (McGinty cited in Emmons 1991, 57).

At the same time, the visible failure of the T-46, which had been a "developmental" program, led the Air Force to determine that its replacement trainer would be "nondevelopmental," a "commercial-off-the-shelf" (COTS) modification of an existing airplane—even though the final product had fewer than 10 percent of its components in common with the Pilatus PC-9, the airplane it was based on (Sladek 2014). The attempt at a replacement trainer required nimble justification, because the robust T-37 service-life extension program (SLEP) offered decades more life from the aircraft while avoiding the new program's cost. A second fear arose: the Air Force would end up stuck with a T-37 for several more decades, an option that was "perfectly fine" in the eyes of budgeting officials but did not provide the training desired for up-and-coming combat aircraft. ATC's planners therefore had to show the T-37's qualitative unfitness to continue as a primary trainer, and did so by emphasizing its lack of power, safety deficiencies, and outdated avionics (Chiabotti 2014). The document that ultimately led to acceptance of SUPT was the 1988 USAF Trainer Masterplan, which laid out alternative courses of action in an easy-to-follow parallel format, while providing definitive, long-term cost and quality analyses that made SUPT an easy choice (US Air Force 1988, iii, 29, 35).

Once the SUPT decision became final the Air Force was able to engage the Navy on an apples-to-apples comparison about primary aviation training. With the T-3 satisfying concerns about seating configuration for beginning aviation students (the T-3 was a side-by-side trainer like the T-37 had been; the T-6 was a tandem configuration) and the T-1 relieving pressure on the T-38 fleet, the Air Force had maneuver space in the realm of primary training. The Air Force was much more comfortable partnering with the Navy on equipment for primary flight training. Partnering in advanced flight training would mean purchasing the expensive T-45 and trying to fit it into an Air Force training scheme. For the Navy, a more modest shift that involved increased simulator time and moving from a three-phase to a two-phase training syllabus was already underway (Boyington 2002, 23); having a trainer in common with the Air Force only helped.

One of the first biservice documents to appear on joint aviation training was a 1988 memorandum of understanding on the development and acquisition of trainer aircraft. Noting the Air Force's "major change in philosophy" on specialized training, it speculated on benefits of common acquisition, identified "commonality in primary trainer aircraft" as the option most likely to bear fruit, and dismissed as "unfeasible" (Hall, Gressy, and Boyd 1988, 1–2) the Air Force's use of the T-45. Thus the Air Force early on in the JPATS process engaged the Navy to manipulate the joint mandate for a common trainer in a direction away from the T-45 advanced-trainer nudge that Congress had provided. Lower-ranking action officers who were ghostwriting for the signatures of general officers pushed for a primary-trainer project that was palatable to both services and preferable to the Air Force (Chiabotti 2008, 266; DeGarmo 2014). This became the first of many compromises the Air Force brokered as it led the march to a joint primary trainer while it kept its dearest institutional goals intact.

Agreeing on equipment was a serious hurdle, but once overcome a follow-on question about jointness arose: Why not have the services conduct joint training with each other's students? Ultimately, DoD would mandate such a course of action. There were reasons why it was not a natural tendency among the services, even with common equipment. An ATC study revealed several substantial differences between the ways the two services trained. From the way they viewed airspace to the control exerted over syllabus timing to grading students, the Air Force and the Navy differed in so many respects that they did not themselves determine cooperation to be beneficial (Bartholomew 2014). This required outside actor orders, with which the services duly complied but never with as much enthusiasm as the joint procurement.

With the unlikelihood of Air Force-Navy cooperation setting the backdrop, a remarkable sequence of activities by a diverse cast of characters was necessary to overcome competitive and dissociative tendencies, but they combined to result in an acquisition program that was joint in fact as well as name.

Acquisition of the JPATS Aircraft and Establishment of Joint Training Programs

As we know, Congress pushed the Air Force to change its acquisition strategy for replacement of the T-37 and T-38. The demise of the T-46 was a bitter event—to the point of "nasty" words being exchanged on the Senate floor—and legislators wanted to control subsequent iterations (Walters 2014). Explicit guidance came in the 1989 NDAA, which directed reports to both Armed-Services Committees outlining "DoD's plans for future training aircraft for the Navy and the Air Force" (Emmons, 2004, 2). The bill's conference report recommended that the Air Force reverse its replacement strategy, completing T-38 recapitalization first, so that it could take advantage of "a warm production line" that was making T-45s for the Navy (Emmons 1991; US Congress 1988). The Air Force realized that it had to comply with Congress's broad intent for jointness and perceived attendant savings, but feared getting burdened with an expensive airplane unsuitable for future pilot training. It also inferred that outside actors' patience for developmental programs had been exhausted and the same COTS approach that saw the Mitsubishi *Diamond* become the T-1 *Jayhawk* would be required for the new primary trainer (Manning 1991, 219). The service was politically informed and incentivized to pivot its strategy toward a difference direction while still complying with legislative intent.

In the dire straits of threats to its trainer-procurement autonomy, the Air Force moved forward from the smoking hull of the T-46 program. The April 1988 *U.S. Air Force Trainer Masterplan* contained a series of options the Air Force could pursue in training new pilots, focusing attention within the service as well as in external organizations. The plan was so effective that it evolved into a DoD-wide product. The Department of Defense 1989 Trainer Aircraft Masterplan served to answer Congress's call for joint trainer acquisition, but it also made a distinct bid for the Air Force to take its own fate in its hands with regard to T-37 replacement. The focus on a primary trainer required the Air Force to sell the program to the Navy, who was not then considering such a move (Walters 2014). Congress's desire for visible joint procurement, combined with the Air Force's fear of losing self-determination of its trainer fleet, set the stage for a long season of compromise that constituted the most successful joint aspect of the JPATS—aircraft procurement. A second joint aspect—that of joint training programs—came about more as a result of outside influence from the DoD. Since it did not require the production of new equipment, the services were able to implement it immediately. As it addressed neither a pressing service problem nor congressional concerns, though, this initiative quickly faded.

Joint Acquisition of the T-6 Texan II

THE WINDING ROAD TO REQUIREMENTS

Stephen Chiabotti (2008, 265) argued that the leadership required for significant changes in military technology lies neither with flag-officer leadership nor with the acquisition specialists supervising the procurement of the new technology. Rather it lies with the planners who define the requirements for the new system, balancing performance, safety, and operator preference against the inevitable cost that each specification adds. When we process-traced JPATS development, we agreed and had an additional observation pertaining to jointness. The people who shepherded this program did all of the following: (1) developed requirements that accommodated a sister service, (2) built political enthusiasm for an attendant joint outcome, and (3) adopted a long-term planning perspective. The effort built significant cooperative momentum. Absent this effort, the joint undertaking may have easily fell victim to a raft of dissociative forces (Kross 2014; Walters 2014). As the 1989 DoD Masterplan for training aircraft stated, "joint specification of requirements and timing are key to the process" (US Department of Defense 1989).

How to Buy a Joint Airplane

Although the JPATS story now spans decades, its joint lineage resulted from intense efforts sandwiched between the T-46 failure and a biservice agreement about what a primary training aircraft needed to do. The formal agreement was an Operational Requirements Document (ORD), a standard format that the military uses to convey to industry what they would like designed and built. The first draft was for an Air Force program known as the Primary Aircraft Training System (PATS), the service's fallback program to try again after the T-46 failure (O'Keefe 2014). The sleight of hand that turned PATS into JPATS was making the ORD a document that enjoyed the full institutional buy-in of both the Air Force and the Navy. Once the ORD existed with joint sanction, several years of difficult procurement work remained, but the program would succeed or fail from that point forward as a biservice entity.

Though Congress has often encouraged joint acquisition, it has never interfered directly in individual services' roles to organize, train, and equip themselves. Therefore, to order up identical training platforms, identical ORDs had to be issued simultaneously from the two involved services. The means used to attain coordination was a comprehensive plan for all military aviation training. The Air Force, putting to use some insight its leaders had gained fighting other Pentagon political battles, assigned some of its best and most motivated people to draft and incubate a series of so-called aviation "masterplans." These comprehensive plans, originating as Air Force–specific documents, came to encompass other services' roles in aviation training as well, and, after some politically astute socialization in the upper echelon of the defense establishment, came to compose the "DoD" plan for creating new military pilots. After the master plan was thus imbued with jointness, it became the basis for a 1990 joint statement of operational need (JSON) for a common primary trainer.

Air Force Chief of Staff General Larry Welch gave then Brig. Gen. Robert Delligatti a personal mandate before Delligatti took over plans and programming for ATC. Welch told him not to rush back into the T-46 program, but rather to see what it would take to get future training back on track. His only specific guidance was the direction to avoid "wide-body" trainers. General Delligatti worked first to strengthen Air Training Command's (ATC) requirements division, which had earlier failed to limit requirements growth—and hence cost increases—for the T-46. He brought together a team of handpicked officers with strong analytical skills. He brought previous planning-and-programming experience from the Pentagon, and in Tome Walters he had someone who could turn the joint-trainer vision into an actual item in the Program Objective Memorandum (POM), the documents the services use to lay out their five-year spending intentions. Delligatti's team reanalyzed pilot training from the foundation up. Studies concluded that the T-37 was not the most pressing problem, but that the Air Force "was flying the wheels off the T-38" (Walters 2014), and would require a replacement for that aircraft first if it did not return to its historical roots and adopt a specialized pilot training system.

Col. Willard Grosvenor, another plans-and-analysis expert who had previously distinguished himself in the Pentagon, led the drafting of ATC's internal master plan in 1988. As mentioned earlier, Colonel Grosvenor's plan became the Air Force plan and then grew "joint legs" in 1989, enabled by the central premise of SUPT. Specialized pilot training made the Air Force's training system mirror the Navy, making joint procurement of the primary training vehicle possible as long as naval representation joined the process without delay.

We interviewed this project's participants as they looked back on a successful endeavor, and their recollections reflect the magnanimity of a winning team, eager to share credit. Thus, the notion to add jointness to what was then the Air Force PATS has a wide attribution, but Col. Stephen Chiabotti received undisputed credit from all for authoring the 1989 version of the trainer aircraft master plan that captured the concept. Then Brig. Gen. Walter Kross (2014) brought his experience in writing master plans to take control of large programs that were foundering in the Pentagon, using them to focus interested audiences toward forward progress. Kross, who went on to lead the US Transportation Command as a general, exhorted his subordinates working on JPATS to "cloak ourselves in the mantle of jointness" (Chiabotti 2008, 268). Having recently completed the new "Capstone" training for flag officers that the Goldwater-Nichols Act had mandated, he believed strongly in the benefits of cooperative training and acquisition, and urged many of the more significant compromises that kept the JPATS trainer appealing to Navy sensibilities (Kross 2014).

Enriched by several ATC-commissioned studies, the master plan gained momentum within both services. Having a focal point for the new direction being pursued for military aviation training for the Air Force and Navy, the regular work of aircraft acquisition could begin again. The military services, led by Air Force representatives from ATC and the larger acquisition enterprise, gathered information from potential competitors for the JPATS (Manning et al. 1991, 222). ATC pilots flew dozens of candidate aircraft that hopeful competitors made available to them—a critical step for a nondevelopmental program. One of the first joint aspects of the program was in screening candidate aircraft. Naval officers detailed to monitor the work the Air Force was doing to research the PATS became de facto members of a team that shared workload and opinions. In 1989, for example, ATC sent a team to Europe to evaluate five possible JPATS candidates. Led by the affable Brig. Gen. Delligatti, the group also included Capt. LynnAnne Merten (the first JPATS program manager) and Lt. Cmdr. Clay Umbach, who represented the Chief of Naval Air Training (CNATRA, pronounced like the famous crooner). Shared perspectives and information allowed the quick release of a Joint Statement of Need that, along with Air Force guidance directing ATC to work with the Navy, conveyed to industry and Congress the seriousness of the two services' intent to cooperate (Manning et al. 1991, 129, 242).

The services continued to develop and refine their aircraft requirements together, culminating in a joint release at a summit on October 18, 1991, that both the Air Force chief of staff, Gen. Merrill McPeak, and the vice chief of naval operations, Adm. Jerome Johnson, attended (Manning 1991, 242). The services regularly and intentionally updated their requirements documents together thereafter. Alterations to these documents, though many were significant, never came close to jeopardizing the joint acquisition. The program's architects sealed this surety during the requirements-definition process, and they were proud of their work. Walters asked and answered himself, "Why was the program successful? Because the requirements were scrubbed, scrubbed, scrubbed, and scrubbed. We had thought deeply and long and hard about each requirement" (2014). A fly-off evaluation led to a June 22, 1995, selection naming the Raytheon-Beechcraft PC-9 derivative the winner (Carroll 1995, 85). Six months of bitter protests followed—standard fare in any military-acquisition process—but a Government Accountability Office (GAO) ruling dismissed the protest, allowing an official contract award on February 1, 1996. Though influenced for five years by service inputs, Congress, DoD staff, legislation, regulations, and industry protests, the joint requirements finalized in 1991 never wavered.

To what may we attribute this enduring joint cooperation? A good measure came from individual military officers who saw intrinsic value in jointness. Then Brig. Gen. Walter Kross, still under the spell of his Capstone experience, "preached jointness" (2014). Though Kross modestly passed credit to his predecessors for making the trainer joint before he joined the project, others interviewees acclaimed him for establishing its joint focus during the late 1980s. He also helped hone the master-plan effort that became one of the more effective mechanisms for ensuring JPATS developed and sustained its cooperative momentum. The team effort created an unimpeachable Air Force plan that grew to become a DoD plan, and then it coopted the highest available level of outside support from Congress.

A benefit of the master-plan construct and its thirty-year time horizon was the leverage it gave programmers and planners to argue for long-term budget decisions. It is difficult to overcome short-term thinking in military budgets. Program cancelations offer tantalizing cost savings in the near term; defense appropriations nominally happen according to five-year plans, but Congress authorizes funds for two years at best and, to maximize its control and oversight, usually appropriates spending on a one-year leash. Fighting to keep the T-1 procurement alive when it faced cancelation in the fall of 1990, ATC leaders were able to point out the multibillion-dollar expense that would result in canceling a then-$910 million purchase for a short-term gain (Emmons 1991, 57). The master plan facilitated similar long-term arguments about operating and acquisition costs for the JPATS, again giving the Air Force some intellectual maneuvering space as it argued its case against impatient budgeters.

So far, we have concentrated on the aircraft, but the "system" in JPATS alludes to the comprehensive training services the winning contractor would arrange. JPATS included "not only the aircraft," but also a range of training aids, simulators, and day-to-day maintenance by civilian contractors (Siuru 1994). The original contract award stipulated that the winning aircraft supplier could not supply the integrated program for simulation and flight-training pedagogy centered on its airframe, and that it must conduct a competition among separate contractors for that portion ("Spring Takeoff for JPATS Ground System" 1997, 14–15). This grew to become a contentious matter later in the process, offering DoD another chance to influence the program, something that we describe later in the chapter.

Perhaps the biggest benefit of the master-plan concept was the easily digested package it provided to JPATS advocates facing Congress. Trainer aircraft do not get the same attention from the officers dedicated to liaison actions with Capitol Hill, so the JPATS team had to engage in self-help actions (Walters 2014). Helpful staffers taught Air Force and Navy action officers to socialize their intent and plans among the "big four" congressional committees—the Senate Armed Services Committee (SASC), the House Armed Services Committee (HASC), the Senate Appropriations Subcommittee on Defense (SAC-D), and its cousin in the House (HAC-D)—that have the most impact on military funding. Skip Ringo was a member of the SASC professional staff. A former Marine, he was "a friend at court" who coached the JPATS team that Congress's mood was right to see the good in joint programs, and that by engaging in the politics of the day the services could gain support for joint acquisition initiatives (Chiabotti 2014). In Walter's phrasing, the joint aspect of JPATS was as appealing to Congress as "apple-hood, mother-pie, and baseball," a sensible political alignment that pleased legislators and ensured there "were no natural predators for this program" (Walters 2014).

FIGHTING TO THE POINT OF COOPERATION

Although the services garnered external support, there were roadblocks and compromises along the path to a JPATS trainer. As participants recalled, the biggest technical-specification hurdles the JPATS faced were the propulsion system and seating arrangement. As we discussed with the requirements-definition process, one of the things that facilitated critical compromise was a near-constant participation of representatives from both services. This lubricated jointness, but had even more effect in addressing intraservice issues inside the Air Force. Most participants recall that the biggest battles were within the Air Force, mostly disputes among parts of the ATC headquarters staff. Had they not been settled in a way that met basic Navy preferences, though, they would have probably scuttled the chances for joint cooperation, which means that the intraservice issues were in fact joint issues.

Many of the obstacles that required attention were at their root simple matters of preference. The Air Force had developed a tradition of flying trainers with two jet engines, a convention that the ATC Operations Directorate guarded jealously. Lt. Gen. Walters (2014) recounted a story about Gen. Kross getting "ambushed" by an ambitious colonel from this office at one of his first meetings with the sitting ATC commander, Lt. Gen. Robert Oaks, over the issue of dual engines. Jet preference could also be expressed in terms of airspeed, another tactic used by the "jet-trainer mafia" in the same office. O'Keefe recalled, "an Air Force Colonel asking, 'What is the top speed for any of the turbo-props?' His subsequent direction was to 'add thirty knots' to that speed as the aircraft requirement" (2014).

Aircraft speed *is* a relevant trainer parameter; new pilots have to learn "airmanship"—the act and art of balancing safety concerns, procedural rules, and mission accomplishment—while flying at a certain pace commensurate with combat-aircraft velocities. A one-hundred-knot crop duster that flies low and can turn in a matter of yards does not create the same decision-making challenges as a jet that cruises at three hundred knots and whose turn radius is measured in miles. Yet structuring a primary-trainer requirement as a function of what a given propulsion system could or could not do was disingenuous and revealed institutional biases at work. The joint advocates had to uphold principles in the face of significant professional pressures to fudge requirements. General Walters (2014) helped General Kross out of the verbal trap by holding forth on the realistic state of the art in engine technology and establishing that catastrophic engine failure had become a rare occurrence, one battle in a campaign of arguing against the ATC Operations Directorate that would mark his tenure at ATC. Observers credited Brigadier General Delligatti as the one who brokered requirements specific enough to allow for quality primary flight training—but general enough so as not to drive a certain type of airframe or propulsion method.

The open-ended approach toward requirements that General Delligatti imparted to his plans and programming team won the day and kept the possibility of joint cooperation alive (Emmons 1991; O'Keefe 2014). The mantra "we don't care what moves the air over the wings, so long as it moves it" became the philosophical basis of the acquisition documents, referring in particular to an agnostic approach to JPATS propulsion but reflecting a broader spirit of compromise that would make joint effort possible (Chiabotti 2014). Media reports in turn noted that the Air Force had lead responsibility for procurement of JPATS, but that evidence of joint accommodation was apparent. Articles that mentioned naval-training traditions allowed that the Air Force remained open to the idea of a propeller-driven aircraft even if it had been "historically less apt to choose a turboprop trainer" (North 1992).

Performance-based requirements also kept the program in line with emerging contracting laws and rules. Ongoing acquisition reforms directed requirements that did not make arbitrary limitations on *how* contractors met the training objectives of the new aircraft. Areas of emphasis were minimum speed, flight deck size, resistance to bird strikes, spin training, ejection seats, and cockpit pressurization (Aris 2003, 148). The simplified requirements aligned the Air Force with the Navy's customary practice of writing a broad specification and reaping a large number of proposals from interested contractors (Laymon 2014, 148).

Even though Air Force students were then flying the turbojet *Tweet*, reporters writing about the replacement program recognized that ATC supervisors were giving clear signals that a jet was not a foregone conclusion, and that there would be room in the competition to consider the benefits of the ruggedness and reliability of turboprop aircraft (Fulghum 1991, 66–67; Sweetman 2000, 18–19). This approach in the end settled the matter of propulsion systems and overcame the Air Force's predilection to use turbojets for trainers. If an airplane can fulfill its training mission with a top speed closer to three hundred than four hundred knots, it becomes very difficult to argue with the low costs associated with a turboprop. It is cheaper to buy outright and sips about half of the fuel of a jet for equivalent flight time.

The other serious area of contention was the airplane's seating arrangement. In the T-37, Air Force primary training was a side-by-side arrangement, the instructor to the right of the student, where she monitored movement of eyes, hands, and feet and could intervene with verbal or visual inputs. The Navy trained in tandem in the T-34, the instructor sitting behind and above the student, able to interact solely via intercom. Tandem seating gives the student pilot and her instructor roughly the same visual references, facilitating the tactile-descriptive knowledge transfer that typifies pilot training. In a side-by-side trainer, the instructor adjusts for his student's view in a different hemisphere of the cockpit; the chirality of maneuvers thus increases the time it takes to produce a proficient student. Finally, a tandem cockpit seems to convey a sense of early pilot-in-command responsibility to students. Without body language from an instructor influencing their decisions, students own their environment earlier, breeding the confidence desired for military airmanship (Manning 1991, 129).

Seating became a matter of interservice dispute, and was settled with theatrical flair. At an early joint-planning meeting, Lieutenant Commander Umbach interrupted Gen. Joseph Ashy when the latter briefed a "requirement" for side-by-side seating. After General Ashy ignored his first comment, Umbach stood and said, "'General . . . as the senior Naval officer present, I must protest. The Navy requires tandem seating and without that requirement, the Navy will not participate in this acquisition'" (Umbach quoted in O'Keefe 2014). The room became silent, so out of routine was it for a junior officer to not only interrupt a four-star general (especially one as notoriously volatile as Ashy), but also defy his instructions. Umbach had direction from his superiors at CNATRA to stand firm on the seating issue. Palpable tension built in the room—at least among the uninitiated—as the exchange unfolded, and then vented away just as quickly when a smile crept across Ashy's face and he said, "I always knew I liked you, Clay" (General Ashy quoted in O'Keefe 2014). Those interviewed on this exchange corroborated it, but the speculation about Clay Umbach's bravado before Ashy ranged from their common matriculation from Texas A & M University to General Ashy requesting Lieutenant Commander Umbach help him with some interservice pressure to solve a political issue that was a tough sell to other Air Force leaders. Whatever the mechanism, individual personalities helped work through the toughest parts of a joint acquisition, all the while holding it together as a concept.

In its early life, the survival of a joint primary trainer was not a given, in part because senior Navy officials were concerned about expense (O'Keefe 2014). The Navy was in the middle of its own acquisition process to require a new advanced trainer, the T-45 *Goshawk*. While the T-45 was not the abortive disaster that the T-46 had proven to be, the Navy was still dismayed to see its procurements cost spiral up as "unchecked requirements" and complex modifications to make the airplane carrier-worthy piled on to the program (Sladek 2014).

Amid the tension and compromise that marked requirements definition, there was an additional subtext of external influence. Later DoD direction regarding joint fixed-wing training programs also included a specific mandate to consider the effect of consolidation on the progress of the 1995 Base Realignment and Closure (BRAC) commission's activities (Pratt and Hayden 1993). Since BRAC is politically charged—experts in the Pentagon speculate that another round will never happen again—the DoD direction was driven by congressional attention. A common theme that remains intact is that both services' primary source of motivation to act jointly was a result of action from the legislative branch, not their own druthers or those of the DoD.

The other observation about interservice interaction to define requirements was that it was marked by robust debate—some would even call it fighting—leading to apolitical discussion of the merits of different approaches. Compromise happened in the conference rooms of the training commands and in the Pentagon; it was not inflicted in a top-down fashion. Perhaps one of the greatest gifts Congress gave the services in its firm mandate for aviation-training jointness was the encouragement to argue with each other to the point of creating a solution. While critics of the approach feel it breeds false unanimity and argue it rarely achieves the best defense outcome, there is little doubt that it forced a form of jointness. At times, participants from the Navy even aligned themselves with Air Force factions who were trying to win an intraservice debate. For instance, ATC planners coopted the Navy in bureaucratic campaigns against the ATC Operations Directorate, again exhibiting the planners' political shrewdness.

Observers allowed that the Air Force made the most significant concessions throughout the entire process. Engines and seating, the two biggest deviations of perspective, went the Navy's way. To describe a problem simple enough for bidders to understand, the Air Force also had to abandon its efforts to study training "footprint," the percentage of the pilot-training syllabus a given aircraft can accommodate. A more capable aircraft can take a student further in the program, which offers a way to relieve pressure on follow-on platforms like the then-overtaxed T-38 (Chiabotti 2014). Giving up this flexibility was probably necessary anyway, as Air Force acquisition experts believed that it to be too complex for competitors to grasp—or at least that it would open up the process to interminable protests (Sladek 2014). Regardless, footprint studies received an early heave-ho to get the two services' primary training syllabi in lockstep for the new trainer competition.

As requirements coalesced, the services prepared to make the training programs employing the new aircraft joint endeavors as well. The direct impetus was direction from the secretary of defense, who ordered consolidation of fixed-wing primary aircraft training (Aspin 1993b). The services established small exchange programs, mingling instructor pilots before educating each other's students. Even without common aircraft, the initial students and cadre indicated that the shared training mission was aligned well enough that there were no significant obstacles to success, though initial rosy assessments proved to be fleeting (Hughes 1994, 40–42).

From Requirements to Source Selection

Once a coarse definition of the requirements existed and had garnered acceptance across both services, the coalescence to a physical system was straightforward. Air Education and Training Command's (AETC) official history covering 1993 through 1995 waxed petulant about the JPATS, complaining about the uncertainty induced by fluctuating, outside influences. The record noted Deputy Secretary of Defense John Deutch's triad of stipulations that the aircraft would accommodate 80 percent of eligible women, that the program would incorporate proposals to reduce acquisition risk and cost, and that contractor recommendations for streamlining the process would be folded into the acquisitions process before the program could proceed (US Air Force Air Education and Training Command 1999, 142). It captures AETC's (ATC had become Air Education & Training Command in July 1993) exasperation at seeing its plan delayed by Johnny-come-lately outside actors' interference. Despite these setbacks, the JPATS program lost no momentum. Having fought for a robust set of aircraft requirements to meet Congress's mandate for jointness, the services' enthusiasm endured in their fight for their shared trainer on its well-vetted merits.

The winning entry, like most of the competitors, was a complex venture that paired a foreign aircraft manufacturer with domestic partners in the aircraft engine and airframe industries. Raytheon structured its proposal using the aircraft manufacturing facilities of its Beech subsidiary to manufacture an airframe based on an existing design, the PC-9, made by Swiss aircraft manufacturer Pilatus (Carroll 1995; Sweetman 1997). The configuration and technology of the winning entry, though, were a product of the necessary steps that had driven jointness; it was not a factor driving cooperation. Once the Air Force made the litany of compromises needed to bring the Navy onboard and "not buy a trainer we couldn't afford," a single-engine, tandem turboprop was likely to be the winner based on cost, which made the Beech-Raytheon partnership centered on the Pilatus design well positioned to win (O'Keefe 2014).

The complexities of acquisition strategies and programs make the outcomes unpredictable, even unintelligible to outsiders. Raytheon's selection announcement came just weeks after the *Ranger 2000* aircraft, a venture of Rockwell and German aircraft manufacturer RFB, had been announced as the best performer in a test-pilot evaluation of competing JPATS systems (Fulghum 1995, 24). While the test-pilot evaluation is a portion of the competition, it is not a decisive factor. With the advantage of hindsight, though, much of the smokescreen of corporate obfuscation has dissipated, and it seems unlikely that anything other than the Raytheon-Beechcraft could win (DeGarmo 2014). It was the only competitor to meet all flying requirements, and, as a single-engine turboprop, was among the choices with the lowest life-cycle costs.

Protests of awards are part of the process for large US defense contracts. Rockwell, Lockheed Martin, and Cessna all lodged GAO protests ("Rockwell Protests" 1995). Cessna's protest was the most earnest and engendered the most congressional attention. The protest was nominally founded on optimistic estimates of low maintenance costs that their JPATS entry would enjoy by being based on the company's newest corporate jet, but in actuality it was based on Senator Robert Dole's clout and disbelief that the Air Force would select a turboprop as a primary trainer ("Thrown Out" 1995; Chiabotti 2014; "GAO Rules" 1996).

Although Raytheon prevailed, delays induced by the gender-accommodation questions and contract-award process protests paused its start on manufacturing for seven months. The 1995 award was a blessing; Raytheon's defense-electronics business was in a slump, and it had lost a bid to build a follow-on lot of Sea Sparrow missiles. Other than JPATS and its recent acquisition of surging E-Systems, Raytheon's defense-related businesses were the company's "weakest" (Johnson 1995, 21). Allowing thirty-three months of government postaward testing, Raytheon estimated it would have "sales off the line" to foreign buyers within four years ("Raytheon Sees" 1995). The pain of the protest and the spoils-to-the-victor outcome exhibit the zero-sum nature of defense bidding, one of the factors that has drastically reduced the inventory of eligible competitors in the United States (Sapolsky, Gholz, and Talmadge 2009, 69).

Beginning Procurement and the Air Force Rollout

The initial production models of the T-6 had to overcome significant hurdles. For all the emphasis placed on the program being nondevelopmental, the resultant aircraft had less than 10 percent of its components in common with a PC-9. Pressurization, instrumentation, ejection seats, harnesses, landing gear, and a requirement for a canopy that could withstand a four-pound bird at 270 knots provided plenty of engineering and manufacturing challenges. During the early years of the JPATS program, both the trainer aircraft and the ground-based training system experienced production delays, funding shortfalls, rising aircraft cost, unforeseen maintenance headaches, and significant production delays.

Naming the aircraft went more smoothly. The name *Texan II* resonated with all concerned audiences, both because the original, World War II–era T-6 trainer was nicknamed the *Texan* and because both services had training bases in the state of Texas (Nolan 1997, 12). The JPATS team continued to display the same political savvy toward outside actors that had provided a solid foundation in the requirements formulation.

Despite the design-and-engineering growing pains, the aircraft enjoyed early success and good user reviews. In 1999, Raytheon attained Federal Aviation Administration type and production certifications for the *Texan II* (Goyer 1999, 44). Multiservice testing to certify the aircraft for student flight operations began in early 2000 (Phillips 2000, 37–38). The first production T-6A arrived at Randolph Air Force Base in March 2000, where pilots from both services flew a series of operational tests (Sligh et al. 2003, 206). The first Air Force students began training in the aircraft at Moody Air Force base in Valdosta, Georgia, in October 2001 (Heines 2004). The initial joint pilot training (JSUPT) class, comprising thirteen Air Force and two Navy students, completed the JPATS phase of training in late April 2002. While the class trained successfully in the *Texan II*, its instructors used a T-37 vintage system for the grading of students and tracking of training progress, a prelude to an issue that would plague the early years of the program and *did* impact its joint staying power (Mason et al. 2005, 195).

By 2007, the aircraft established a safety record that compared favorably with primary trainers overall and the T-37 it had replaced (Orton 2006; Groendyk 2007; Nunn and Arnold 2006). Commanders of fixed-wing training squadrons praised its performance and compatibility with the tactical fleet (Sherlock 2003; Vandiver 2014). Although JPATS achieved initial operational capability (IOC) at Moody Air Force Base, joint testing and full acceptance of the T-6 remained. The services' respective operational-test organizations completed their work in January 2003, releasing mixed results. Both agencies deemed the aircraft well suited to the role for which it had been designed, but discrepancies in reliability rates earned it an "unsuitable" rating overall. Their report went to great effort to ameliorate the harshness of that rating, and pleaded that "it [was] almost there" (Mason et al. 2005, 198). A similar much-ado-about-nothing "rejection" had occurred earlier; though OSD found the aircraft "unreliable" in a publicized report sent to Congress in late 2001, the Air Force certified Raytheon for full-scale production less than two weeks later ("Raytheon Plane" 2001; "Air Force Approves" 2001).

The Navy again benefited from the Air Force's initial struggle with the ground-training systems associated with JPATS, waiting to begin its initial test of the Training Information Management System (TIMS) at NAS Corpus Christi, Texas, until September 2004 (Mason et al. 2005, 200). TIMS, complex even by Air Force standards, was a quantum leap forward for the Navy, who had used a grading system "that could be done on the back of a napkin" for decades of aviation training; instructor pilots "were not thrilled" with the extremely capable, but complex, new computerized system (Bartholomew 2014). Despite the early growing pains, in its role as a primary trainer the *Texan II* has been, with few qualifications, a success for the US Air Force. The Navy was next in line to see if it deemed the aircraft, which it had followed the Air Force's lead on, adequate.

CONTINUED PROCUREMENT AND THE NAVY PURCHASE

Part of the Navy's easy cooperation in the JPATS program arose because of the in-turn nature of the program, meaning that the Navy would buy its aircraft *after* observing the Air Force take control of its own. Such an arrangement is beneficial to the second service. As Lieutenant General Walters summed it up, "It *was* joint, but it was joint in the best of all possible ways [for the Navy], which is to say joint-sequential. If it had been joint-concurrent, the Navy might have argued more violently than it did" (Walters 2014).

The Navy's O'Keefe highlighted the difficulty of procurement for an out-of-cycle procurement that a service has not planned for; the Navy was "told to participate and find a way to pay for the program" (2014). The cost to the Navy POM was two F/A-18s and one T-45, explaining in part why the Navy showed a tendency to delay purchasing its own fleet of T-6s. The Navy was not as desperate for a replacement primary trainer as the Air Force had become in the 1980s, and its budgetary concerns over its aviation fleet are of a different tenor than the Air Force's. The Navy has to protect a whole aviation budget from the imperatives of the larger fleet, whereas programmers in the Air Force almost always give first priority to aircraft.

In 1997, the Navy's procurement goal was 339 aircraft, not including twenty-nine needed for biservice navigator training ("Air Force, Navy Name" 1997). By 2011, the Navy amended its JPATS requirement to 297 total aircraft (US House of Representatives 2009). Congress had to encourage the Navy to buy the T-6 at its planned rate; as early as 2001 it began to indicate that it would delay procurement from what it had originally stated (Roughead 2001; Wolfe 2001). Prior to passage of the 2002 National Defense Authorization Act, Congress added language to force a Navy buy even though the service had requested none for that year; the motivation was to keep unit price and maintenance costs lower than they would be if fewer aircraft were purchased ("HASC: Air Force May Need" 2001). Congressional pressure balanced any Navy reluctance to continue purchasing JPATS, and the service bought at a predetermined or even accelerated rate, as evidenced again by 2003 NDAA markups (Wolfe 2002).

Congress continued to prove unabashed in weighing in on JPATS through specific markups, legislative addenda that prevented the Navy from spending a portion of its budget on anything *except* the JPATS. Markups are effective since no service likes to willingly leave money on the table. Congress used this tack several times in the early 2000s to keep the T-6B on track; the Navy bought to plan, though it curtailed its buy in successive POMs and ended up with just over 250 in total (Castelli 2010; Castelli 2011).

The brightest facet of the JPATS program was the aircraft acquisition, so we turn now to the lesser success: JPATS's attendant joint-training programs. An afterthought from the very start, the impetus for joint training was different, and its outcome less enduring, though in practice it may have yielded the best dividends in terms of improved interservice relationships.

Building Joint Primary Aviation Training Programs

It was clear that Congress desired to see a joint acquisition during the T-6 era. The services, seasoned by experience, knew to acquire a training system accompanying the aircraft being bought, either by having the airplane manufacturer deliver that system or—as in the case of the JPATS—by acting as the contract-awarding body (Chiabotti 2014). JPATS evolved to encompass a third aspect, one that appears to have been outside Congress's original mandate and the services' initial conception. That aspect was the creation of fully joint training, i.e., bases and programs that accommodated students from multiple services in the same facilities and flight-instruction units.

This systemic integration went above and beyond the original congressional mandate to consider a common acquisition of an aircraft and training system that both services could use for their existing, independent training programs. Congress planted the idea of joint acquisition in legislation, the Air Force watered it with a trainer aircraft master plan that brought onboard the Navy, and Air Force and Navy action officers made it grow through diligent framing of requirements and sparking interest across Washingtonian officialdom. Another entity, however, introduced the concept of joint training, placing the Navy and Air Force together in the hothouse of training each other's student. Complementary to Congress's original intent and prevailing wisdom that military training should be joint on every possible facet, the idea carried the day—even though the services probably would have never suggested such a scheme on their own.

SERVICE PREFERENCES

The Navy paid close attention to the Air Force's effort as the acquisition process moved forward, and its interest redoubled again when it realized that the outside attention—this time from DoD—toward the matter was serious and would lead to normative decisions about acquisition. More plainly, according to one of the key Navy figures involved, "Without pressure coming out of the

Pentagon, JPATS [training] would never have happened" (O'Keefe 2014; Bartholomew 2014).

The DoD directive induced perturbations that, had the upfront requirement vetting by the services not been so solid, might have put the program at risk. The AETC history office observed that changes in acquisition plans reflected a decreased procurement goal for both services (a result of the overall defense drawdown of the 1990s) and a change to the planned initial joint training bases (an outcome of Secretary of Defense Les Aspin's directive to "consolidate fixed-wing aircraft training across the board and to get started right away" [Manning et al. 1991, 128]).

Earlier we mentioned a realization that decisions made about primary aviation training would affect BRAC assessments and the basing options of both services. As O'Keefe (2014) recounted, the Navy realized early on that participating in joint training as well as joint procurement would provide additional leverage in bargaining over BRAC decisions. The services believed that having visible joint aviation training programs at certain bases would strengthen their resistance to BRAC closure. Since BRAC's attendant politics is a black box masquerading as a transparent, rule-based process, the assertion is impossible to prove, but none of the joint aviation training bases (Vance Air Force Base, Oklahoma; Whiting Field, Florida; and Pensacola Naval Air Station, Florida) succumbed to BRAC closure or significant curtailment of their training missions (Sapolsky, Gholz, and Talmadge 2009, 67).

There is ample evidence from early on in the development of JPATS that the services perceived external demand for visible cooperation. Evidence that the concept of JPATS as a fully joint undertaking had firmly established itself came in deliberate efforts by the services to ascertain how jointly administered pilot training would work well before the request for aircraft proposals was complete. In 1993, Navy instructor pilots arrived at Reese Air Force Base in Texas. Two joint primary pilot training squadrons were established with joint leadership (squadron command rotated between Air Force and Navy personnel) and instructors trained a pool of students without regard for the service from which they came (Hughes 1994, 40). Although it was a DoD mandate they might not have preferred, the services complied promptly.

The Navy and Air Force also quickly consolidated their training of non-pilot aircrew. "Almost before the ink was dry on Defense Secretary Les Aspin's April 1993 memo mandating consolidation of fixed-wing aircraft training, the Air Force and Navy agreed that the idea of joint navigator training held considerable promise" (US Air Force Air Education and Training Command 1999, 167). Suffering BRAC outcomes that had shuttered key training bases, the two services emerged in 1994 with a plan for all primary navigator students to complete training at Pensacola Naval Air Station. Some Navy intermediate students trained at Randolph Air Force Base, and both locations had squadrons that were fully joint. By the end of 1995, joint navigator training—including the production of Electronic Warfare Officers at the Navy's Corry Station, something the Air Force had not done for more than two and a half years—was in place at joint-service locations throughout Florida and Texas.

Joint training for navigators proved to be faster to establish than that for pilots. Specialized undergraduate pilot training (SUPT), the original nucleus for joint training, threatened a specter of unifying all fixed-wing training under the Air Force, which had a larger, below-capacity training infrastructure. DoD in that era was cost-conscious and not shy about inserting itself into decisions. The Navy instinctively resisted such a loss of control of its aviation-accessions training—not only would it mean a loss of shore billets coveted by officers eager for a break from sea duty, but there is a fundamental resistance to turning over control of the system that generates a service's newest trainees in any discipline, especially aviation (Webb 1996, 37). These sentiments caused the Navy to pursue status as full partners in the incipient system. Joint training, though external stakeholder input inspired it, did not have the same internal spark that illuminated acquisition. It thus emerged and existed in a constant state of discomfort within the services.

Fixed-wing joint training programs are not the only ones that have showed signs of strain. The Air Force was caught unawares in 2003 by an Army announcement that it would be shutting down its UH-1H training fleet in late 2004 despite the Air Force's reliance on that fleet since 1971 (Mason et al. 2005, 1, 168). The ensuing scramble to provide an alternative means of training fifty pilots a year demonstrates the risk of outsourcing segments of specialized aviation training to sister services, justifying the Navy's suspicion of being cut out of primary production as a result of JPATS's joint-training facet. We conclude that without long-term external actor oversight and enforcement, joint-training programs that affect core service identities (as does aviation training in the Air Force and Navy) are unlikely to survive.

Building toward a Framing Theory of Joint Military Cooperation

Considering the narrative of the JPATS development against the sample of theoretical questions raised in chapter 2 helps to identify some of the most important dynamics at play. As in the case of AirLand Battle, many of the theoretical considerations are relevant to defining the space in which jointness was affected, and these speak chiefly to reasons why jointness is difficult to attain. A smaller number have explanatory power for how tendencies against jointness may be overcome, and we concentrate on those in this section.

Theories of Organizational Interactions

PUBLIC GOODS

The magnitude of the primary-trainer problem, viewed from a national security lens, is of much smaller scope than the AirLand Battle concept of a Soviet invasion addressed in the preceding chapter or of the counterinsurgency warfare addressed in the next chapter. However, even in this peacetime acquisitions case study there are pieces of theory that apply to all these cases. For example, the Air Force tacitly recognized the appeal of free-riding to the Navy and structured the acquisition program so that it could initially do just that—in this case by being the second service to procure the T-6 after some of the initial-run kinks had been smoothed away. Once the problem was reframed in a way that stated it in terms of mutual biservice interest and established its appeal throughout all layers of the defense establishment, free-riding became an accepted feature of the creation of a greater public good. The training units of the Navy came to view the T-6 as vital to their interests and congressional pressure was able to ward off any larger institutional temptation to curtail or slow down the purchase of the Navy's share.

ORGANIZATION THEORY AND OBSTACLES TO MULTILATERAL NEGOTIATION

The JPATS saga has many of the characteristics of a multilateral negotiation in which no formal regime is available to impose certainty and order. We found that the successful acquisition effort traced back to a series of master plans, authored first to articulate the Air Force's strategy to recapitalize pilot training, and subsequently offered as an answer to Congress on behalf of all services and the DoD. Fen Hampson summarized the chief barriers to successful negotiation in these circumstances as "complexity created by the large number of parties to the negotiation and issues on the table, uncertainty heightened by the difficulties of communicating preferences and exchanging information among a large number of participants" (1999, 23). This description affirms the nature of public goods in the hands of organizations described in chapter 2—an overwhelming number of dissociative forces make the likelihood that large bureaucracies will act to preserve or advance public goods seem unlikely. Given that at least four identifiable large bureaucracies were in play in the JPATS process (the Air Force, the Navy, DoD, and Congress), one could easily expect Mancur Olson's warning to apply: "the larger the group, the farther it will fall short of providing an optimal amount of a collective good" (1971, 35).

Our story revealed, however, that individuals and institutions overcame these barriers to cooperation in the case of the JPATS. The trainer-aircraft master plans played a pivotal role because they accomplished so many of the functions essential in multilateral negotiation. They served to assist with identification of the problem, the search for options, agenda debate, issue definition, and details for a future agreement—all of which are in the purview of "experts . . . and the bureaucrats" (Hampson 1999, 24, 26) who figure in the three-phase framework that Hampson described for multilateral negotiation. Several items could have derailed cooperation since compromise meant that one side or the other would not get its desire. However, the master plan (1989) and the associated vetting dialogues at all times kept the service figures focused on the point at which their interests converged rather than diverged, i.e., satisfying Congress's unambiguous call for jointness in acquisition. The influence of the master plans, as well conceived and shrewdly constructed as they were, underlines the power of the legislative branch in joint military matters, a factor that we reenergize across later cases studies and in our conclusions to this book.

CRISIS COOPERATION: FIGHTING AS AN INDICATOR OF COOPERATIVE STRATEGY AND INTENT

The Navy and Air Force have a history of public conflict, often airing their disagreements at the expense of any meaningful compromise. Despite many opportunities to disagree on and fall apart over JPATS, though, the services argued their way to effective compromises. We lead this back to the positive correlation between decision-situation "fighting" and an overall negotiation strategy of "signaling trustworthiness" in crisis-cooperation literature (Svedin 2009, 128) If not the sole deciding factor in cooperation, fighting over contentious issues in the short term can be beneficial to a longer-term relationship based on trust. As the Air Force-Navy history and our interviews suggested, overcoming institutional distrust was a necessary step for working together on JPATS.

Given the reputations of several key individuals involved in JPATS requirements negotiations, we speculate that "violence or coercion" was used as a tack to prevail in some decision situations. A summary of crisis-cooperation data also reveals that an organization that uses these behaviors also tends to "make voluntary concessions," a behavior that participants from both services readily admitted that the Air Force did (Svedin 2009, 127). We may question whether this case study approaches the status of a crisis, but given the uncertainty of final outcome, the services' unease about losing aviation-training autonomy, and the time pressure of acquisition, procuring a new trainer fits within the malleable bounds of the definition. More useful to the understanding of joint cooperation, however, is a basic appreciation that it is helpful to talk about differences rather than act as if they do not exist, even if an argument occurs along the way. The alternative might have been a breakdown in interservice trust, and a hiding of real preferences and intent over the project's multiyear lifespan.

PROFESSIONS AND PROFESSIONALIZATION

Abbott's (1988) view of professions—suggesting that they work to control the work done and claim abstract knowledge necessary for the job—helps us understand the desire of both the Air Force and the Navy to have autonomous control of their primary training programs. Professionalism, in Abbott's account, also entails dominating outsiders who attack the control of the work or the abstract knowledge claimed. Pilot training is one of the fundamental mechanisms the services use to shape new members in each aviation group. Few cultural artifacts are as important to the Air Force as its "pilot culture," explaining the reflexive move away from Congress's T-45 nudge and instead toward building an airplane of its own choosing.

Ironically, the Air Force was so effective in using the leverage of joint procurement for a *platform* that it appeared training *programs* would also move toward increasing commonality, intruding on both services' autonomy. However, over time the Navy moved back to a point of equilibrium on its own. Painting its airplanes orange and specifying slightly different requirements for the T-6B seem like small matters, but in concert with obstacles like increased training expenses and syllabus harmonization, enough friction between the services existed that a slow, steady move back to de facto autonomy for both services resulted. Separate training pipelines allow distinct service cultures to develop and support the idea that each service contributes a separate professional capability, when really the baseline education and experience of Air Force and Navy aviators are very similar.

PRINCIPAL-AGENT RELATIONS

Feaver's (1999) model of military agency resonates in T-6 development. Feaver suggests that military behavior is a function of its interests, the degree to which the state as the principal monitors the military agent's behavior, and how much more information the military has that the principal does not have. While the Air Force did not ignore a *formal* legislative directive to buy the T-45 in the wake of its failed T-46 project, the service did read the urgency of that intent. By dint of Congress's own enthusiasm for commonality and "joint" labels, the Air Force was able to exploit a monitoring-and-punishment gap with respect to the T-45, turning that previous support into congressional support for JPATS. Such an outcome would have been impossible without dedicated and politically savvy officers who were able to provide all levels of the defense establishment with an acceptable alternative to the first legislative "suggestion." The mechanisms by which the Air Force effected this flying change are discussed in the next section.

Military-Specific Theories

From a perspective of military-innovation theory, the JPATS trainer does not lend much substance. The changes required and introduced were incremental ones related to the in-cockpit presentation of information, flying pedagogy, and the economics of operating large fleets of airplanes on a limited budget. Implicit guidance that the aircraft would follow off-the-shelf commercial designs further reduced the incentive for innovation. JPATS has satisfactorily readied pilots and other aviators for the tactical platforms they would fly later in a military career, but without a leap forward in airplane technology per se.

On the other hand, civil-military relations theory fits this case well, but it is most effectively employed in concert with a discussion of service cultures and institutional response to bureaucratic threats—in this case, the threat was the Air Force's perceived loss of autonomy with respect to its selection of primary-training aircraft. We see that Donnithorne's discussion of the way service cultures interact to shape responses to legislative and other outside influences is relevant in the JPATS case. We point to a few additional structural and actor-based factors that significantly influenced the JPATS case and thus inform our framing theory of jointness.

SERVICE CULTURES AND FEAR

The Air Force's push to comply with Congress's guidance about joint procurement while burying the specific recommendation to buy the Navy's T-45 recalls Donnithorne's (2013, iv) observations about service agency: namely, that we should expect the services to comply with external guidance, but that they will attempt to spin *how* they comply in a way that favors their existing institutional preferences.

The services, sensing congressional anger and reading implied direction, acted out of fear. For the Air Force, the single biggest worry was its loss of autonomy to determine its own training fleet. Entering the 1980s with a plan to replace its oldest trainer, at the end it had nothing to show for its effort except language in a defense bill to consider buying the same airplane the Navy was procuring. The Air Force seemed eager to turn this situation around and hopeful to avoid having to buy the T-45, not to get stuck with the T-37 for several more decades, and not to "fly the wings off the T-38" (Chiabotti 2014) by continuing generalized UPT. Perhaps the Air Force, a service originally focused on aviation, retains a subconscious conceit that it should control its own flying-training destiny. Whatever the driving motivation was, it turned unilateral direction from Congress into a joint plan in which the Air Force defined terms, through which Congress's requirement for joint participation was met, and with which the Navy went along because the Air Force compromised with the Navy on all of its key requirements (Donnithorne 2013, iv). Although the choice of a primary-trainer aircraft may seem a minor issue within the defense establishment, to Air Training Command it was a major concern, and received attendant senior-officer attention.

While some Air Force participants characterized the Navy as "always behind" in the T-6 joint acquisition process, some of the Navy's fears about what might happen in a joint acquisition process have explanatory power for its involvement throughout (Sladek 2014). The Navy, just like the Air Force, realized the prevailing importance joint acquisition had in Congress and feared getting stuck with a trainer that was far from its needs. For the Navy, who sends more than 80 percent of its primary students to a helicopter or a turboprop school for operational flying, one of the least desirable outcomes would have been a high-cost, high-maintenance, high-performance jet. The Navy was happy with its T-34C *Turbo Mentor* and not quite ready to replace it, but perturbed by the unexpected development costs that had gone into procuring its new advanced trainer, the T-45.

A second motivating factor with a foundation laid partly in fear was the ongoing BRAC process. The Navy realized that pilot-training bases were among those to be affected by directed infrastructure closures. A slide deck prepared for Pentagon briefings clearly anticipates the potential effect of BRAC, and explicitly ties the design of joint primary aviation training to its outcomes (Pratt and Hayden 1993, slide 3). The Navy and the Air Force both became aware that they had equities at stake in the BRAC process—namely, those training bases they wished to close and keep open—and recognized the need for a unified front before the congressionally appointed commission who made final recommendations (O'Keefe 2014).

Next, the Navy had an interest in closely following and being a part of both the aircraft requirements and the training system in which it was used because of its preference to run its own helicopter training system. For years there had been calls for consolidation of naval helicopter training under the Army's massive umbrella at Fort Rucker, Alabama, which had significant excess capacity. Some asked for the upcoming JPATS purchase to be reduced commensurate with a Navy curtailment of separate fixed-wing training for helicopter pilots, starting them directly in rotary-wing platforms (Webb 1996, 37). The Navy's resistance to helicopter consolidation proved foresightful; the Army's early-2000s transition to a new syllabus and aircraft mix at Fort Rucker left the Air Force flat-footed. In the end, the Air Force acquired Vietnam-era UH-1 *Huey* aircraft from the Army and established a separate training system at Fort Rucker, which was a dissolution of a joint aviation-training program that foreshadowed the split of similar Air Force-Navy programs (Tiron 2004, 38–39). The Navy's diligent JPATS participation kept pressure off to further consolidate rotary-wing training until that attention faded.

Despite the benefits that accrued to the Navy through joint cooperation, the Air Force's leadership and management of the overall decision stream are a classic example of what William Riker (1986) termed "heresthetics," the act of manipulating a "situation in such a way that other people will want to join them—or will feel forced by circumstances to join them—even without any persuasion at all" (1986, ix). By presenting the issue of JPATS first as an easy ride for the Navy, and then building external enthusiasm for jointness so that it started to threaten the Navy's autonomy to control its aircraft budget and basing decisions, the Air Force found itself with a willing partner without having to argue on the merits very much at all. Willingness turned into a sense of urgency—even apprehension—as the Navy realized the political momentum of the joint project to which it had agreed. This momentum included entailments, such as JSUPT and DoD efficiency management, not conceived in the earliest meetings.

A Diminished Role for the DoD

JPATS is analogous to the 1960s TFX program because it involved an external influence that caused the Air Force and the Navy to jointly define the requirements for a system that both services might have preferred to design on their own. The Navy demonstrated its distaste for the TFX result by never buying its planned variant of the F-111 (Lorell et al. 2013, 9). However, there are significant differences that speak to the relative influence of the type of external influence at play. The dominant force acting in the two most important TFX decisions—the decrees that it would encompass "commonality" between the two services and the direction to select General Dynamics's design over the joint-service choice of Boeing's submission—came from Secretary of Defense Robert McNamara (Art 1968, 158).

In contrast, most of the defining external inputs for the JPATS came from Congress. From the idea that the Air Force should consider a joint procurement for its next trainer, to aiding the services' effort to build momentum for their jointly developed aircraft requirements, to ensuring that the Navy purchase its agreed share of aircraft, Congress was the political force of reckoning. Although DoD directives mandated some aspects of jointness, notably to establish joint training programs in primary aviation training, these appear to be bandwagoning responses to the congressional directives that had set the overall tone of the program, and were less enduring than the legislative direction. The Pentagon protested the requirements Congress put on JPATS in 1992 as one of a handful of model acquisitions programs, saying they would place unnecessarily complex and expensive reporting burdens on manufacturers ("Pentagon Says JPATS Plan Would Burden Contractors" 1992, 29). Congress reinforced its own decision with additional legislation, and by 1996 DoD had changed its refrain, saying that reform had positively impacted the program ("DoD Says Acquisition Reform Pilot Programs Making Progress" 1996, 1).

DoD, in shaping JPATS with respect to the matter of gender accommodation, acted with significant impact as an agent of the executive branch. The matter received Pentagon attention, rallying women's advocacy groups, and attracting notice of Assistant Secretary of Defense Edwin Dorn. Nina Richman-Loo and Rachel Weber (1996, 140; see also Weber 1997, 239) traced the development of the JPATS through a lens of gender, noting that the system's original sitting-height requirement of thirty-four inches would have excluded 50 to 65 percent of the female population. This would have undermined Secretary of Defense Les Aspin's April 1993 directive that said, "the services shall permit women to compete for assignments in aircraft, including aircraft engaged in combat missions" (1993a, 1).

Excluding half of the nation's female population via anthropometric-ergonomic standards derived from the military's standard male population measurements had an appearance of deliberate resistance (Rosser 2001, 170–174). The under secretary for defense for acquisition directed the assistant secretary of defense for personnel and readiness to develop new sitting height requirements that would accommodate at least 80 percent of women, a requirement that became, according to the aircraft manufacturer, "the single greatest challenge in the JPATS program" (Aris 2003, 148). Weber (1997, 243–244) showed that a debate internal to DoD—chiefly about whether to justify change on the basis of gender equality or foreign military sales potentials—over the proposed alternatives ensued, but ultimately resulted in a recommendation to alter the sitting-height criterion.

The DoD initiative in directing gender accommodation, a direct extension of White House policy, had comparatively little joint effect on the services' acquisition programs, though Cessna did hold up the contract with a protest on gender-accommodation grounds (Mintz 1996, E1). Given executive intent and legislative mandate, all services would have had to comply with the instruction. The JPATS trainer aircraft did, however, give DoD a convenient single entity through which it could impose the directive. The existence of JPATS and its ongoing requirements development provided a host of outside actors a vehicle through which policy could be shaped. This demonstrates the wide array of actors who may exploit joint initiatives that carry sufficient, self-sustaining momentum. One need not act to *force* jointness; an able practitioner can use its existence and the dynamics it fosters to further other political ends.

To the services, DoD input on major acquisitions programs spans a spectrum from well-intended but tone-deaf advice to punitive encroachment.

An example from the former was an OSD recommendation that contract award be delayed from February 1995 to February 2002 (Emmons 1991, 30). Then the director for program analysis and evaluation under the Secretary of Defense William Lynn's mid-1994 "budget drill" was used as a buttress to a Congressional Budget Office report recommending T-37 life extension by relying on the Navy's T-34 fleet for a higher percentage of primary training (Fulghum and Morocco 1994, 23). Nine senators, led by Dole, wrote to Defense Secretary William Perry, reminding him that "Congress has been deeply involved in and guiding the JPATS effort since its inception" and chiding him for "effectively canceling the program" (US Air Force Air Education and Training Command 1999, 147). Underlining the relative strength of Congress over DoD, the two reports (one released by a congressional entity) served as a warning to the services that being *too* joint invited unwanted interference, as it allowed outside entities to identify potentials for efficiency that might not align with service preferences or autonomy.

We can picture acquisitions programs as an object being acted on by myriad external forces. The services generate some, DoD others, the White House and Congress still more. In the JPATS instance, the weaker but steady service forces seem to have worked mostly in line with congressional intent, with the DoD and the larger executive branch only occasionally expressing perturbations. Given that JPATS led to actual joint acquisition—where TFX and several other programs like it did not—the question arises as to whether Congress's external influence is qualitatively different and more effective on the service than that provided by DoD (Lorell et al. 2013, 18, 20).

STRONGER CONGRESSIONAL INFLUENCE

We have seen that external influence exerted significant force on JPATS. The most obvious effect from the highest strata of defense decision-making is an unbroken thread of interest, legislation, hearings, and directives related to the program. From the moment the T-46 failed, members of Congress made clear their displeasure at the Air Force's requirements creep. The Air Force clearly received this communication and the formulation of an earnest response was coded in the organization's DNA for the next several years (Morgan 2006, 40–41). The words "directed by Congress" appear in early program briefings and the acquisition documents that followed. A conference-committee report harmonizing the House and Senate versions of the 1989 National Defense Authorizations Act endorsed the Air Force's plan to return to specialized pilot training, directed the Air Force to consider procuring the T-45 as its advanced trainer, asked to examine the possibility of procuring the PATS aircraft in concert with the Navy, and directed DoD to submit a report to the HASC and SASC on its plans for joint aircraft acquisition (US House of Representatives 1989, 1–2).

Congress also reflects and magnifies public debates onto the defense establishment. JPATS received extra attention because it developed during an era of significant military cultural change—opening a large number of military positions to women—and the legislative branch refused to let the executive branch have the only say. Legislative intervention imparted final resolution to the matter of gender accommodation, with a Senate amendment threatening nearly $40 million of the Air Force's $41.6 million trainer budget unless the Pentagon revisited the cockpit design to accommodate more women ("Senators Want Training Aircraft to Accommodate Female Pilots" 1993, 6). The Pentagon complied, relaxing the sitting-height requirement (Secretary of the Air Force n.d., 9). More significantly, the controversy thrust JPATS into the public arena, making it a proxy for the social issue of women's combat roles (Weber 1997, 243–244; Banks 1994, 20; "Sit Tight" 1996, 329–330). Once the controversy was settled, the services spotlighted the airframe's inclusivity by having well-known female aerobatic pilot Patty Wagstaff fly the aircraft at large military air shows (Goyer 1999b, 36). As in the case of DoD concern over this matter, gender issues did not greatly impact jointness, but rather witnessed Congress using a strong, ongoing joint project to demonstrate proactive control of a politically significant issue.

Whether it appears explicitly in the Congressional Record or not, individual delegations have an outsized effect on aircraft procurement. While Beech, as a partner of Raytheon, was a winner in the JPATS contract, it did not fare as well in a later competition for a large Air Force contract, and reacted in a way reminiscent of its competitors' challenges in 1995. Action officers at the Air Force's Office of Legislative Liaison would regularly take phone calls from Kansas protesting the "outsourcing of national defense to Brazil" and other unpatriotic, if not treasonous, allegations during initial selection of the Air Force's Light Air Support (LAS) platform (Biass and Fox 2012, 23). The LAS was a counterinsurgency aircraft concept to be sold or given to countries that receive US military assistance. Had it progressed, it would have had a potential contract value of approximately one billion dollars. Hawker-Beechcraft undermined a contract originally awarded to Brazil's Embraer and Sierra Nevada Corporation in 2011 with a protest to GAO (McCoy 2012). After restarting the competition, the restyled Beechcraft again lost a 2013 competition to the Embraer–Sierra Nevada partnership and was again rebuffed through a subsequent GAO protest (McMillin 2013).

During the JPATS contract fight, Senate Majority Leader Dole was in the awkward position of "supporting one constituent [Cessna] and slamming another [Beech]" when he wrote a letter criticizing the Air Force's award to the Raytheon-Beech proposal—Beech and competitor Cessna are neighbors in Wichita, Kansas, and the Raytheon-Beech team had recently attained a significant victory by winning the competition to provide the Tanker-Transport Training System (TTTS) aircraft in the form of the T-1 (Mintz 1996, E1). Kansan advocacy later shifted back to Beech, which benefited by being able to develop weapons demonstrations for the AT-6—an LAS candidate—"funded by earmarks from the Kansas congressional delegation" (Dorr 2012, 64). Roy Braybrook (2013) noted this external influence in highlighting Pilatus's development of a high-performance turboprop trainer out of a desire to not "compete directly with the PC-9-derived, *Washington-backed Beechcraft T-6*" (Braybrook 2013, 47). Beechcraft was always ready to line up at the trough of defense appropriations with congressional help, and its frequent bouts with bankruptcy preceding its acquisition by Textron in 2014 suggest it was less able to manufacture economically viable aircraft. The JPATS legislative history suggests that a practitioner of jointness who knows acquisitions processes can exploit domestic politics to help cause an interlevel coalescence of interests. JPATS was artfully conceived and forcefully marketed on Capitol Hill, mostly by its Air Training Command advocates, granting it enduring legislative attention. With the initial procurement questions settled and a viable airframe ready for manufacture, JPATS remained popular with congressional appropriators. The SASC, for example, directed additional procurement in the 1997 National Defense Authorization Act, marked up a 2000 appropriations bill with an additional fifty-four million dollars to buy twelve more JPATS aircraft than the administration had requested, and recommended forty-six million dollars in advance procurement (Foote 1999, 1). The congressional markups admonished DoD for attempting to delay trainer procurement ("Senate Appropriators" 1996, 1). Even when JPATS ran afoul of Nunn-McCurdy requirements for cost overruns in 2007, it received prompt recertification and continuation as a program "vital to national defense" (US Department of Defense 2007).

The extensive political attention heaped on JPATS, if read alongside some of the harsher criticisms released by the test community during early development, might lead a cynic to believe that the T-6 was a suboptimal product. In fact, the T-6 is quite good at accomplishing the mission for which it was designed. The lack of bells and whistles further suggests that designers limited the requirement creep that had plagued the T-46 and so many other military aircraft. As Sapolsky, Gholz, and Talmadge noted, "Projects require more than desire and friends to succeed. The process cannot be only about politics. No matter how many congressmen feel that their districts got a big chunk of the work, they, too, will abandon the project if it cannot serve its assigned missions" (2009, 94). The *Texan II*, with a foundation of a well-designed COTS airframe and buoyed along by a rigorous requirement process that held the line on essentials, retained support throughout its political half-life on the basis of its technical merits.

Things to Move Forward With

A summary of how the JPATS case fits with the variables of the framing theory appears in table 4.1. The biggest appreciation we gained here is for the outsized effect of external influences, which at once seem to be the impetus and continuance for the successful parts of joint cooperation. A lesser outside directive from DoD to conduct joint training seems to have faded. Thus, the organizational, collective-goods, and professions barriers to cooperation are in full view via the failed TFX initiative, while the effects of outside influence are apparent because the Air Force and Navy overcame all of these things with JPATS. As in the case of AirLand Battle, JPATS also showed the need for a strong service leader or leaders to carry the banner for jointness

Table 4.1: Theoretical Observations on the JPATS Acquisition and Associated Cooperation

JPATS Observations	
Theory	*Observed Outcomes for Institutional and Operational Jointness*
Public Goods	Any temptation for free-riding quickly overcome by mutual service interests
	Acquisition jointness pursued in good faith, overcome by service preference
	Training jointness most likely a forced endeavor; did not endure
Organizations	Fear of losing acquisition autonomy motivated the Air Force to lead process
	Effective heresthetic approach by the Air Force got the Navy onboard; good faith partnership and negotiations kept it there
	Continual congressional interest applied enduring pressure for cooperation
	Outside influence and occasional crises kept the program in the spotlight
Crisis Cooperation	Services exhibited fighting decision–situation dynamics at times on the way to a signaling trustworthiness cooperative interaction strategy in the project overall
	Examples of coercion counterbalanced by concessions from both services
Professions	Desire to shape service identity through service-specific control of primary aviation training
Agency	Services possessed an information asymmetry with respect to Congress, but rather than use it to mask intentions, early participants waged an aggressive public affairs campaign to build support for the desired program, which was in line with the overall legislative intent

continued

JPATS Observations

Theory	Observed Outcomes for Institutional and Operational Jointness
Military Innovation	(Little substantive innovation observed. JSUPT enabler was a return to an earlier training method. Candidate aircraft were modified to win contract, but using common aircraft technology.)
Civil-Military Relations	Congressional dissatisfaction with Air Force program sets the stage for JPATS
	Goldwater-Nichols Act and jointness zeitgeist are a subtext for all cooperation
	The services used the momentum of congressional enthusiasm for joint projects to build support for their preferred brand of jointness
Service Cultures	Service history over aviation and contrasting training cultures made JPATS cooperation without exogenous influence unlikely
	Training preferences were never completely resolved; joint training pipelines were relatively short-lived as a result
	Sequential acquisition strategy helped to sell Navy on program, but also allowed the loss of economies of scale in production
Defense Department & Joint Staff Structures	DoD figures make several interventions in the program, reflecting their understanding of its prominence and symbolism to Congress
	DoD interventions, while garnering publicity, threatened or delayed cooperative programs without substantially improving them
	JROC process a key part of approval for the program in the Pentagon, but the responsible service commands again used it to advance their desires, which had been settled among endogenous-level figures
Other Exogenous Factors	Congressional interest is far and away the biggest cause, driver, and shaper of the JPATS process, interacting at all phases of the program with service preferences

without tiring. Outside influence can provide momentum and support, even the initial imposed requirement, but service action is ultimately required for something joint to happen. Finally, theoretical sources predict the temporary nature of the kind of cooperation exhibited in the instance of JPATS. When coordination emerges to achieve "well-defined packages of issue-specific goals," coalition-like behavior that exhibits strong appearances, teamwork, and solidarity accompanies the effort, but "the unity is focused narrowly both *substantively and temporally*" (Najam 2001, chap. 2, 39). Having used jointness to see off the threat of greater congressional or DoD interference in their aviation training programs, the Air Force and Navy were free to drift apart when other issues absorbed the outsiders' attention and lessened the threat of outside involvement.

Conclusion: Satisfactory Cooperation

Though a peacetime acquisition project for the most foundational military-training requirement, the JPATS program was a successful example of joint cooperation. It produced an acceptable result for both services and an efficient acquisition that neither service would have done alone. The Navy felt that the "Air Force contracting expertise" yielded the Navy a "better training system than we would have obtained on our own, particularly in regard to the simulators" (Vandiver 2014) and other ground-training aspects. Both services sang its praises as a suitable trainer, easy enough to fly so as not to overwhelm students but with power and aerobatic capability sufficient to prepare them for advanced training.

The JPATS case had many of the expected advantages of joint cooperation within reach but the case also illustrated that compromise often means neither side is truly happy but both feel that they have gained enough. We concede that its pursuit of a joint program robbed the Air Force of some flexibility in deciding what balance to strike between primary- and advanced-training platforms, and that killed potential savings. Factors like this are one reason why joint programs tend to cost more than comparable single-service efforts (Lorell et al. 2013, 17). Though neither a tactical fighter nor a 'complex' aircraft that tend to be the worst offenders, the T-6 nevertheless exhibited the higher cost growth typical of joint programs (Lorell et al. 2013, 17).

We consider JPATS a success case because the T-6 was well suited for its mission requirement. Even forceful critics of the services' joint training programs conceded that the aircraft itself was sufficient (Sirak 2007, 1). Once satisfactory requirements emerged from the rigorous biservice process of the late 1980s and early 1990s, the process of identifying the airplane and its manufacturer was by military standards of experience smooth. Joint aviation training, a DoD-directed addendum, was less successful. Though yielding interservice exposure for hundreds of young officers, it was not an original motivation for JPATS. Despite surviving for a time in the spirit of jointness that characterized the 1990s, it lacked the congressional imperative assigned to the platform acquisition. The services identified reasons why it would be easier to revert to independent programs and, lacking enforcement or oversight from DoD or Congress, slipped back into old, independent training habits. Airspace, efficiency, and BRAC concerns subsided over time, removing external forces that kept the most delicate part of the experiment intact.

Case Epilogue: Nostalgia for Jointness

By 2014, consensus within both services' pilot-training enterprises held that the "Joint" in JPATS was an artifact of a bygone era. Despite that, the joint acquisition program remained successful, though that is an unsurprising outcome. Once purchased, services are committed to the platforms they buy through their design life and very often well beyond that. Congressional attention ensured that the Navy completed its total procurement, and auditors' reports reflected a mature acquisition program with neither significant risk nor outlandish cost increases (US Government Accountability Office 2014, 161). How far did the joint-training aspect of the program regress? Air Force and Navy instructors no longer teach students in common units where primary leadership positions rotate among the services. There are no Navy or Marine Corps students at Vance Air Force Base and no Air Force students at Whiting Field. On the installation where both services do maintain primary aviation training (at Pensacola Naval Air Station, where Air Force navigators and Naval Flight Officers perform initial training), a previous effort at jointness and integration has disappeared, though the services do at least share the same runways.

Reflecting a strong sense of ownership, those involved in the incipient JPATS program exhibited some regret for the current vector of primary aviation training. Overall, a sentiment of bittersweet nostalgia for the jointness manqué pervades conversations. Those who laid the foundations for JPATS and the multiservice training that grew out of it do not grasp why successors have diverged from a road map that provided economies of scale in aircraft production and early-career exposure to sister-service perspectives. Instructors who taught in those joint programs are not as nostalgic; they emphasize the difficulty associated with producing a quality graduate from the mixed lines, going so far as to refer to joint training as "a controlled failure" (Bartholomew 2014; Armstrong 2014). It does not seem that there are any significant political forces pushing the services back together on this front, and rumors of a congressional investigation into why the Air Force and Navy needed to construct completely different facilities at Pensacola when they had shared buildings for so long never materialized (Bartholomew 2014).

It is possible to trace the people and technology that led to the construction of JPATS through the lens heterogeneous engineering. It is more difficult to chronicle its dismantling, because that responsibility goes unattributed. Official histories rarely record who made a decision to end participation in a joint program. The contracting officer who signed the order to paint T-6Bs according to a different scheme than the Air Force has his name recorded for posterity, but the decision was not his and responsibility is untraceable behind a screen of several committee meetings. It is difficult to assign responsibility for the undoing of jointness to specific people. The loss of jointness in primary aviation training resulted not from a flag officer's willful declaration in the wake of an interservice controversy or substantive reorganization; rather it bled out from a thousand cuts. Where "jointness" had been a rallying cry of the early program, service culture started to hold sway as aircraft began to arrive on the ramps.

Different training base commanders began to put their own stamps on the program (O'Keefe 2014). A Navy decision to adopt its traditional orange-and-white paint scheme for trainers began to differentiate the aircraft, and the Navy's T-6B variant created two separate fleets with minor but divisive maintenance differences (O'Keefe 2014). Early syllabus disputes that had seemed minor in the early 2000s became a major focus of interservice dispute, and both services involved began to focus on the extra training (each service found it necessary to institute "top-off programs" for students training at a sister-service base before integrating them into advanced flight training) and additional expense (the Navy had to move students to and from Vance Air Force Base, Oklahoma, and the Air Force students had a diversion to Pensacola, Florida, for example) that inhered in joint training (Armstrong 2014; DeGarmo 2014; O'Keefe 2014; Vandiver 2014).

We cannot help wonder what the state of JPATS would be today if joint training had retained priority in the external defense establishment. The 1993–1994 DoD directives to execute training jointly were not rescinded; they just faded from memory. Chiabotti (2008, 272), describing the benefits of a turbojet over a turboprop, wrote that "the physics of rotational mechanics are unrelenting." Spinning components must be tightly bolted together, and the pilot at the controls must adjust for forces pushing the system apart. The same analogy exists for joint programs and JPATS. Pushing services to do something together when there is no internal motivation precipitates a tendency for those programs to fly apart. If no one acts as custodian, jointness ends. Where was the caretaker of joint-aviation training after the early 1990s? A major shift in the world security climate occurred in September 2001, and DoD's focus went more to combat than training just as the first students started training in the T-6. Previously a guarantor of jointness, DoD had its attention and gaze fixed first on Afghanistan, and then on Iraq. Urgent counterterror focus siphoned off the energy available to keep JPATS joint. Our next chapter returns this investigation to the realm of combat and partially helps explain why JPATS, like ALB, only had temporary sway in creating jointness within its realm of influence.

Chapter Five

Air Support in Counterinsurgency, 2001–2012

One fact of life is not likely to change by the year 2000:
the Soviet Union will still be the threat—
either directly or through surrogates.
—General Wilbur Creech, US Air Force, October 1981

The Navy and Marine Corps fighter pilots routinely flew as
low to the ground as they could to achieve the effects, even
when it was below what was deemed minimum safe distance.
They were terrific. The Air Force had to work through airspace
management—aircraft were stacked up to the ceiling and
could only be flown in, in a few numbers.
—Major General Franklin Hagenbeck,
US Army, June 2002

Introduction and Background

Counterinsurgency (COIN) and Direct Air Support to Ground Forces

Thus far we have examined two instances of interservice cooperation that occurred during peacetime. This chapter examines the phenomenon during combat operations. In contrast to AirLand Battle (ALB) in chapter 3, the context for cooperative effort is not a hypothetical battlefield facing the Soviet Union. The focus was instead on fighting together in a type of combat labeled variously as "asymmetric war," "low-intensity conflict," "small wars," "counterinsurgency" (COIN), "security and stability operations," or "guerrilla war." We focus on specific air-power capabilities that support ground combat.

The epigraph from Gen. Wilbur Creech, who for six years supervised the Air Force's battlefield support provided to the Army under AirLand Battle, is representative of US military strategic preferences. Senior military leaders of Creech's era, and the American military services in general, tend to concentrate planning efforts on responses to an advanced, technologically comparable enemy (a contemporary term is "near-peer competitor"), who puts essential US interests at risk. Anticipated conflict involves "major combat operations" (MCO) and avoids uses of force characterized by greater political restraint (Builder cited in Brzezinski 1979). Creech's prediction, inaccurate though it was in describing the conflict the US military would wage for the first two decades of the twenty-first century, reflects a sentiment shared by many civilian government leaders and strategists (Carver 1986, 786–787). The military services, forced against expectations to perform COIN warfare in Afghanistan and Iraq, had to rediscover long-forgotten habits and competencies in order to be relevant and effective in the context of these conflicts.

Corum and Johnson (2003) drew the following conclusions about COIN conflicts:

"The support role of air power is the most important and most effective mission in a COIN conflict";

"COIN wars are intelligence-intensive, meaning that knowledge of where small bands of guerilla fighters hide and the coordinating of such information among different military and civilian agencies is essential"; and

"Effective joint operations are essential for the effective use of air power; again, the cooperation must occur among both the military services but must also involve other agencies involved in the overall COIN effort" (2003, 427, 34, 33).

The characteristics of the Afghan and Iraqi conflicts underline these conclusions. Both wars witnessed a heavy reliance on "close air support" (CAS) and airborne surveillance to support ground forces, and they featured military strategies enabled by detailed intelligence of the type described by Corum and Johnson. These two specific air power applications played important roles in these COIN wars, and their effectiveness was dependent on the degree of joint cooperation that occurred. CAS and reconnaissance enabled by "unmanned aerial vehicle" (UAV) are thus indicators of cooperation and germane topics for exploring jointness.

Direct air support to ground operations has a checkered history of joint cooperation in both COIN conflicts and conventional wars. CAS and the provision of aircraft-gathered intelligence to ground units have suffered similar maintenance neglect historically. The specific capability of drone-enabled reconnaissance is a relative latecomer to combat, but the Air Force, in the development and fielding of drones (remotely piloted aircraft, or RPAs in Air Force lexicon), exhibited the same reluctance as it has toward supplying CAS to ground troops. There exists a repeated pattern of Air Force-Army interaction: the Air Force has resisted supporting the Army by apportioning aircraft directly to ground commanders, instead favoring centralized control whereby an Air Force commander with responsibility for an entire region sets priorities for the use of air assets. This trend sets the context for the cooperation—and occasional lack thereof—that is this chapter's focus.

Themes in This Chapter

Our broad research finding is that the military services increased their use of air power in direct support of ground forces during the campaigns in Afghanistan and Iraq, albeit in a way that ran counter to services preferences, especially for the Air Force. Though such cooperation was an example of combined-arms jointness, the extent and quality of cooperation fell short of ideal, facing cultural and bureaucratic obstacles in the case of the Afghanistan and Iraq conflicts. Where our first two case studies looked at institutional joint cooperation for which a need was identified by and pursued by service leaders acting jointly (level b in military-innovation parlance), the pattern is markedly different for COIN air power. Significant (but historically consistent) battlefield failures (a failure of operational jointness) preceded a "decapitation" of Air Force leaders. The removal of both the secretary of the Air Force and the Air Force chief of staff by the secretary of defense sent an urgent message about that service's interest in COIN, and constitutes a rare imposition of jointness by echelons above the military services (level c) *in response to battlefield conditions.*

Initial overtures toward institutional jointness rose because of the security crises the US military faced in Afghanistan after September 11, 2001, and then later after invading Iraq in 2003. Both wars began as a wounded and enraged America struck out with allies against entities it believed supported terrorism. The conflicts followed a pattern that witnessed the dismantling of an entrenched regime as the first stage of a conflict; opening battles marked the closest approach to major combat operations on the spectrum of conflict. Afterward, however, both conflicts witnessed the United States and its allies contending with power vacuums in the countries they had invaded. After the ousting of the Taliban government in Afghanistan and the de-Ba'athification of Iraq, nascent insurgencies began to threaten US goals and the stability of these new protectorates.

In addition to the first-order interest of removing terror organizations' safe havens, a secondary concern became the establishment of peaceful, democratic governments in the wake of deposed regimes. To address burgeoning sectarian anarchy and tribal warfare in both countries, military efforts later focused on destroying the elements deemed harmful to stability and rebuilding national institutions capable of maintaining order. The plan for operations in both countries attempted to isolate and reduce the influence of terror organizations and uncooperative warlords, struggles that typify COIN warfare. While operations at this low end of the spectrum of combat have not been historical favorites of either service, in this instance the Army embraced the concept more quickly than the Air Force did (Beebe 2006), and the Air Force's recalcitrance to do so quickly became of concern to a Pentagon leader who was postured to address it decisively.

Two of the most notable advances during this period were air power specialties of particular use in COIN warfare: CAS and RPA surveillance. RPAs, whose nascent capability had shown promise in previous conflicts and yet struggled for decades to find a suitable home in the US military, became an entrenched part of military air power during this case. A confluence of internal service factors (level a), among services factors (level b), and external stakeholder relations factors (level c) contributed to overcome this and other barriers to jointness. For the Air Force, this provision of support conflicted with its most dominant doctrinal concepts, what the service calls "centralized control" organizing all air power capability under a single commander. The services and DoD had to confront these preferences to realize the level of joint cooperation eventually attained, with the Air Force going through the highest amount of institutional trauma—via Secretary of Defense Gates's removal of its senior leaders—to get on the path to jointness. In the examples we examined, individual actors were decisive in overcoming obstacles, although revising uncooperative habits occurred gradually over the long years of conflicts. Yet because the push for jointness came from the exogenous stratum and not among the services themselves, advancements toward jointness consisted of what could be grown organically (on the battlefield in terms of tactics and what was already in the development pipeline for military contractors in terms of equipment) rather than the truly novel, generation-skipping systems and doctrines that emerged from AirLand Battle.

The increase in COIN air support capability was evident in areas of materiel, technology, and training. During the two extended major combat operations (MCO) combined with COIN conflicts during the first decade of the 2000s, substantive improvements in CAS and, later, RPA-reconnaissance capacity developed during these persistent armed conflicts. It was motivated in part by the threat of failure, injury, or death that confronted the soldiers, sailors, airmen, and marines charged with executing US security strategy under conditions of enduring combat. A principal observation from this case study is that the urgent and high-threat nature of war drives cooperation where many other mechanisms fail (Svedin 2009, 19–20). The characteristics of military service at war and the challenges it faces are, in the parlance of technological history, a perfect opportunity for "heterogeneous engineers" to disrupt the status quo and create new technological systems (Law 1989, 113, 131; Hughes 1987, 64–66).

Even with the forcing function of combat creating positive inputs, significant service-level obstacles to jointness remained in place, not the least of which was the Air Force's predilection to eschew CAS. Command and control also remained problematic, especially in transitioning from an MCO effort to a protracted COIN struggle. The Army and the Air Force, after a tragic breakdown of joint coordination early in the Afghan conflict, reacted swiftly to correct battlefield coordination measures that had fallen fallow during peacetime. The Air Force, despite a jarring blow to its autonomy, continued to exhibit a reluctance to support ground missions, and interpersonal dynamics often interfered with efforts to make air power truly responsive in a COIN fight. Prospects for air support of a ground force that marks a successful COIN effort thus remain in doubt for future conflicts.

Differing views of the nature of the armed conflicts in Afghanistan and Iraq competed in the minds of component commanders, creating discontinuities of strategic aim. The split occurred between the external and internal parts of the defense hierarchy, with the services (level a) and interservice structures (level b) at times out of sync with larger the Department of Defense and higher executive branch (level c). The services, Joint Staff, and combatant commanders at the first two levels of analysis viewed the armed conflicts as significant crises, while the Department of Defense under Donald Rumsfeld initially adopted a business-as-usual attitude, only later ascribing more urgency to the conflicts (Jumper 2013). Gates's direct intervention to remove the top echelon of Air Force senior leadership in 2008 marks a distinct inflection point as the conflicts ground on. Congressional attention throughout mostly went to technology and acquisitions programs, with little focus on military cooperation per se or even warfighting effectiveness. External actors' dithering and inconsistency with respect to strategy resulted, presenting a paradox that affected jointness. The paradox was rooted in US preferences about the use of force, which may have exacerbated the conflict between the Air Force and Army about how best to support ongoing operations with air power.

In addition to strategic inconsistency, the refusal of senior military leaders to compromise on doctrine further hindered jointness. Failure to adapt command-and-control structures to the military operations occurring in Afghanistan and Iraq caused high-level interservice strife. A second paradox arose, this one with regard to how best to structure command and control of joint forces to meet the operational requirements of COIN. These disagreements in turn made the application of air power in combined-arms efforts less efficient. That these threats were later overcome by different individuals occupying the same offices as the conflict wore on speaks to a major conclusion of this chapter and of this study: the role of individual actors—at *any* level of the political-military bureaucracy, and especially at a relatively low level of the service strata when institutional leaders cannot find common ground for cooperation—is important to the success or failure of joint endeavors. "Who leads matters" in crises (M. Hermann et al. 2001, 84), whether fighting on a battlefield, establishing command-and-control systems, or providing clear direction about military strategy.

Counterinsurgency (COIN) Warfare and the Role of Air Power in COIN

COIN AND ITS APPROACHES

Since 1945, the US military has faced a blurry muddling of wartime and peacetime considerations. Previous work on this topic summarizes in the following points (Birch 2014a):

> The US military, contrary to its cultural preferences and broader American strategic preference, often finds itself involved in conflicts that fall short of full-scale war;
> COIN, including the operations waged in Afghanistan and Iraq beginning in 2001 and 2003, respectively, mostly inhabits this realm of low- to mid-intensity conflict;
> There are two approaches to waging COIN: the *direct approach* is annihilative and aligns better with US strategic preferences; the *indirect approach*, a more subtle strategy of separating combatants from a neutral civilian population, has met with more historical success.

During the limited time frame that this chapter focuses on, the US military found itself fighting conflicts of a type that it does not relish, while debating the philosophies and means best used to wage them.

CASE STUDY FOCUS ON COIN AIR POWER

The Afghan and Iraqi conflicts from the early 2000s constitute an extremely large body to draw observations from. To limit the scope of material to a tractable level, we have focused on the air-support missions used in the COIN conflicts in Afghanistan and Iraq that began for the United States in October 2001 and March 2003, respectively. The capabilities selected for attention are CAS and drone operations. With respect to UAV operations, the study focuses on the support they provided fielded ground forces, which started out mostly as observation and intelligence, growing over time to include kinetic strikes, usually with the same types of missiles and bombs that are associated with CAS.

While we emphasize the role of specific air power capabilities that often appear in COIN conflicts, these capabilities are useful in armed conflict of any intensity. "COIN air power" is convenient shorthand that captures the missions in their most recent context of the Afghan and Iraqi conflicts, but applies to all COIN conflicts. In describing the development of these COIN air power missions, we differentiate their application in "regular" and "irregular" wars. Trends observed across the spectrum of conflict are important in assessing the likelihood of interservice cooperation in the COIN context. We limit our case-specific discussion in this chapter to the CAS- and reconnaissance-specific application of air power, even though other types of air power like military airlift are essential to both conventional war and COIN (Corum and Johnson 2003, 427; Lorenz 2013).

CAS: CLOSE SUPPORT TO GROUND TROOPS FROM AIRCRAFT

CAS is a military term of art that refers to delivering bombs, bullets, or even just intimidating noise from aircraft flying close to friendly ground forces—so close that their operations need to be tightly coordinated as a battle unfolds (Joint Chiefs of Staff 2009). In contemporary conflicts this coordination comes from expert tacticians with detailed training in helping pilots find and attack their intended targets, while avoiding unintentional harm of friendly forces. These experts are called "Joint Terminal Air Controllers" (JTACs) in US and NATO military language, and they communicate with airplanes while embedded in ground units or while they themselves fly over the battlefield in an aircraft with special sensor and communications capabilities.

Close air support performed by the US military sets the pattern for the inconsistency observed across the military in all proficiencies relevant to COIN: the services, having let training lapse during peacetime, are generally weak at the beginning of any new conflict, and then go through a period of development that causes proficiency and technology to rise to an acceptable and effective level. Later in this chapter, we examine how CAS—regardless of where on the spectrum of intensity a war falls—conflicts with the Air Force's culture and foundational doctrine.

ISR: OBSERVING THE BATTLEFIELD FROM ABOVE

Another mission that waxes in wartime but in which air forces otherwise lose institutional focus is the act of observing the close battlefield. Gen. Robert Marsh (1999) said of tactical reconnaissance that, "When we're at war, everyone wants it, but in peacetime, nobody wants it." The military services collectively refer to information gleaned about the areas in which they are operating or will operate as "intelligence, surveillance and reconnaissance" (ISR). A ubiquitous military acronym, "ISR" refers to the information itself as well as the means and acts of gathering it. The scope of the enterprise is vast, encompassing several subspecialties and tens of thousands of system operators and processors of the data those systems collect (Callahan 2013). Our discussion focuses on a narrow band of ISR activity that deals with gathering information from airborne drones, most often referred to as "UAVs" by the military (Ehrhard 2000). Originally the first military application of aircraft (and the reason why the first airplanes belonged to the US Army's Signal Corps), observation, reconnaissance, or other forms of intelligence gathering have long been conducted from dedicated aerial platforms (Hurley and Heimdahl 1997, 4–5). We focus on the use of unmanned platforms to accomplish these tasks. More importantly, in the Afghan and Iraqi conflicts, ground forces grew increasingly dependent on the ability of UAVs to provide a real-time picture of the territory for which they were responsible.

Although serious work on unmanned aerial platforms was underway as early as the end of World War II, the role of UAVs in accomplishing battlefield reconnaissance in the form it exists now did not evolve until Vietnam, and was then quite primitive (Whitmore 2021). Ehrhard's (2000, 494) history of UAV development recorded approximately thirty-five hundred US Air Force combat UAV reconnaissance sorties in Vietnam. After a period of innovation, institutional forces and technological limitations conspired to stop UAV development until the mid-1990s. Since then, however, the platforms and their capabilities have pervaded battlefields around the world, with an undeniable effect on combat (Whitmore 2021).

Our investigation found that UAV-ISR support to ground forces shares many of the same characteristics that make CAS a source of tension between the Air Force and ground forces in general. The tension, as with the CAS mission, is especially palpable with the Army, for which the Air Force has historic—so traditionally grounded as to be inescapable—ground-support responsibility (Pyatt 2014). As with CAS platforms, the Army has pursued its own UAV-ISR fleet after complaints that the Air Force inventory and apportionment schemes did not meet combat needs in COIN wars. The Army, understandably, would like to have a complete view of any battlefield, down to the last square inch of terrain. As with CAS aircraft allocation, the demand for ever-more streaming UAV "feeds," able to show enemy movement near remote outposts in the Afghan wilderness or around the next corner in a hostile Iraqi city block, led to an insatiable appetite for the aircraft and systemic support that provide full-motion video.

Before beginning the historical narrative, it is worth noting the overlap between aircraft described as "CAS platforms" and those described as "ISR platforms." Though distinctions were fairly clear even as late as the early 2000s, the increasing ability of military aircraft to perform multiple missions clouds the matter and can lead to confusion. For example, since 2001, US UAVs have had some capability to fire missiles or drop bombs, a "hunter-killer" capability that complements their role of gathering ISR over the battlefield (Keller 2013). Fighter aircraft that perform CAS have developed more elaborate and capable suites of sensors that allow them to observe areas of interest for ground commanders, sharing their live video feeds. The Air

Force labeled this role "non-traditional ISR" (NTISR), a term that now covers the many thousands of hours aircrew have spent performing it.

Blurring the line between CAS and UAV capabilities, in 2008 the Army created an aviation brigade cobbled together from its own UAVs and some modified Beech C-12 aircraft that provided live video (Shanker 2008). The unit's aim was to counter the plague of improvised explosive devices that made the two theaters of war so deadly for COIN ground forces. A similar effort started in Afghanistan, and the Defense Department *mandated* that the Air Force purchase and operate fifty MC-12 *Liberty* aircraft. The *Liberty* was a manned, fixed-wing, propeller-driven ISR platform that increased the full-motion video feeds UAV orbits provided while the Air Force continued to grow its capacity in the UAV area (Gates 2014, 131).

The Relevance of CAS and ISR in COIN Warfare

In brief, CAS and airborne ISR—whether provided from UAVs or a manned platform—have a special relevance to COIN warfare. Part of the reason COIN is an intelligence-intensive endeavor is that success requires detailed knowledge of local populations. Ground forces speak of "patterns of life," a normal routine of village or city life that, when interrupted, gives US forces warning of nefarious activity or impending attack (Pincus 2009).

COIN, if pursued through an indirect approach, must also be discriminate and precise in its application of firepower. COIN ground forces should be close enough to the population to determine good from bad intent, in clues sometimes as nuanced as mere body language or tone of voice. When a reason to use force arises, ground forces must use force with a great amount of care not to hurt or punish an innocent person or family. Getting it wrong can lead to a growth of proinsurgency sentiment. The inhospitable terrain of Afghanistan and Iraq, as well as the unfavorable impression of occupiers using heavy ground-based and less precise weapons, makes air power a preferred choice to serve as a source of precise, on-call firepower.

These conditions mean that "always-on" ISR systems, coupled with CAS, gave a COIN ground force commander the best chance for success with an indirect strategy. Appetites for full-motion video, both to analyze patterns of local activity and to ensure the safety of ground forces against attack, grew alongside a larger deployed COIN force. The desire to have air power overhead, even when no specific operation was ongoing, reached a similar fervor because of the unpredictability of insurgent attacks.

Air Support to Ground Forces: A Historical Dilemma

The use of air power for military purposes has evolved as technology and capabilities have become available. For example, soon after military aviation gained an offensive capability, aircraft from opposing forces began to attack each other, both in the air and on the ground. The concepts of pursuit, escort, and other types of "counterair" missions thereby evolved along with the technologies that made them practicable. Another mission that developed of direct benefit to ground forces was airlift, i.e., the aerial movement of personnel and supplies to and around the battlefield. As rotary-wing aircraft became practical "air-mobility," a hybrid of airlift closely linked to the tactical maneuver of ground forces rounded out US air power capability in war. Air power, from this perspective, has enabled ground combat since the inception of flight.

Our interest in this study is the ways US military services collaborate. The US Air Force's CAS and ISR support to the Army and Special Forces on the ground in conflicts is one example of that. However, the Air Force's historical record of and affinity for close air support is undeniably weak. Applying air power in support of ground forces conducting COIN operations is one of the US Air Force's least-favorite, least-practiced, and lowest-priority missions. The ability to apply air power elements, both CAS and ISR, to COIN conflicts tend to atrophy during periods of comparative peace. These capabilities then exhibit rapid improvement during times of war, catching the US military ill prepared at the outset of a conflict but witnessing great proficiency as conflicts draw down (Birch 2014a).

USAF COIN air power capability was not as flat-footed as historical trends might have suggested when OEF and OIF began because of the Balkan conflicts in the 1990s, but the challenges identified earlier still set the context for the state of COIN air power capability at the beginning of the Iraqi and Afghan conflicts. Gen. Robert Foglesong (2004), who had an insider's perspective as the Air Force vice chief of staff and later as the commander of the US Air Forces in Europe, offered this commentary on the state of affairs between the Air Force and Army when operations in Afghanistan began:

> In a sense, the Air Force and the Army had . . . drifted apart over the years in close air support. And it wasn't because somebody, years ago, made the decision that we wanted to drift apart, it was just that that had happened. . . . So our cultures had moved away, and . . . so had our dedication to a couple of things. . . . Now, we had a new form of close air support that was being delivered from 30,000 feet. . . . So it was an uncomfortable thing, to a degree, for the ground forces, that all of a sudden have to accommodate this change in culture. . . . So we . . . now have remarried, I guess. But we still have work to do. (Foglesong 2004)

General Foglesong's description outlines a historical pattern; the atrophy of joint capability and the need to improve cooperation; the typical state of affairs at the beginning of a new military campaign that requires joint-arm application by air and land forces.

Afghanistan and Iraq: Cooperation in the Post-9/11 COIN Conflicts

The Nature of Air-Ground Cooperation in Afghanistan and Iraq

Following a typical period of peacetime neglect, the US conflicts abroad during the early 2000s witnessed key developments in military organization, training, and equipment that enabled air power to better support COIN fighting. Military activity in Afghanistan immediately after September 11, 2001, was, like Desert Storm and the Balkan conflicts of the 1990s, an air power–intensive undertaking most closely resembling MCO, albeit between poorly matched enemies. Six weeks after an "easy" bombing campaign began on October 7, 2001, the Taliban regime had abandoned Kabul and run with Al-Qaeda forces to Tora Bora. Happily, "not a single American soldier had been killed" (Guest 2013, 62). Enabled by the US Central Intelligence Agency and significant air power support, a Special Forces contingent numbering in the mere hundreds had helped topple the Taliban (Naylor 2005, 14). Interdiction missions by land- and carrier-based aircraft were able to "flip all of Afghanistan in three days" (Gersten 2013). This meant destroying, with use of the air-strike assets available to the coalition arrayed against the Taliban as well as a remarkable special-operations effort to identify them, all designated targets in the country within seventy-two hours (Johnson 2007, 139).[1]

The campaign that followed this early success lasted much longer and required more manpower, consistent with senior military officer predictions about successful indirect COIN strategies (Shanker 2007, A13). A steady-state peak of more than 130,000 American and NATO allied military personnel deployed to OEF, waging a campaign to remove remaining insurgent forces and rebuild Afghanistan's national security forces. When the Global War on Terror expanded in 2003 to include Iraq, a maximum of 207,000 Western military personnel (United States and its allies) deployed. US involvement increased to a maximum of 187,900 military personnel present in either Iraq or Afghanistan, with as many as 294,000 US military personnel supporting the operations in the US military's Central Command's (CENTCOM) area of responsibility (Belasco 2009). The upshot was a huge, ultimately failed attempt to rebuild Afghan government security structures to resist Taliban influence.

Both the Afghan and the Iraqi conflicts involved enemies classified as "nonstate irregular" organizations. The Afghan Taliban and Al-Qaeda in

Iraq consisted of fighters lacking formal discipline and employing dispersed, cellular structures of small formations. The use of short-range small arms, mortars, incendiary explosive devices (IEDs), and mines as the sole sources of firepower made air power largely uncontested above three thousand feet. Furthermore, ISR for overhead observation and signals intelligence was among the most important uses of air power, along with air mobility and CAS (Johnson 2011, xxii–xxiii).

In spite of, or perhaps because of, failed attempts during the 1990s, rhetoric from 2001 suggests that the Air Force and Army were confident in their ability to execute CAS in a joint fashion entering 2002. Interest in CAS did not diminish as rapidly as it had after previous conflicts. The recent Balkan conflicts and a transition of the Air Force leadership mantle to fighter pilots, who gave more focus to tactical air power, may have contributed to this enduring CAS attention. Despite conditions that might have been cause for optimism, we must acknowledge that evidence of interservice jointness is sorely lacking at both the beginning and end of this case study—the jointness observed bubbled up from the battlefield (showing the strength of existential threats to driving jointness) and was imposed by the Defense Department (evidence of the strength of this seldom-used avenue to jointness). Both services retained liaisons in major headquarters and planning staffs, although they were not as pervasive as they needed to be given the Army's devolution of combat autonomy to lower echelons (Jacobs et al. 2009, 62). Gen. John Jumper, in his roles as the commander of US Air Forces in Europe and Air Combat Command, and eventually the 2001–2005 Air Force chief of staff, continued a trend toward CAS proficiency that General Ryan had advanced during his tenure as Air Force service chief following his European war experiences, but his efforts did not overcome the Air Force's overall institutional inertia against CAS and his successor amplified this friction.

Jumper also made a significant commitment to building further UAV capacity in the Air Force, committing in 2005 to buying "every *Predator* General Atomics can produce," promising to quadruple the number of UAV squadrons in the service, and dismissing out of hand the idea that pilots devalued the role of unmanned aircraft in combat (Roughton 2011). In spite of the existence of a degree of CAS proficiency and the relative ease with which the US-led coalition dispatched the Afghan government, the transition to steady-state COIN air power applications was difficult.

Operation Anaconda: An Inauspicious Start

Even though sufficient fixed-wing capability existed to give special operators air-strike support at the beginning of the Afghan campaign, it became clear that CAS for conventional Army units was not a well-honed enterprise. Operation Anaconda, executed in the first half of March 2002, was the first major conventional battle for the US military in Afghanistan. Because it led to the first casualties that received significant media attention, the operation inspired a great deal of soul-searching within the Air Force and the Army. Initial reports from Combined Joint Task Force Mountain, the organization with nominal command and control of Anaconda, claimed a tactical success based on the number of Al-Qaeda and Taliban fighters killed in the operation (Geibel 2002, 75). However, the casualties, including eight Americans killed in action, told of a confusing scheme of operational control and botched coordination, and a protracted interservice argument about the planning and intelligence analysis prior to the operation imbued it with controversy (Bochain 2014; Neuenswander 2003, 3; Davis 2004, 16).

Fortunately, a good historical record and after-action reviews allow us to extract lessons about cooperation from Anaconda (Neuenswander 2014; Donnelly 2014; Jordan 2014). The US-led coalition forces for Anaconda included about twenty-five hundred soldiers. They were comprised of two hundred US Special Forces, fourteen hundred US conventional forces from the 10th Mountain and 101st Airborne Divisions, and a remainder of Afghan fighters. This force, commanded by Maj. Gen. Franklin Hagenbeck, planned a hammer-and-anvil attack against an enemy force that his intelligence section estimated to be between 125 and 200 persons. Once the attack began on March 4, however, it became apparent that the enemy force was much larger. Closer to fifteen hundred to two thousand enemy personnel (the original US CENTCOM estimate), the Al-Qaeda resistance was well concealed in high terrain, holding significant small arms and several large mortars. A friendly-fire incident by an AC-130 gunship with an errant navigation system inadvertently repelled the hammer portion of the attack, which was to be executed by US Special Forces led Afghan troops. This left entrenched Al-Qaeda fighters unchallenged for a time, allowing them to attack the US conventional and Afghan force–backed anvil. The ensuing battle raged for two weeks instead of the expected three days, and took a continuous emergency CAS effort to drive back enemy attackers (Davis 2004, 123; Loeb 2003).

Shortly after Anaconda, an accidental bombing of Canadian paratroopers at the Tarnack Farms Range near Kandahar, Afghanistan, suggested that airspace management and CAS procedures were not yet perfected for the war on terror (Friscolanti 2005; Gersten 2013). Lt. Gen. T. Michael Moseley, then in charge of air operations across the Central Command region, authored a memo directing his subordinate air commanders to review rules of engagement with aircrew. It reflected mounting frustration with the friendly-fire incident, the overall CAS difficulties emerging in the theater of war, and the lack of situational awareness the aircrew who would be called on to provide CAS had with respect to the location and intentions of conventional ground forces. But one can also read it as an operational commander pushing down responsibility for coordination that a lack of institutional interservice focus on jointness in this area had failed to deliver during the march to conventional armed conflict at a scale beyond special-operations forces:

> I need everyone's head in the game—we cannot afford another tragic incident. Commanders will sit down with their aircrews and "chair fly" [the rules of engagement] to ensure a complete understanding. Before aircrew step to the jet, they must have a solid understanding of the ground situation and the ongoing dynamics. I cannot overstate how fluid the ground environment in Afghanistan is and the challenges this creates in identifying friendly vice Taliban and al Qaeda forces. Friendly forces on the ground are lightly armed and working in difficult terrain against an elusive enemy. These troops have come to rely on airpower when they become engaged. We need to be there and do it right in accordance with the ground commander's priorities and deconfliction plans among the many teams throughout the area. (Scarborough 2002)

The aftermath of Operation Anaconda had ramifications across all levels of the interservice relationship. In media reports, the incident became a distraction, delaying cooperative efforts among flag officers while both sides dug in for an ideological battle. Most of the subject-matter experts we interviewed argued that Anaconda put "laser-like focus" on CAS and other air support for conventional ground forces (Bochain 2014). Reality was in between. As General Jumper remarked, "you have to gather the facts before you can discuss what happened" (2014, 2). Fact gathering for Anaconda took almost three years, culminating in authoritative books about the Anaconda incident and an Air Force report that was defensive in tone (Naylor 2005; Davis 2004). The Air Force's corporate answer to Hagenbeck's criticism can be summarized the following way: if CAS planning had been incorporated into the Task Force Mountain's plan from the inception of the operation instead of waiting until less than a week before its initiation, a more adequate theater air-control system could have provided adequate CAS, without the mid-air conflicts and confusion observed in the event. As it was, the report argued, CAS was responsive and delivered thousands of precision munitions to good effect; shortcomings in Anaconda were mainly due to poor planning and inadequate intelligence about the number of enemy fighters present in Shahikot Valley (Davis 2005). Such equivocating looks predictable in hindsight from an institution that had not wanted to do and so did not do the more difficult work of pursuing jointness via an investment of senior leaders' time in developing the appropriate command-and-control doctrine, organizations, and infrastructure that enable these kinds of operations.

While Anaconda revealed personal and organizational shortcomings that we will analyze later, for now, we can say that it was yet another inauspicious debut for US CAS in a conventional conflict. Against a massed but concealed enemy force, the United States had once again entered major combat operations without a satisfactory plan for close integration of air power with conventional ground forces. Exuberance over the success of interdiction and the ability of air power to aid in the defeat of a badly overmatched conventional enemy gave way to dismay when the remnants of that enemy dug in and concealed its true strength. The rest of the Afghan campaign would involve a close-fought battle with insurgents concealed among the population, a manner of fighting consistent with classical Maoist warfare, and the measures to integrate air power into that war had to be refined to a higher degree than during the first two months of the operation.

As in previous periods of peace, a pattern of recognized CAS weakness (and amelioration of that weakness) emerged as the defining context from Anaconda, but the proper lessons were slow to emerge. Given this work's interest in service-driven jointness (vice operational or meso-organizational levels), we would contend the services missed the lessons entirely. Although Hagenbeck's post-Anaconda comments suggested a lack of ground-based air controllers, the truth was that there were thirty-seven in the battle. This meant that a density of about one terminal air controller per square kilometer existed throughout the battle. Not only were JTACs present for Anaconda, they had been integrated all the way to company-level echelons of conventional Army forces and were well known to the commanders of those ground units (Kraft 2014)

Having many coordinators in the fight, however, did not automatically lead to a responsive tactical air ground system. In fact, the density of autonomous air controllers in Operation Anaconda was far too high for effective fire control, lacking an overarching organization. Such organization is supposed to originate from a higher-echelon Air Force organization that was then integrated with an Army Corps or, alternatively, from the Air Operations Center commander (Neuenswander 2003, 3). Instead, independent controllers directing fires on a rugged, hostile battlefield crowded with friendly forces among a hidden enemy led to a disjointed "pick-up game" (Gersten 2013). Some CAS pilots withdrew from live firefights on the ground after airborne near misses; they realized that there was substantial danger having so many aircraft close together and controlled by ground parties who could not talk to each other on the same radio frequencies (Neuenswander 2003, 2). Since Lieutenant General Moseley was not at his headquarters during the bulk of

Anaconda planning, it is entirely plausible that the scope of Anaconda eluded him, just as the size of the enemy force eluded the 10th Mountain Division (Naylor 2005).

Ample fodder exists to assign blame to both the Air Force and the Army for the setbacks and casualties of Operation Anaconda. On the one hand, Army planners did not involve the Air Force until the last minute and then underestimated the CAS required, a symptom of faulty intelligence estimates. It is also true that the air liaison squadron (which would have been capable of establishing an ASOC and an overall air-power application plan) assigned to the 10th Mountain Division was denied passage to Afghanistan, a move the division's chief of staff blamed on troop-level limits imposed by the secretary of defense (Donnelly 2014; Bochain 2014). Erroneous comments by the Army fire-control officer for Anaconda about which units took enemy fire revealed a startling lack of situational awareness at the headquarters that persisted even after the operation was complete (Naylor 2005, 273).

On the other hand, the Air Force paid lip service to Anaconda from the Air Operations Center (promising "the world" with respect to air support), but did not realize how dense the calls for CAS might become (Naylor 2005, 137). Only at lower levels within the air-ground coordination hierarchy did remarkable persistence result in an inordinately high ("eight times" the normal) number of CAS aircraft being scheduled and airborne during the opening hours of Anaconda (Donnelly 2014). Although admittedly caught flat-footed during planning, in the aftermath of Anaconda Air Force generals lamented the "CNN image of the [Army] commanding general running around saying, 'Where's my air? Where's my air?' when he knew damn good and well where it was—he had left it behind" (Jumper 2013). On top of this, an alleged focus of attention on the soon-to-open Iraqi theater and a sense of complacency that developed as the Central Air Operations Command (CAOC) became "used to controlling air strikes from their base in Saudi Arabia" (Naylor 2005, 272), rather than providing responsive CAS to the shifting needs of a ground commander in battle, drove the Air Force into a defensive crouch over Anaconda (Call 2007; Lemann 2002).

Unfortunately for the forces involved, the CAS "lessons" of Anaconda did not prevent further tragedy in Iraq about a year later. In March 2003, US Marines called in air strikes from A-10s against a sister company after confusing their relative positions during an advance on two bridges in Nasiriyah. Eighteen Marines died and another seventeen were wounded. Because the fight was fierce and enemy fires intense, the friendly fire contribution is still unknown, but Central Command's final report about the matter raised familiar issues that have plagued CAS efforts as long as the mission has existed: lack of trained air controllers and difficulty with communications equipment. The report added a new wrinkle of difficulty, mentioning that "lack of training on their new 'Blue Force Tracking' situational awareness systems" ("CENTCOMM Commander Urges Close Air Support Examination" 2004) had contributed to the Marines' mistaken air strike requests.

The Blue Force Tracker (BFT) is not specific to CAS missions; it is a system that uses radios to create a data-link network among friendly forces. The intent of BFT and other situational-awareness technologies enabled by local wireless networks is to enable friendly forces to better track where nearby units are, in part to prevent friendly fire incidents. The system, relatively new to most ground units at that time, failed to prevent friendly fire in the incident. Whether a failure due to training or simple friction of war, the incident is a reminder of the persistent difficulties that arise in integrating air power into a close ground battle, one of the most fluid, and therefore unpredictable, scenarios in the scope of military operations.

Advancing from this series of uneven starts, we now addresses efforts to improve the speed and lethality with which CAS could impact enemy forces, making it easier and safer for friendly forces to request and use.

CAS and UAV Reconnaissance "Pinnacle" of 2008

ROUTINE, MULTISERVICE, MULTIPLATFORM PRECISION ENGAGEMENT

In addition to the controversy and defensive interservice dialogue it inspired, Operation Anaconda demonstrated room for improvement in capacity and capability to perform CAS in rugged terrain. Iraq provided similar lessons with respect to CAS executed over cities—"urban terrain" in military jargon. Improved capabilities indeed emerged as soldiers and airmen continued to articulate requirements for equipment, training, and organization, leading to more agile fire support for land forces involved in the COIN fight. In turn, the military services funded, trained, and fielded these capabilities via the forces they provided to joint-service combatant commanders.

This evolution unfolded over several years and witnessed a shift toward air-power competencies that favored direct ground support in lieu of other capabilities. In the priorities for equipping and training the fleets and aircrew of fixed-wing fighter aircraft, CAS took priority over other mission sets, such as deep interdiction and air superiority (Johnson 2014). During this period, the technology that enabled CAS grew rapidly, and training opportunities to put it into practice increased as well. At a tactical level, coordination between fielded Air Force and Army units became closer, more routine, and more responsive, with a subsequent growth in mutual trust among tacticians. The relationships between senior commanders that enabled this cooperation, though, marked an uneven trail of progress, which is characteristic of jointness that grows organically on the battlefield rather than being driven by service-level interests.

Despite a patchy record, progress did occur in Afghanistan and Iraq. An unclassified summary of aerial engagements conducted on a single day in July 2008 provides a representative sample of extensive combined-arms co-operation. The technologies found in the weapons systems it references and the command-and-control relationships that existed to allow this collaboration reflect a myriad of the most pertinent CAS advances that had occurred since 2001. Ten different types of Air Force, Navy, and Marine aircraft from a wide-ranging international coalition employed at least twelve different types of ordnance, half of which had been developed or refined for use in CAS since 2002. The account reflects command-and-control mechanisms that allow the apparently seamless use of multiservice and coalition assets to complete direct-support CAS and ISR missions, as well as a wide range of precision-guided munitions technologies. As an almost throwaway aside, it casually mentions 142 airlift sorties delivering 514 tons of cargo (including 34,000 pounds of troop resupply air-dropped in Afghanistan), and fifty-nine air-refueling sorties providing 2.6 million pounds of fuel to 190 receiving aircraft (US Air Forces Central 2008). This level of integrated aerial activity is a marked departure from the chaos of Anaconda and represents a significant level of normalized joint cooperation. This routine press release mentions more than a dozen significant advances allowing for better integration between air and land forces. Along with the manned, fixed-wing air power capacity mentioned, the quantity of Air Force UAV support available to provide real-time pictures to ground commanders of the threats and terrain they faced had grown more than twentyfold by 2008.

In addition to the technology and weapons systems apparent in this description, several other command-and-control changes allowed these operations to proceed safely and with close coordination among the appropriate component forces. To explain the changes that made such reports routine, many of which would have been unattainable in 2001 or even 2005, we have arranged the following thick description using an "organize-train-equip" rubric, paralleling the mandates found in law and joint doctrine for the individual services to train, organize, and equip their forces. As discussed in chapter 2, services provide trained and equipped fighting and support units, organized into domain-specific components, which they believe are best suited to give joint commanders the required combat capability. It is up to joint force commanders to employ the resultant forces. This section deals with those three fundamental service responsibilities, leading to discussion in the third part of the chapter about how well the joint-force organizational construct worked in its employment of COIN air power. Table 5.1 summarizes changes across the case study, grouped into the categories of "equipment" (mostly improvements in airframes, weapons, electronics, and other technology), "training" (the repeated practice military personnel, especially soldiers and airmen, received prior to deploying), and "organization" (the way air and ground units were assigned by commanders to provide mutual support to each other).

Equipment

CAS and ISR improvements after 2001 occurred across a broad spectrum of technological innovation, improved tactics, and refined command-and-control procedures. The most tangible and numerous improvements in the ability of the US military to successfully execute CAS and ISR missions in direct support of COIN ground forces are weapons and sensor technology, now in routine use, that did not exist, or existed in much more primitive forms, in 2001.

MANNED AND UNMANNED AERIAL RECONNAISSANCE

One of the biggest changes over the battlefields of Afghanistan and Iraq was the increase in persistent air surveillance from manned and unmanned aircraft. To start with a reference point of 2004, there were an equivalent of three MQ-1 *Predator* combat air patrols (CAPs) in the Central Command region, two in Iraq and one in Afghanistan (Gambold 2013). Several hundred MQ-1 *Predator* and MQ-9 *Reaper* aircraft plied a total of sixty-two full-time UAV-ISR CAPs at the end of 2013, a twentyfold increase (Lee 2013). The organizational change necessary to support such growth is remarkable, especially given the Air Force's institutional reluctance to provide tactical ISR support, and the fact that it came from a platform that was a technology demonstration in 1995.

A marked increase in capacity in no way indicates that the Air Force completely met joint force commanders' desires for increased UAV reconnaissance capability, however. A measure of how far short of the ground commanders' requested support level the Air Force fell comes from counting the UAVs acquired by the Army. As of October 2013, the Army possessed seventy-five *Gray Eagles* (a larger *Predator*-like UAV with similar capabilities), with plans to acquire a total of 152 aircraft and thirty-one ground control systems (US Government Accountability Office 2013, 101). The Air Force bought just over seven hundred RQ-1, MQ-1, and MQ-9 *Predators* and *Reapers*. However, the Army and Marines in 2013 operated more than six hundred *Hunter* and *Shadow* aircraft, and the Army alone had more than six thousand of the hand-launched micro-UAVs, including *Raven*, *Puma*, and *Wasp* (Dunnigan 2012). Numbers like these bear out the Army's near-insatiable appetite for organic UAV capability ("Army: Tremendous Demand for UAVs" 2010).

ADVANCED TARGETING PODS

Prior to the beginning of the Afghan conflict, the Air Force had recognized the need for better targeting pods for its multirole fighter aircraft. Existing infrared pods were suitable for finding large, fixed targets like large weapons depots or military headquarters—the kind encountered in deep-interdiction missions. For concealed targets and urban areas, however, the limited acuity of these pods was a liability. The Air Force began to acquire the Sniper, a combined television and infrared targeting pod with up to five times the resolution of legacy pods, in August 2001 ("USAF Orders Sniper Targeting Pods" 2002). The first Sniper pods went to the F-16 to enhance its air-to-ground capabilities, and the system debuted on the F-15E in early 2005 ("F-15E Debuts with Targeting Pod" 2005). Eventually, all Air Force CAS platforms received the capability to carry the Sniper, giving capability to generate target coordinates precise enough to employ GPS-guided munitions against stationary targets. At about the same time, acquisition of the Israeli-designed Litening, with capabilities similar to the Sniper, commenced in the United States ("USMC Orders Litening Pods" 2001). Air National Guard, Navy, and Marines Corps missions tended to incorporate the Litening pod, and the addition of two advanced pods increased the capability of CAS platforms to identify features of interest to ground forces and target them when appropriate.

In 2010, the Air Force began to upgrade all of its Sniper and Litening pods with improved-resolution sensors and two-way data-links, demonstrating that capabilities essential to CAS over the past decade had become fully entrenched in the tactical fleet (Gething 2010). Software upgrades to the pods provide video overlays, showing unique symbols for entities designated friendly or hostile over tactical data-link networks. They also provide the capability to display range rings on target pod video, a quick way to determine if ground forces are too close to the intended point of impact for requested CAS munitions. New targeting pods act as a data-link relay for ground forces, allowing units to send each other video imagery by routing signals through pods without necessarily communicating with aircrew at all (King 2013). The Navy began similar initiatives for its advanced targeting pods, determining that a long-term investment in CAS capability was appropriate for its tactical air fleet (Maloney 2013). While the capabilities of advanced targeting pods are indisputably useful for providing faster, more precise CAS, their ability to interface with one- and two-way video terminals carried by CAS controllers provides an integration capability an order of magnitude better than anything enjoyed in previous CAS-intensive conflicts. One might hearken back to AirLand Battle and expect to see four-star generals from several services leading the charge for this new equipment and the means of integration it provided. On the contrary, most of it seems to have been the natural progression of defense-contractor development arcs given outlet by battlefield demand. However, there are notable exceptions where service-level leaders, responding to the plaintive cries of experienced battlefield veterans, increased the uptake of technology.

Air-Ground Video Interface Stations and Imagery

The first linking of full-motion UAV video feeds to other airborne platforms occurred in November 2001, when specially designed antennae and equipment installed by the Air Force's "Big Safari" office allowed AC-130 gunships to see *Predator* video feeds over Afghanistan (Grant 2013). Late in 2001, a request from a motivated soldier, Chief Warrant Officer Christopher Manuel, led to testing of video receivers for use by ground units (Barnes 2007, 1). Remarkable efforts in 2003 by Air Force Lt. Col. Gregory Harbin continued to propagate the system through Afghanistan and Iraq, making it a "must-have" for more and more ground commanders (Bochain 2014). By 2005, the Air Force recognized that the wide introduction of lightweight, man-portable Rover video terminals, which allowed a JTAC to see on the ground exactly what aircrew saw in their targeting pod displays, would provide a leap forward in CAS capability.

The success of integrating demand signals offered by battle-tested warriors gained cachet in the Pentagon, and General Jumper insisted on including a feature on follow-on Rover systems that allowed ground parties to draw on their screens, highlighting to aircrew points of interest. JTACs in Iraq and Afghanistan "used the system just about every day" (Krooner 2006). With Rover, JTACs diminished uncertainty whether they and the aircrew with whom they worked had located the same target. By extending the visibility of the targeting pod to the JTAC, CAS platforms allowed ground commanders a view of the battlefield without the risk of exposing themselves to enemy fire on high ground vantage points. Integration of Rover to all CAS platforms and UAVs happened quickly, finally expedited by enthusiastic support from the Air Force bureaucracy.

By 2010, fifth- and sixth-generation Rover terminals could provide communications links among tactical UAV users, as well as full-motion video feeds to the strategic ISR network of UAV ground stations ("Rover Develops" 2010). Rover, unimagined in the 1990s, solved many of the frustrations of CAS as well as some of the precision-target location issues associated with UAVs, although it still involved stitching disparate systems together rather than compatibility off the assembly line (Harvey 2014).

The burgeoning growth of CAS platforms (manned and unmanned), along with advances like Rover and the ability to provide feeds to any ground control station, created a watershed moment for the amount of ISR data available to commanders. "With the arrival of digital imagery from photo-recce, plus the fallout imagery from other applications such as surveillance and targeting . . . the potential raw material available for exploitation has mush-roomed exponentially" (Gething, Gourley, and Janssen Lok 2006). The intelligence-intensive nature of COIN warfare that Corum and Johnson de-scribed was in full view in Afghanistan and Iraq, and the new technologies used to gain an advantage fostered change within the military organizations that exploited them (Hughes 1987, 52–53; Law 1989, 112–113).

WIDE INTEGRATION OF LASER DESIGNATORS AND RANGE FINDERS

Devices that can send a beam of laser energy to a target ("designators") or detect reflected laser energy ("seekers," or, if they can also determine dis-tance, "range finders") have become an essential part of CAS capability, both for airborne platforms and as part of the equipment kit that JTACs carry. Only weeks into the Afghan conflict, General Jumper frequently shared the story of "the young Special Forces troop, out there in the hills of Afghani-stan, riding a horse and carrying a laptop computer hooked up to a satellite, using laser goggles to put a precise designation on targets (Dougherty 2002). Jumper's first experience as a senior leader integrating laser designators and range finders into new combat systems was with *Predator* in 1999, giving the platform the ability to provide target coordinates to fixed-wing strike aircraft in addition to full- motion video of areas of interest (Anderegg 2001, 29–31).

But airborne-laser designators have been in use since at least 1968 (when Jumper was still a lieutenant), when the first "smart" guided bombs struck targets in North Vietnam. Initially, a laser designator shone a beam on the target from a pod attached to a "guider" aircraft. An accompanying "mule" dropped the bomb in a precalculated envelope that gave the weapon and its guidance apparatus the best opportunity to acquire the reflected laser energy and follow the beam to the target. Improvements led to gimbaled targeting pods that allowed a systems operator to "self-lase" a target throughout the time of flight of the weapon until impact (Gillespie 2006, 66–95).

As with much of the other military hardware described in this section, handheld systems of this type predated even the Afghan conflict. The Hand-held Command and Control Wireless Communications (HC2WC) system, a development of the Joint Expeditionary Digital Information (JEDI) system, existed in mid-2001. The soldier's pack complemented a base station and included "a rugged palmtop computer, GPS receiver, laser range-finding binoculars, and a satellite handset"—it is precisely the kind of system that Jumper was so enthusiastic about (Hewish 2001).

Born of battlefield necessity rather than the vision of any high-level ser-vice leaders, OEF and OIF witnessed the first widespread use of handheld laser designators by the Joint Tactical Air Controllers (JTACs). In concert with GPS and a computer display that provided a "moving map" of the

JTAC's position and surrounding terrain, the handheld designators were used to generate target coordinates. As advanced targeting pods became more common, their ability to lock on to reflected laser energy meant that a JTAC could simply point his designator at a target and aircrew would quickly acquire it if environmental conditions permitted. Designators evolved additional capabilities, including infrared beams that were compatible with the night vision goggles aircrew wore. Again, the effect was to reduce the time required to talk an aircrew's "eyes on" onto a target or friendly forces. As the wars continued, the capabilities of these multipurpose systems increased, and the batteries required to operate them became a bit more efficient and lighter to carry.

ADVANCED TARGETING POD INTEGRATION ON HEAVY BOMBERS

Air power employment in COIN conflict demanded targeting pods providing improved battlefield resolution for aircrew and tactical air control parties. Prior to 2006, targeting pods only operated from multirole fighter or attack aircraft. After Air Combat Command received authorization for advanced targeting pod integration testing on the B–52 in 2002, the first combat use of the pod occurred in April 2003. There was a lag in the full implementation, but the first active-duty B–52 unit completed the training procedures for combat use in July 2006 (Edwards 2014). For the B–1, which carried the largest and most diverse weapons load in the Air Force inventory at the time, an urgent need statement in 2006 led to the first successful test of the capability in 2007 and combat capability by 2008 (Bates 2009). Although bomber aircraft had been involved in providing CAS since the beginning of OEF hostilities, aircrew had been reliant on outside sources for precision coordinates or laser designations to deliver accurate GPS- or laser-guided munitions. Having onboard targeting-pod capability allowed for bombers to complete the CAS targeting cycle autonomously.

GPS-GUIDED MUNITIONS

Like *Predator*, Joint Direct Attack Munition (JDAM) GPS-guided precision munitions saw initial combat in the Balkan conflicts of the 1990s. The two-thousand-pound version of the JDAM was developed starting in 1997 and first used in Kosovo in 1999. GPS-guided weapons offer all-weather capability, as they hit targets under any atmospheric conditions, provided a satellite signal is present. The other requirement is that the weapon be given accurate coordinates to which it can guide, and no weapons system is immune to fratricide-inducing or collateral damage–inducing user error. A notable instance occurred in late 2001, when a JTAC, while calling in air strikes in support of US Special Forces and Afghan troops under the leadership of Hamid Karzai, sent a B–52 his own coordinates. The two thousand-pound JDAM dropped in response killed three US soldiers and five Afghans, injuring forty others. The JTAC had in error changed a battery in his GPS receiver, causing it to default to displaying its current position rather than the enemy position he had identified using a separate range finder (Burgess 2002). CAS tactics soon incorporated steps to crosscheck friendly positions against proposed weapons delivery locations.

HYBRID (GPS- AND LASER-GUIDED) MUNITIONS AND LOW-YIELD MUNITIONS

Included in the abundance of new munitions developed since 2001 are many specifically adapted to meet the challenges of COIN warfare. Since intelligence about elusive insurgents is often fleeting, the need to quickly locate, track, and destroy moving targets spells the difference between success and failure in targeting an enemy's network. GPS weapons allow precision strikes beneath the weather, but they are almost always ineffective against moving targets. Laser-guided weapons, in concert with some of the training methods discussed later in this section, provide an opportunity to hit moving targets, but exclusive use of laser-guided weapons removes some targeting flexibility in unpredictable weather.

Hybrid weapons, which can use GPS or laser guidance to find a target, offer the flexibility of two separate guidance methods. A JTAC can provide precise coordinates for a given target, and aircrew can in turn use the accuracy of a GPS weapon to hit it. However, if that target subsequently becomes a "mover," or if a higher-priority target emerges in close proximity, the crew has some latitude to guide the weapon to a different terminus, including a target moving at a high rate of speed. The first combat use of the GBU-54, the Air Force's five-hundred-pound hybrid laser-GPS bomb, occurred in Iraq in 2007 ("Laser Guided JDAM Debuts" 2007). The Navy accepted its initial delivery of the weapon in 2008, and designated it to fulfill its requirement for attacking moving targets in 2010 ("Navy Awards Boeing $23 Million for Laser JDAM" 2012). In tracing the development of these weapons, however, the path ends at the program offices of various platforms and types of weapons. These exist at relatively low levels of the service hierarchy, and are heavily populated by tacticians who rotate between these offices and flying operational sorties. One could say that the battlefield was simply recognizing what it needed and buying it with a little help from the American military-industrial complex.

Corum and Johnson (2003, 429) stressed the point that bombing civilians while attempting COIN is particularly damning to the effort; collateral damage caused by errant or oversized weapons tends to increase the number of people devoted to the insurgent cause. Several efforts to decrease the likelihood of unintended civilian casualties ensued. Even with the utmost in precision, some weapons are simply too large to be used against a legitimate target without the unintended consequence of destroying or killing something or someone nearby. To this end, the services have developed smaller bombs with fuses and other modifications that provide limited blast and fragmentation effects (Hewson 2008).

The five-hundred-pound JDAM (GBU-38) moved rapidly into use in Afghanistan as an alternative to the two-thousand-pound GBU-31 when COIN-focused efforts required limits on collateral damage. When *Predator* received the ability to strike targets in 2001, the reason for arming it with *Hellfire* was that the laser-guided missile's small size made it the only weapon the UAV could reasonably carry (Boyne 2009, 44). While the 109-pound weapon allowed for precise engagement of moving targets, an unforeseen COIN advantage was that its twenty-pound warhead also prevented excessive collateral damage, a fortuitous side effect.

Another low-yield weapon developed during the era was the 250-pound GBU-39 small- diameter bomb (SDB), a GPS-guided weapon that also increased standoff capability with its glide characteristics. While the F-15E was the first US CAS platform to integrate the SDB in 2006, the weapon's first iterations saw limited use in combat, primarily because its long glide times made it undesirable to ground commanders who wanted to strike potentially fleeting targets without several minutes intervening between weapon release and impact. Future variants solved that problem and provided multimodal terminal guidance methods, adding three means of moving-target engagement to the basic fixed-target GPS capability.

Precision Visualization, Mapping, and Targeting Systems

The individual tools such as laser designators, GPS receivers, and video receivers are of scant value to JTACs if they cannot assemble the various pieces of information to create a contextually meaningful representation of the battlefield they want to influence. Mapping systems, usually mounted on a rugged laptop computer, meet this need by showing pertinent data on a static or moving display. There is a tradeoff between advanced technology and the urgency of war. As one JTAC remarked, "You don't want to be typing into a terminal when you are being shot at, nor do you want to have to stick your head above the parapet to point a laser designator" (Pengelley 2004).

Improved Communications Systems

Reflecting a challenge that has existed since World War I, communication with ground forces via radio links proved difficult in Afghanistan and Iraq. In Afghanistan's topography, rugged terrain often denied the unbroken "line-of-sight" required for air-ground radio communications to work. Several means of overcoming this difficulty arose, the simplest being the adaptation of airborne relays: an aircraft (including a UAV) airborne over a ground party's location could relay radio transmissions from the ground to a more distant aircraft that would otherwise not hear the transmission. This adaptation occurred as a matter of necessity during the first air-ground battles fought in both Afghanistan and Iraq.

The next increment of adaptation came with the installation of radio transceivers more capable of communicating on the frequencies commonly used by ground units and their associated air liaisons. Functioning as a bellwether that indicated the Air Force's embrace of the CAS mission across all platforms, the F-15E from 2006 received upgrades to carry VHF-AM and VHF-FM radios, allowing improved communication with JTACs and Army units, respectively.

Afghanistan's rugged terrain also inspired the installation of satellite communications (SATCOM) terminals in many different CAS aircraft and the allocation of sufficient bandwidth to allow its routine use. Since SATCOM signals effectively go straight up and come straight down, they are not subject to the same interference faced by UHF or VHF signals in a steep mountain valley or urban canyon. This was the final major communications innovation realized, with testing beginning in late 2008 and installation to the fleet in 2009; the F-16 received a similar upgrade in 2009–2010 (Zessin et al. 2008).

DATALINK INTEGRATION

Known by jargon-rich and acronym-heavy names including Link 16, JTIDS (Joint Tactical Information Display System), SADL (Situational Awareness Data Link), and Blue Force Tracker, among others, wireless data-links provide friendly force position, direction of movement, weapons inventories, and other useful information. They also allow shared targeting across multiple platforms and the "cross-cueing" of sensors to rapidly home in on unknown entities and ascertain hostile or benign intentions. They have steadily improved since 2001 and have undeniably made the sharing of data, with or without voice communications, easier among the US and coalition forces fighting in the CENTCOM area of responsibility.

New technology does not serve the cause of jointness if military personnel do not understand how to employ it. Having described some of the equipment and technology innovations that have increased CAS and UAV-ISR capacity and capability in COIN warfare since 2001, we turn next to the training schemes developed to make operators proficient in their use.

Training Personnel: Terminal Air Controllers

Far from the neglect characteristic of peacetime years, the CAS training of both terminal air controllers and aircrew witnessed significant effort across the services as they waged COIN- intensive war in Afghanistan and Iraq. The now-defunct US Joint Forces Command acted to harmonize CAS procedures across all services, worked with the Air Force to create a simulator that could train terminal air controllers faster to the level of desired proficiency, and certified all controller training schools to a common standard (Pengelley 2004). While somewhat quotidian work, this effort at standardization is critical to making COIN CAS lethal, scalable, and repeatable, and is a rare example of institutional-level joint success in this case study.

While increasing the number of ETACs and, later, JTACs remained a priority, their integration into meaningful predeployment training meant that CAS-capable aircrew also had more opportunities to hone the skill sets required for the urban and rough-terrain environments encountered in the countries of interest. Aircrew completing training to prepare for CAS-intensive deployments could expect to receive several dozen chances to interact with JTACs in live training scenarios, perfecting the difficult art of identifying a target or friendly formation to an aircraft through verbal communication. As the equipment and ordnance available for executing CAS improved, these training opportunities offered chances to improve procedures for using video data-links by which JTACs saw the same targeting pod video that aircrew monitored, greatly increasing the speed and verifiability of CAS targeting.

Affirming an oft-expressed Army sentiment that "there can never be enough JTACs," the Army's artillery branch began training forward observers (FOs) on joint CAS procedures. While these soldiers were not fully certified as terminal controllers, they attained a familiarity with CAS procedures that allowed them to interface with JTACs, extending the range over which CAS assets could operate with a single controller (Pengelley 2004). Maj. Gen. Mike Maples, a commander of the Army's field artillery branch, made a call for the Air Force to reciprocate and train its JTACS to control the Army's long-range artillery systems, but proposals for a "universal observer" did not gain the same traction that improved JTAC training did. The Marines made similar efforts, expanding training opportunities for FOs and using them routinely to extend the range at which JTACS could provide "Type II" CAS control—a practice where the JTAC controlling a strike observes either the aircraft performing the strike or the target, but not both at the same time (Harvey 2014). All in all, there are fits and spurts of institutional jointness throughout this era, but absent the force of aligned senior leaders or service culture, they never take off in a way observed in the first two case studies.

UAV TRAINING PIPELINES

The need for additional personnel to operate the twentyfold increase in UAV orbits realized after 2003 was significant. The Air Force had fifty UAV pilots in 2001; by 2013 it had over thirteen hundred (Peck 2013; Church 2014).

The intelligence tail required to support the constant data feeds pouring in from each CAP, estimated at twenty to forty personnel per orbit (Deptula 2013), was another significant burden. Two operators—a pilot and sensor operator—manned each remote UAV control station; additional personnel ensured airspace coordination. Adding to this manning was a requirement for roughly twenty intelligence personnel to process the sensor data coming from the aircraft at distributed common ground system (DCGS) facilities scattered around the globe. With so many people directly supporting each CAP, plus additional administrative and maintenance requirements, the manning bill for ISR operations grew precipitously (an average of 162 people per orbit for twenty-four-hour operations) as theater demand grew (US Air Force 2010). This in turn required the rapid buildup of training units to graduate personnel qualified in all facets of the UAV mission. Though the intelligence personnel requirement was significant, both subject-matter experts and official Air Force reports hold that pilot training was the biggest inhibitor of faster growth (US Air Force 2009; Campo 2013). The center of the UAV training universe—Creech Air Force Base, Nevada—addressed these challenges.

In spite of the dramatic increase in UAV CAP capacity, by 2014 the Air Force had not "normalized" the manning ratios for the aircraft it operated (Church 2014, 36). In plain language, this means that crews did not get to practice noncombat practice skills training expected for a sustainable long-term force (Campo 2013). This growing pain accompanied the breakneck pace of UAV capacity development and continual surge operations in Iraq and Afghanistan. It may also reflect an enduring Air Force predilection to move away from the tactical UAV enterprise, though a consistent Air Force message during this era was that the UAV career field, along with an accompanying path to career advancement and command, had arrived (Church 2014, 37–38).

The US exit from Iraq and ever-impending wrap-up in Afghanistan provided transition opportunities, as do predictions of warfare against peer enemies in expansive, highly contested theaters. Those knowledgeable about ISR capability interviewed for this study suggested that the only way to keep the ISR platform infrastructure viable within the Air Force is to rethink the platforms as part of a strategic, non-COIN air war. A strategy like this anticipates the Air Force's historic postconflict treatment of COIN-specific airframes. Such thinking reflects an instinctive grasp of institutional culture by those who are the most ardent advocates for keeping a robust UAV enterprise extant in the Air Force. Inasmuch as a former aide-de-camp to the Air Force chief of staff derided the peak of UAV ISR capability as "65 soda straws" (Bell 2014), there is evidence that the Air Force never believed this to be part of its core mission.

Moving-Target Tracking and Strike

The capability to hit moving targets existed at the outset of both the Afghan and the Iraqi conflicts. It takes focused and repeated training to master the art, however. Capability improved throughout the Air Force as aircrew—especially those who had not performed extensive CAS previously—practiced to become proficient at this unique skill. Prior to the development and fielding of hybrid GPS-laser-guided munitions, the only options for striking moving targets from the air were with a laser-guided munition or an aircraft's internal cannon. Strafing "movers"—shooting a CAS platform's twenty- or thirty-millimeter cannon at a relatively small vehicle—is an art that requires repetition to master. Even when aviators perform without fault, direct hits from weapons smaller than thirty-millimeter caliber can fail to stop some vehicles.

A more effective and widespread way to hit moving vehicles is with laser-guided munitions, but effective use of these also requires training with the technology. The Air Force spent years developing techniques, procedures, and rules of thumb for targeting movers using laser-guided munitions. CAS units spent extensive time practicing the art of tracking vehicles, using advanced targeting pods and data-link systems that allowed two-ship formations to share information and have at least one aircraft ready to make a successful attack. The fielding of hybrid weapons, improved targeting pod software, and evolved laser seekers that compensated for target movement made the task easier. Over time, aircrew became more proficient at engaging moving targets, even those they had to follow for minutes or hours, biding time until they moved away from areas where collateral damage was a serious concern.

National Training Center and Joint Regional Training Center Integration

Yet another notable change in close support training habits for COIN occurred at the Army's national training centers, where combat-bound units underwent certification prior to deploying. The Air Force repurposed its Green Flag exercises (traditional electronic-warfare training opportunities) as CAS practice. Green Flag replaced and rejuvenated the existing Air Warrior exercise, aligning participating Air Force units with Army maneuvers taking place in the training centers at Fort Polk, Louisiana, and Fort Irwin, California (Magnuson 2006). Training outcomes depended on the commanders who put their units through the paces in these exercises, but the simple existence of the opportunity showed that an institutional barrier to effective combined-arms training fell in the wake of Anaconda and the initial days of the Iraqi invasion. When NTC–Green Flag training exercises worked well, they addressed General Jumper's complaint that CAS integration in the Army's NTC exercises trained ground commanders to say, "I've got those airplanes up there but they're not doing me much good" (Jaffe 2003).

Even more symbolic was the Air Force's revamping of its vaunted Red Flag exercise, traditionally styled as MCO rehearsal for multiaircraft formations that simulate the opening hours of full-scale, conventional air war. After 2006, parts of some Red Flag exercises included elements that more closely resembled the COIN fights occurring in Afghanistan and Iraq, with smaller sections of aircraft pursuing fleeting targets with the assistance of ground controllers operating in and around urban terrain. The new side of Red Flag did not lose its MCO focus completely, though. The CAS-like exercises remained, per the vision of Red Flag, "more intense for pilots than actual air combat" ("US Air Force Eyes Better Integration with Army" 2004), with the pursuit of high-value targets happening in contested airspace, something that never occurred in Afghanistan or Iraq. The shift of focus, though, leaves little doubt that the Air Force reallocated significant training resources to making sure its aircrew were better trained to support COIN missions. On the flip side, the fact that as of this writing all of these changes have reverted to status quo ante or become even more focused on high-end, near-peer warfare demonstrates how countercultural this whole era was for the Air Force.

LESSONS-LEARNED ORGANIZATIONS

Directly affecting the usefulness of predeployment training is its pertinence to ongoing combat operations. Units that went to Afghanistan and, later, Iraq would return with dossiers of information about preferred tactics, techniques, and procedures for providing CAS in the unique terrain and combat conditions of Afghanistan and Iraq. Trusted instructors, usually graduates of the Air Force's Weapons Instructor Course, would be charged to share the information with their counterparts in other squadrons. A network of well-qualified instructors with high peer credibility in their units thus developed predeployment training plans that incorporated the latest data, enabling squadrons to deploy better prepared (Magnuson 2006).

This widespread, informal practice led to the Air Force's reactivation of the 561st Joint Tactics Squadron, an organization dedicated to collating lessons from combat and ensuring units preparing to deploy had access to them. The "Joint" in the squadron's name implied a CAS focus, emphasizing the work the squadron did in conjunction with the Joint Air Ground Operations Group at Nellis Air Force Base and the Air Force's Green Flag exercises (Rondeau 2007). Demonstrating the institutional embrace of CAS in its training system, the Air Force added significant discussions of CAS tactics, techniques, and procedures (TTPs) to pertinent tactical publications, in many cases inserting a dedicated chapter where previously had existed a single paragraph or no mention at all. The move is significant since Air Force aviators tend to spend more time reading and studying their tactical publications than any other printed matter that purports to hold sway over flying operations.

The Army, which has maintained a lessons-learned institution in the Center for Army Lessons Learned (CALL) since the mid-1980s, continued to glean and disperse lessons about maneuver warfare as practiced at the national training centers and in the CENTCOM theater. As expected, Army discussion of CAS leans more toward a doctrinal, even philosophical viewpoint, as the typical ground unit views CAS as one of many available "calls for fires" that can put ordnance on a target (Bochain 2014; Kraft 2014). The JTACs who interface with the air-ground control system to provide those fires are generally Air Force enlisted personnel or Air Force–trained special operators, so improved CAS procedures do not impose an obligation for extra training on Army personnel. The notable exceptions are the increase in FOs already mentioned and the manning of more ground liaison officer (GLO) positions, typically filled by Army artillery soldiers in Air Force units.

Joint Air-Ground Operations Office

In October 2004, Air Combat Command established the Joint Air-Ground Operations Office (JAGO) at Langley Air Force Base. The office's first director, Col. Michael Longoria, made reference to the need to improve institutional cooperation between the Army and the Air Force. Colonel Longoria, who oversaw air-ground operations from the combined aerospace operations center (CAOC) during Operation Anaconda, noted "serious deficiencies in this air-ground domain that we can and must fix" ("US Air Force Eyes Better Integration with Army" 2004). Further elaboration of the office's role, though, revealed a concentration mostly on advanced technology that would more quickly provide precision-targeting data to JTACs—information they could subsequently share with aircrew through the various communications systems aboard CAS platforms. JAGO does not seem to have resulted in substantial new doctrinal changes or improved working relationships with the Army in and of itself, however. We found that organization and relationships, discussed next, remained chronic points of friction in spite of remarkable equipment and training progress over the span of this case. A key takeaway in observing the JAGO is that the Air Force finds it more appealing to adapt technology offered by defense contractors in response to its contemporary tactical problems than it does to address its fundamental doctrines or beliefs about the use of air power. And Colonel Longoria, influential though he was in telling the story of CAS applied to COIN, was ill matched to take on the quest of institutional jointness with Air Force culture stacked against the effort.

Organization

BUILDING THE UAV FORCE

The same difficulties that have affected CAS requests in all wars also plagued UAV-ISR normalization in Afghanistan and Iraq. Special-operations forces, whose doctrine, tactics, and equipment all tend to be more flexible than those of the conventional forces from which they are derived, helped compensate for some of the service-level jointness otherwise lacking in this case study. Air planners from the mid-2000s related that Special Operations Forces (SOF) requirements created overwhelming demand for UAV orbits. The Air Force exerted itself to create dedicated special operations support UAV squadrons. The Third Special Operations Squadron (3rd SOS) reactivated in 2005 to fly MQ-1B missions in support of special operations forces, and tripled in size during its first twenty months. Operators spoke of the inflexible grip that special-operations missions had on UAV assets, both those assigned to the Air Force Special Operations Command and Air Combat Command's conventional UAV squadrons. SOF missions usually received comprehensive dedicated air support (both manned and unmanned), and ground commanders were loath to release the support once a mission was underway, regardless of the probability it would be used.

Competing for a scarce resource with special-forces operators who tended to get the support they requested, conventional ground commanders experimented with support requests until they found a combination of metrics and language (phrases like "dynamic targeting," e.g.) that gained CAOC approval. Sometimes the motivation was curiosity or jealousy rather than true operational need. As one pilot recounted, some conventional Army units "didn't know what to do with you once they had you" (Gambold 2013), consigning the aircraft to be a flying observer for hours in an area devoid of movement or threat. In describing the relationship at the CAOC between the "A3" (an operational manager responsible for assigning *Predator* support to priority missions such as high-value individuals or ground troops taking fire) and the "A2" (an intelligence manager responsible for prioritizing ISR collection), interviewees often resorted to analogies of a dysfunctional family, a theme captured in Callahan's (2013, 1–2) description of Air Force UAV command and control.

The reestablishment of the 3rd SOS as an MQ-1 squadron is illustrative of the influence Air Force Special Operations Command (AFSOC) and SOF support had in growing the UAV force. The squadron did not put existing SOF aircrew through a training program. Instead, it divided an existing conventional intelligence squadron, shunting the majority of its members into the AFSOC UAV force, and then regrowing both squadrons using novel accession and training plans (Campo 2013). According to James Gear (2013),

"They wanted to grow ten CAPs indigenously in AFSOC" (Gear 2013), which spurred ever-faster growth within the conventional Air Force UAV-ISR world.

Special-operations missions were not the only driving force, though. As with manned CAS assets, conventional Army ground commanders, learning of the situational awareness and security that having a full-time surveillance feed gave them, became more vocal about the need for more ISR orbits. Once ground commanders learned to use tailored verbiage in air-support requests, they could claim priority more often in the CAOC's prioritization queue. As they learned how to manipulate the system, more "must-do" missions entered the air-ground tasking system. A positive-feedback loop ensued, and a cry for UAV orbits reverberated all the way up the Army chain of command, through the Joint Staff, and up to the Department of Defense.

During the height of operations in Iraq and Afghanistan, UAV CAPs continually increased. DoD's UAV action plan in 2010, which led to the 2011 "ISR surge," specified a final target of sixty-five orbits in 2014, and the Air Force reached a steady state of sixty-two orbits in 2013 (Lee 2013). However, this end state belied the reality of ever-moving, ever-increasing operational-capability targets that continually frustrated Air Force planners after 2005. The arbitrary increases vexed plans to stabilize training systems, procurement of airframes and ground-control facilities, and key components of the organizational architecture, including the final number and disposition of UAV squadrons (Peck 2013). Even the secretary of defense, by his own admission, overreached on occasion, stipulating a number of Air Force UAV-ISR orbits that would have "eclipsed the sun" (Gates 2014, 131). The Air Force's organizational effort to enable the sheer increase in the size of the military UAV force was a struggle. As the next section will describe, it at times led to strife throughout the US defense hierarchy, which became a significant threat to jointness.

OPERATIONAL-LEVEL COMMAND-AND-CONTROL STRUCTURES

Operation Anaconda demonstrated that operational-level command-and-control structures were not in place for effective CAS in the early days of Afghanistan. Improving these structures and honing them prior to Iraq received significant attention. The Air Force realized its provision of CAS had not been convincing simply because ground commanders did not see airmen very frequently. Gen. Ronald Keys (2004) commented, "We need to refine the air component coordination element concept, which is air power's representation at the table in the land forces headquarters and command posts." Keys's comments also made reference to a "General Custer–type of scenario" wherein a land force overextends itself and gets "wiped out" by a hostile force. Demonstrating that Army preparation for Anaconda and subsequent complaints about Air Force CAS still struck a nerve, his remarks also indicated the service had plans to address perceived shortcomings. Accordingly, the Air Force established command-and-control elements within both Iraq and Afghanistan to ameliorate that shortfall.

Command and reporting centers (CRC) and deployment of the Airborne Warning and Control System (AWACS) aircraft facilitated a link between the tactical-level air-ground systems and the CAOC. The theater-wide focus on air support to ground forces was unmistakable, as was an overall eagerness among senior leaders to hone joint operations in MCO. From a firsthand witness: "When we went into Iraq for major combat operations, Moseley, Keating, and McKiernan [the air, maritime, and land component commanders, respectively] were like a band of brothers. All the relationships were there for MCO. Where it got difficult was when MCO ended" (Rew 2014). The reluctant transition to COIN warfare in Iraq witnessed an end to warm interservice relationships, which gave way to an extended period of parochial interservice doctrine debates.

The Air Force's Air Component Coordination Element (ACCE) construct demonstrated the on-again, off-again nature of air-ground jointness in Afghanistan and Iraq. According to Lt. Gen. William Rew (2014), General Moseley conceived the ACCE construct during the ten months he worked on the battle plan for Iraq. "He wanted to keep the land-component relationship close, modeled after the one he enjoyed with Lieutenant General McKiernan" (Rew 2014). The ACCE concept placed a general-officer airman with each service component in both Afghanistan and Iraq. It was similar to the arrangement between Lt. Gen. Omar Bradley and Maj. Gen. Pete Quesada in World War II, who, as the respective commanders of 1st Army and 9th Tactical Air Command, remained in the same place at all times, a hedgerow alone separating their sleeping quarters to mitigate the risk that a single attack would kill both commanders (Hughes 1995, 156). If the formal arrangement survived Moseley's tenure, the spirit of its implementation did not. Rew (2014), who went in 2007 to monitor command-and-control arrangements in Iraq and Afghanistan, perceived "visceral hatred" of the air component by the ground component, and attributed the animosity to a CFACC bent on asserting the independent, centralized control of the air arm within a theater of operations that included two very significant COIN operations being run by two different four-star ground-component commanders.

LIAISON BILLETS AND ORGANIZATIONAL ALIGNMENT

Whereas the ACCE concept—at least during times when it worked well—sealed a gap at the operational level of warfare, demand for low-level tactical air-ground liaison seems to have been adequately met, even at the outset of Afghanistan. Even so, tactical-level capacity begged for expansion in the opening years. By early 2004, the idea that all services should be able to control CAS provided by air assets of any service or a participating coalition member had momentum within the US military and the NATO alliance, with training mechanisms to match (Pengelley 2004). The missing piece was a reliable organization to coordinate among controllers at the maneuver-unit level. One of the shortcomings evidenced in all CAS efforts from World War II to the present is an initial lack of personnel adequately qualified to control air assets integrated with the ground units they support, but in Afghanistan this came in a slightly different form of difficulty than it had in the past.

Although a dearth of adequate control plagued the early years of Afghan and Iraqi CAS, simply producing more qualified terminal air controllers and putting them with ground units was not the remedy. The issue in Afghanistan that glared in Operation Anaconda was the lack of coordination one level up from individual maneuver units. In changing alignment of Air Support Operations Centers from the corps level (which served only as administrative organizations in this era) to divisions (which served as the top level of the actual fighting force), the Air Force started to address the real issue (Neal, Green, and Caraway 2012). Along with trying to transfer a portion of CAS capability to the reserve component, the Air Force made a commitment to improving CAS (Jinnette 2014). Once a capability is in the reserve component, it is much harder to reduce or eliminate than if it remains in the active-duty force. The political dynamics of the reserve component, particularly the National Guard, mean that local political considerations have more decision-making sway than a singular service. The remaining step was to field the appropriate command-and-control structures, such as the Air Support Operations Centers left behind by the 10th Mountain Division, and allow them to act as a liaison between the CAOC and maneuver units for the provision of COIN air power. This remained a shortcoming—Monte Cannon (2012) derided it as "pasting over seams with still more liaisons"—and remains an issue the US military has never brought to an appreciable point of closure (see Dixon 1976).

NAVAL INTEGRATION

This case study has focused on the Air Force. As the lead proponent of air power in the joint realm, one might suppose that the Air Force might have led interservice integration during this era. We have examined how doctrinal and cultural predilections impeded that development. Throughout both conflicts, the Navy played an increasing role in CAS and ISR missions. Adapting the carrier-aviation role (which was, by naval doctrine, primarily focused on fleet defense) in the early interdiction campaign, naval aviators matured in the joint CAS arena to a level of precision and flexibility ground troops expected from Air Force and Marine Corps air platforms (Lambeth 2007, 45).

As in the Balkan campaigns of the 1990s, integration of naval aircraft into the ATO was at first vexing for both planners and the naval aircrew supporting those missions. Frustrations with the Air Force's system that arose in Operation Desert Storm and the Balkans still plagued efforts to integrate the Navy into an air war fought over land. Though difficulties remained in using a centralized command-and-control system headquartered in a CAOC, the efforts to coordinate made in the Balkans continued to translate into better relationships in Afghanistan and later Iraq.

Naval aviators reported that integration of carrier-based air power since 2001 has improved steadily. One noted that "CAOC planners had a better appreciation" of the carrier deck cycle and "carriers came to understand that they could not operate in a completely rigid manner" (Maloney 2013). Inability of the carrier battle group to receive and interpret the ATO—a common problem in Desert Storm—resolved during the 1990s. The Navy increased its use of ground liaison officers (GLOs) over the decade beginning in 2001; these Army personnel embedded in Navy units provided insight about planned air support requests. Rather than simply fly to an assigned area and wait for a tasking, CAOC planners made deliberate use of naval assets whenever intelligence and planning allowed.

Maritime component air capabilities relevant to CAS amplified many of the issues described thus far for the Air Force. Difficulties with communication equipment and designated frequencies to talk to ground controllers were common in 2001–2002. Yet by the middle of the decade, naval aviation communications equipment—including secure, frequency-hopping, jam-resistant radios—operated in the air-ground command-and-control system routinely. The Navy and Marines added additional radio capabilities to all CAS platforms, including Link-16's shared positional awareness via airborne data-link. The Navy's CAS targeting pods had one-way video data-links by 2007, evolving to two-way data-link systems that gave aircrew and ground controllers the ability to see and communicate with each other what they saw on the ground (Maloney 2013). But the Navy's advancement of CAS for COIN support has much in common with the Air Force's: rather than being driven by the vision of a cadre of senior service leaders, it grew up more organically. It was driven by a general understanding at all levels of the service that budgetary relevance in the Pentagon meant contribution to the COIN fights and the improving aviation technology (weapons and sensors) available throughout the era.

Summary of COIN Air Power Advances

Table 5.1 summarized the COIN air power advances realized in the Afghan and Iraqi combat theaters after 2001, revealing these trends:

Materiel and technological advances were numerous, developed rapidly, and exhibited steady improvement throughout the case.

Training programs emerged to provide manning commensurate with huge growth in the number of UAV systems and coordination with ground troops. Within platforms that already existed, new emphasis on CAS-enabling technology and procedures improved proficiency and coordination throughout all services.

The Army and the Air Force developed a system to link close-support operations to the centralized air-control scheme used for MCO after Operation Anaconda. However, command-and-control organizations for COIN air power made slow advances, witnessing notable failures and regression. Relationships among senior officers seemed at odds with technological, training, and coordination advances at lower levels.

Table 5.1 also lists some factors that enabled or hindered joint cooperation. We proceed to examine how interservice relations in this case study overlap with our previous theoretical survey.

Explanations of Jointness

To tie the historical process-tracing narrative back to the foundational theories discussed in chapter 2, a summary of the relevant theoretical observations—table 5.2—summarizes this section and bridges many of the explanations.

Theories of Organizational Interaction

FIGHTING CAN HELP FOSTER COOPERATION

Though the cooperation and the combined-arms efficacy it brought about in this case study are the weakest example in this book, it is nevertheless remarkable in the context in which it occurred. The extensive technological and limited command-and-control advances occurred in areas in which the Army and Air Force find it most difficult to cooperate: limited-scope war and direct air support to ground units. Historically neglected in peacetime and marked by steep learning curves during periods of conflict, this case exhibited the same trends. Where cooperation occurred, it was facilitated by fighting, both actual battlefield combat that revealed weaknesses in combined-arms effectiveness and the resultant interservice conflict that forced leaders to address those neglected specialties. A sharp rap on the knuckles to the Air Force from the secretary of defense certainly contributed to the milieu under which jointness developed by making it generally understood in all services that COIN recalcitrance was unacceptable. "Beating up one

Table 5.1: COIN Air Power Improvements in OEF and OIF

Summary of COIN Air Support Advances		
Area	*Tangible Improvements*	*Enabling (Inhibiting) Actions*
Equipment	USAF medium–altitude UAV capability increased 20-fold	Industrial development
	42 MC–12 *Liberty* aircraft	Congressional markups and lobbying for increased UAV systems
	Advanced targeting pods (ATPs), ATP integration on bombers, Rover video terminals, and data-link systems	SECDEF interest and focus on deploying additional systems (2007–2009)
	Laser designators, range finders, GPS-enabled weapons, hybrid laser–GPS weapons, low–yield (collateral damage-limiting) CAS weapons, and UAV-specialized glide bombs	DoD ISR task force
		Additional UAV squadrons added
		Intel support grew to match CAP capacity
	JTAC position/visualization tools, TACP-compatible radios	JUON & other urgent needs processes
		Interservice acquisition provided evidence of joint need
Training	CAS role for all fixed–wing multirole fighters	561st CTS established; "lessons learned" widely pursued
	AF training for UAV operators	Pilot production surge/WIC shutdown
	GREEN FLAG, RED FLAG revamp, NTC, & JRTC training	
	CAS prominent in tactics manuals, moving target procedures	Air Force unit training aligned w/ accompanying Army units; concentration on COIN missions over MCO
	JTAC integration routine to company level	Development of dedicated, nonaviator ALO/JTAC career field
	AGOS at Nellis AFB evolves to become 57th Operations Group	Increased JTAC/FO manning and increased GLO manning
Organization	**Endogenous**	AF PA organs broadcast "all-in" ethos
	Matching JTACs to ground units, growth of operational control	Expeditionary culture normalized (ILO became JET)
	Meso-Organizational	Institutional history of dialogue (e.g., Army–Air Force Warfighter Talks)
	ACCE/JACCE	
	Naval asset integration	"No more Anacondas"
	JAGOC	Theater air commanders not empowered
		Lack of operational control mechanisms
	Exogenous	Air Force senior leadership scrutiny/decapitation
		Inconsistent strategic guidance (OIF)

inmate to get the prisoners to fall in line" fairly analogizes what Secretary of Defense Gates did to Secretary Wynne and General Moseley.

Several social science theories combine to complement one another and offer plausible explanations for this behavior pattern. One comes from organization theory: war is a first-order threat to the literal survival of military members, who must respond with improved tactics. As in chapter 4, crisis-cooperation theory also offers an explanation of the role that fighting can have in cooperative relationships. In this case, the threat was marked by uncertainty, both the enemy's unpredictability and the uncertain national political aims. The Army and Air Force responded to the perception that they prepared poorly for Operation Anaconda by tackling the ongoing problems of air support in Afghanistan and the upcoming invasion of Iraq with several months of preparatory joint planning. The theory of professions also predicts a similar outcome, at least as it pertains to preparation for the next immediate battle, as neither service wanted to lose more credibility. Organizational autonomy and survival of bureaucratic service interests felt as much under fire as did US and Afghan soldiers on the ridgelines around the Shahikot Valley in early 2002. The services banded together for support against these outside threats.

Although cooperation emerged in response to these forces, it was fleeting. As combat operations reached steady-state rhythms in the wake of the Iraqi invasion and the transition to a COIN-intensive conflict in both countries of interest, interservice relationship stasis set in again. After personalities that had utilized open fighting as a means to achieve the close cooperation of a "band of brothers" exited joint collaboration, the services were lulled into a sense of fait accompli with respect to their combat responsibilities, but without appreciation of the friction that needed to be overcome (Rew 2014). Combining these structural issues with personalities dedicated to fighting for service specific command-and-control ideals made the atmosphere ripe for a breakdown in helping behaviors—though outward relations may have remained cordial, seething civility masked the depth of dysfunction.

Coalescing Interests Helped Solidify Cooperative Gains

Afghanistan and Iraq showed how coalescence of objectives between organizational levels can shape tactics and strategy. For example, in the major combat operations accompanying the Iraqi invasion, Cannon (2012, 271) portrayed Gen. Tommy Franks's alignment of service interests with those articulated to him by DoD and other external actors as an effective means of getting the services to support a hurried invasion timetable. By offering the Army and Marines their own independent routes of attack to Baghdad, by allowing the Air Force its shock-and-awe air campaign, by incorporating the Navy's available special-operations and carrier assets into the attack, and by implying punishment for wayward actors ("The childish behavior we saw in Afghanistan will not be repeated"), Franks brought to heel divergent interests that risk-averse services might have used to threaten his tightly knit plan, hewing closely to the desires of Secretary of Defense Rumsfeld (Gordon and Trainor 2006, 93).

In the realm of COIN air power, subgroups within the services coalesced to accelerate the growth of UAV-ISR capability. The emergent UAV community, which had been treated like a "leper colony" in an Air Force culture dominated by manned platforms, saw its legitimacy solidified by a huge increase in demand for the capabilities it provided (Mulrine 2009). These in turn captured the attention of Secretary Gates, who then pushed the Air Force to increase its capacity for MQ-1s and MQ-9s more quickly than the service preferred (Deptula 2013). Viewed from the perspective of organization theory, the development shows the importance of coalition-building behavior in bureaucratic politics (Allison and Zelikow 1999, 255–258). It also brings to mind the means by which subgroups held in low esteem by their parent organizations facilitate effective cooperation in collective efforts.

Military-Specific Theories

A FUNDAMENTAL CLASH OF SERVICE PREFERENCES INHIBITED JOINT COOPERATION

Viewing how cooperation waxed and waned, one might reasonably conclude that something forced the services to cooperate in ways absent outside pressure. While arrogance and chest-thumping may not further goodwill or cooperation, they are a part of steady-state interservice existence and should not absorb an inordinate share of responsibility for breakdowns. Analysis of the structure of the components of a joint force, along with the cultural preferences of the services that supply the bulk of each component, suggests a more compelling reason for the enduring dissociative forces that plague joint cooperation. Air support to COIN warfare puts the Army's and Air Force's distinct ideals of command-and-control structures in competition with each other, in a way that continually brings them into sharp relief. The Army was committed in its official COIN doctrine to push autonomy to its lowest-level commanders, reasoning theirs was the best picture of local conditions critical to effective engagement of a population. This desire to devolve initiative to the fringes of the military organization conflicts with the Air Force ideal of centralized control and apportionment to meet priority-mission assignments with scarce, theater-level resources.

David Johnson's (2007, xxiv) argument that COIN warfare requires decentralized command and control of air power complements Rew's (2010, slide 8) ideas for improving operational coordination; both realized its reliance on smaller, peripheral units to carry out the overall strategy. These ideas contrast with Jeffrey Vandenbussche's (2007, 68) prescription for balancing the centralization and decentralization of Air Force operations. He posited that political sensitivity toward a given conflict was inversely related to its relative intensity. High-intensity, existential conflicts merit decentralized, mission-type orders, whereas limited-scope conflicts like COIN call for centralized control to prevent the commission of serious errors that might thwart political aims, as they rely heavily on not angering a civilian population (68). The resulting command-and-control paradox is that the Army and Marines try to push initiative and control toward lower echelons in COIN, while the Air Force takes the opposite tack. We conclude here that these service preferences constitute an ongoing obstacle to closer joint cooperation in any conflict, but especially one characterized by COIN warfare and small-unit tactics that rely on air power for fire support.

INCONSISTENT EXTERNAL GUIDANCE INHIBITED JOINT COOPERATION

Our first two case studies found that outside influence can have a positive influence on joint cooperation. We will end up there in this study as well, but the path is rockier. Here, the finding is that inconsistent external guidance inhibited joint cooperation, instead encouraging the services to pursue more competitive and confrontational behaviors. The first example came from DoD, where a sharp difference in the approach of two consecutive secretaries of defense sent confusing messages to the services. In the case of Donald Rumsfeld, an insistence on a small-footprint military force deployed quickly to Iraq was in line with Air Force ideas about the strategic effect air power can have, but clashed with the land services' ideas about the size of the force needed for postinvasion stabilization. The other notable contrast between the two defense secretaries was their focus during the conflict. Whereas one Air Force service chief complained about Rumsfeld's focus on futuristic "transformation" of the defense establishment at the expense of attention on Afghanistan and Iraq, a successor felt that Gates micromanaged the COIN conflicts and sacrificed long-term defense planning (Jumper 2013; Tirpak 2014, 54). Objective civilian control of the military was undisputed throughout the era, but the animosity and distrust it engendered between layers of the defense establishment may have placed additional barriers to future cooperation.

While DoD guidance was inconsistent, it was at least extant and somewhat forceful, particularly in the case of Robert Gates vis-à-vis the Air Force. Less guidance came from other outside sections of the defense establishment, especially from Congress, who in typical form stood clear of questions about military performance until well after conflicts had ended or appeared to be drawing to a close. Congress did enable large amounts of additional military spending during the peak years of COIN conflict though, which solidified some of the technological-materiel gains already mentioned, but also created opportunities to misappropriate so-called "joint" wartime acquisition processes with service-specific preferences. Even if they did advance battlefield cooperation, auditing agencies noted that these acquisition efforts proceeded without substantial oversight or coordination.

One exception to an otherwise general dearth of consistent external guidance to the services was increasing executive reliance on the options afforded by UAV platforms, for both surveillance and strike options. This preference was reflected in DoD direction to the Air Force, but stands as a force in its own right. However, the external guidance and attention that proved so effective in the two previous chapters were generally absent in this instance, and the quality and consistency of joint cooperation seem to have suffered as a result.

IDEAS ABOUT MILITARY INNOVATION

The last major theoretical observation for this chapter involves the mechanisms of military innovation as they relate to joint cooperation. The motivation of Air Force and Army personnel who recognized a need for improved coordination mechanisms led them to document shortcomings, propose remedies, build consensus across bureaucratic levels, and ultimately stabilize the use of new systems within theaters of operation. Greg Harbin's experience in advocating and propagating use of the Rover video repeaters is an archetype of this kind of innovation, and it echoes the constructivism and activities of the heterogeneous engineers documented in the JPATS case study. Again, someone willing to put in energy and effort that went beyond the normal churn of bureaucratic affairs made possible a cooperative system enabled by technology.

None of the competing theories of military innovation emerged with clear explanatory power. Interservice competition à la Coté helped spur the Air Force to develop better UAV technology, as the service watched the Army's relentless deployment of its own systems intrude on its turf. As discussed earlier, though, plenty of outside influence augmented this possible motivation. A stable career path for UAV operators also emerged during the period of conflict. Along with subgroup coalescence, this stabilization of a new professional group—associated with the *peacetime* innovation of Rosen's formulation—provides another explanation for the normalization of an important air-support technology and highlights the difficulty of characterizing these periods of conflict as "war," at least as far as bureaucratic actors were concerned. To round out military-innovation theory, a nod must go to Posen, whose key argument that military failure drives innovation may help explain the fits and spurts of command-and-control and technology advances that occurred after Operation Anaconda's ignoble prelude.

While a great deal of general and military-specific theories provide plausible explanations as to why certain cooperative behaviors emerged, the one thing that seems lacking—seeming to point to why cooperation was so sporadic for well over a decade of armed conflict—was explicit outside influence to force multilevel interests to coalesce into stable cooperation. Absent that forcing function, the external threat of war and the efforts of individual leaders shone through at times, but were never enough to overcome the dissociative forces that service preferences have historically erected as obstacles to joint cooperation where air support in limited war is a concern.

Conclusion

Our examination of COIN air power during recent Afghan and Iraqi conflicts reveals a variety of mechanisms, informed by theory and descriptions of the defense establishment, that can encourage and discourage joint cooperation. Though leadership and coalescence of institutional interests remain important, the simple threat of war seems to offer additional mechanisms for creating combined-arms success. The US military was steeped in COIN for so long that some level of joint institutional cooperation was probably inevitable, no matter how inefficient the machine used to build it. Whether it is the desire to be a member of an effective team facing down a crisis, a passion to avoid failure, a grasp for professional credibility, or the incentive for innovation in the heat of battle, the pressures of war serve to drive joint cooperation even as they remove some obstacles that might prevent cooperation. The obvious shortcoming of this approach is that the pressures dissipate in times of peace, even during times of steady-state conflict. The months and years when the nation is not at war ostensibly constitute an opportunity for undistracted thinking that could more readily build long-term interservice military capability. War is at any rate an expensive way to learn lessons, in terms of both materiel and human life.

As our examples showed, individuals can advance jointness. In the story of COIN air power, people proved their mettle fighting on a battlefield, delivering new equipment to users in the heat of combat, insisting on the betterment of lackluster command-and-control systems, pushing a service to develop capability faster than it might have, and reaching out to mend damaged interservice relationships. As in other chapters, successful efforts spanned all levels of the defense hierarchy, as empowered agents acted in the role of heterogeneous engineers to build consensus and stabilize the "answer" they devised long enough to make a significant contribution to efficacy. The urgency of battle and the greater budget latitude associated with war gave greater leverage to individuals who wanted to advance joint capabilities.

Overcoming bureaucratic inertia and the tendency to seek individual political gain proved to be among the biggest obstacles to jointness. Here the efforts of individuals prevailed at times as well. Where a single person could not overcome difficulties, coalitions of stakeholders striving to increase combat capability formed. Given a suitable amount of time in the crucible of combat, the goals they sought were sometimes realized. And of course, an empowered, decisive meso-organizational leader like the secretary of defense can swing a heavy hammer, but the ensuing organizational change is likely to be more fleeting than if jointness originates through service cooperation.

Our theoretical examination of the causes of jointness suggests equifinality; there is more than one path to encourage, even force, joint cooperation. It also raises a few paradoxes. The unique traits and sources of pride resident in each military service cause it to strive for greater capability, an essential building block of joint-warfare effectiveness. Yet the institutional personalities driving ambition may cause resentment in other services, potentially negating joint benefit. When the ideal command-and-control mechanism to achieve battlefield effects seems evident from one perspective, the immutable political nature of war makes it ill advised from another. An individual, particularly a high-ranking or influential leader, can provide the force and focus to achieve joint cooperation when it seems out of reach. Yet the same force of personality may serve to demoralize a force or intensify its bureaucratic resistance to cooperation, meaning that the use of individual influence must be deliberate and measured. In the end, Margaret Hermann et al.'s (2001) conclusions about organizational leadership proved particularly prescient for organizations at war. People, particularly leaders, matter; and they matter at all levels of armed conflict, from tactical to strategic.

Case Epilogue: COIN Air Power after Afghanistan and Iraq

CAS subject-matter experts interviewed for this investigation (Bochain 2014; Dabros 2014; Donnelly 2014; Gersten 2013; Jinnette 2014; Maloney 2013; Neuenswander 2014; Orchard 2013) generally agreed that Air Force CAS focus and capability reached a temporary peak between 2009 and 2011. Though CAS continued to be an essential part of COIN warfare in the CENTCOM region, the decade after Operation Anaconda delivered the bulk of demonstrable improvements in the ability of all services to coordinate and execute CAS. Low ground-troop concentrations with relatively little organic firepower, dispersed widely over a large geographic area, made the application of precision airborne fires critical to the success of the campaigns. Accordingly, air power presence—along with an ability to coordinate its effects among the services—has only grown since March 2002. ISR capabilities, especially those provided with the persistence offered by UAVs, also became a critical part of the fighting style employed by the United States and its allies throughout the Middle East.

As past history would suggest, with the Afghan war brought to an abrupt and failed terminus and the United States officially out of Iraqi operations, the service's patience with the COIN warfare mission has long been waning. As the conflicts wound down, concern over neglected MCO missions emerged. Attention on CAS has been fading, just as it has after all conflicts in which the Air Force has had to provide significant support to the Army. The Air Force plans a dramatic cut to the TACP enterprise, and while whether it will end its practice of colocating air support units with Army divisions remains unknown, postconflict drawdown of the Army necessarily means that CAS training, familiarity, and proficiency within both services will diminish.

In a similar vein, discussions as distant as the 2013 Army-Air Force Warfighter Talks suggested that the demand for COIN ISR capabilities within the services had reached a maximum (Clark 2013). Military withdrawal witnessed a reduction in ground forces and a commensurate decrease in demand for air missions that support COIN. Most assuredly, the Air Force has returned to its desire to develop UAVs that can operate in airspace contested by advanced air defense threats, and the Navy looks to follow suit. Given the importance of UAV-enabled strikes and ISR to national objectives that reach beyond the Afghan and Iraqi campaigns, there is little chance of these systems suffering neglect from the Air Force. Enduring close integration with ground forces is another matter, though.

Congress is unambiguous in its distrust of the Air Force on CAS and UAV ground support. The report (US Senate 2014, 10) accompanying the 2015 NDAA, for example, is one of dozens of times the Air Force's claim that it needs to retire the A-10 to meet larger budgetary constraints has been questioned. Amid a slew of limitations designed to slow or prevent the Air Force from retiring any type of airframes without a rigorous prior explanation to Congress, legislators also slowed the retirement of the MQ-1 *Predator* and are taking the same tack on *Reaper*. These measures, while they convey the overall displeasure of the legislative branch, are blunt control tools: they focus on materiel procurement but do not address the command-and-control and theory-of-victory problems that lie at the heart of Army-Air Force difficulties observed during these last wars. They are also driven by local political considerations (i.e., jobs in a given congressional district) rather than by any consideration of effectiveness in warfare.

The prediction by General Creech about the nature of twenty-first-century war that opened this chapter was as inaccurate as Maj. Gen. David Baker's comment to an industry group in 1956 (at least up to this point): "We can readily see that except for certain types of missions, the manned combat aircraft will become technically obsolete in the future." There is a helpful parallelism in missed prophesies about the nature of war and the demise of unmanned aviation, though. The Air Force and the other military services looked for peer competition in large-scale conflict and pursued combat platforms, including UAVs, to operate independently in the contested environment characteristic of that type of conflict. Instead, it found itself working closely with the Army and Marines in a COIN conflict of limited proportions; CAS and UAV development accelerated during years of combat to accommodate the ground-force needs in that very permissive air environment, albeit not fast enough to keep pace with the appetite of the ground force for on-demand organic support.

The contrast of reality with expectations says as much about the difficulty of constructing technological systems in anticipation of future combat requirements as it does about the Air Force's perceived reluctance to develop CAS assets or UAVs. Jointness is difficult to construct for the same reasons: the means to build it and the forms it must take differ with external circumstances. The Air Force's institutional culture would have rather avoided the Afghan and Iraqi conflicts and their incarnations as COIN-intensive attrition campaigns. However, finding itself enmeshed in such conflicts, it developed combat systems to accommodate the requirements of the environment, including enhanced CAS capabilities and a great leap in UAV-ISR capacity. In the description of John Law (1989, 111–113), the systems developed were not socially constructed as a result of the cultural preferences of the Air Force and Army. Rather, the systems developed and stabilized around the conditions of the era as these conditions interacted with service preferences.

While anticipatory development of technology is and always will be a difficult proposition, it is also necessary to restate one of the fundamental arguments of this chapter: the Air Force is loath to spend money, time, and people on systems that allow (or force, depending on one's perspective) the service to expend air power capability on missions that are organically apportioned to and executed at the will of ground commanders. The concept is anathema to the service's organizing principles and its fundamental doctrines. We differ with Benjamin Cooling's (1990, 2) assertion that "wartime experience—World War II in particular"—is the "proving ground" for CAS doctrine. CAS doctrine receives appreciable attention *only* during times of active conflict. At other times, only Army complaints, attention from Congress, or Air Force forays into acquiring new aircraft put any focus on the matter. Indeed, the story of COIN air support in post-9/11 conflicts shows that organizational structures and cooperation among senior military leaders remain the aspects of jointness most inhibited by interservice friction.

I. B. Holley's summarizing statement that "the processes and procedures by which success was achieved, usually belatedly, in each war . . . were largely forgotten by the armed forces by the time they again became actively involved in fighting" (1990, 535) is an apt description of CAS. We have outlined, through historical process tracing, some of the mechanisms through which cooperation occurred, despite obstacles—between the Air Force, the Army, and the maritime services on this thorny combined-arms effort. While concrete examples of cooperation exist, they are largely based on materiel and technology and enabled by training; there is little evidence of enduring institutional changes or of flexibility organizing in favor of COIN missions. The willingness of senior military leaders to cooperate in furtherance of joint goals is spotty in this case, although the failure of one leader could be overcome by the efforts and outreach of a successor—or externally delivered blunt-force trauma against a service reluctant to pursue jointness.

Even though by some measures the level of cooperation appears impressive when outlined over two decades, significant shortcomings are evident. The Army remained unconvinced of Air Force dedication to the CAS mission, as evidenced by vocal campaigning against the retirement of the A-10. Nor did the Army appear satisfied with the level of UAV-ISR support it received, demonstrated by the nearly seven thousand UAVs the Army acquired for itself during the conflict. Finally, vitriolic exchanges among senior officers did not diminish as the conflict wore on, leading observers to worry about scar tissue inhibiting interpersonal relations should the services need to cooperate on a new mission before the current generation of senior generals leaves the service (Rew 2014). Jointness, just like new technology, is difficult to construct, especially when external leadership is in short supply. If individual leaders see a necessity to build jointness though, the materials and mechanisms seem always to be available, perhaps even more plentifully during a time of conflict. The uncertain variable is the existence of human will to make an effort.

Chapter Six

Joint All-Domain Command and Control

> *Pale Ebenezer thought it wrong to fight,*
> *But Roaring Bill (who killed him) thought it right.*
> —Hilaire Belloc, "The Pacifist"

We now turn to our final case study, easing into it with martial poetry. Belloc grasped both sides of pacifism, and expressed his thoughts succinctly on the matter. Hoping for similar even-handedness and economy of expression in discussing fighting, we borrow from a better-known American poet to assert that joint cooperation is at the stage where two roads diverge (albeit in the Pentagon, not Frost's yellow wood) for this last topic of discussion. This final case study is contemporary, and so lacks the historical perspective that informed previous chapters in this book. It is too early to say whether the collective US military consciousness will look back in later years on this time to reflect on whether it took the road less traveled. By that we mean whether the US defense institution walked down the much harder path to joint cooperation in a peacetime context or if it took the path of lesser resistance to arrive at "silos of excellence," to borrow a pejorative Pentagon cliché. The more often-traveled way is that the services remained focused on their respective projects, talking regularly about partnering together and all the joint synergy that could result, but never making programmatic decisions that might put progress meaningfully in the hands of another service—or endanger service budgets.

Introduction to Joint All-Domain Command and Control

The context of this discussion is Joint All-Domain Command and Control (JADC2) (Hitchens 2019). JADC2 is at once an obvious concept and quite a lot to wrap one's mind around. A good starting point for grappling with the concept is to call it "the internet of things for war," or what battlefields will look like when big data, ubiquitous networks, machine learning, and automation come to military platforms and the weapons they control (Barnett 2021). We are familiar with this in everyday life. It is why one rarely has to set foot in a bank to handle financial transactions. It allows us to order from Amazon in the morning and have the object of our desire arrive that afternoon. And it is why Meta, Amazon, and Google can show us a targeted advertisement about something we might have only discussed in our living room or walked by in a store the same day. The application of this logic and these algorithms to military matters is quite a different proposition, however, and we will explore why later in this chapter.

Because of the recency of JADC2 concepts, our inability to predict if they will endure, and to what degree they will change the US military, our overarching theme for this chapter is one of uncertainty. In fact, our endeavor here is fraught with peril. It could easily become a headline-chasing exercise that withers quickly; after all, there is no guarantee that the current framing of the problem will endure long enough in the Pentagon that a reasonable test of our pretheory can be derived from it. While we can look back at AirLand Battle and see that it had a profound effect on every aspect of the contemporary military it impacted, from doctrine to organization, the potential for JADC2 to do likewise is there but the outcomes are uncertain. The JPATS trainer resulted in a shared materiel solution for the Air Force and the Navy; however, JADC2 has not yet resulted in any meaningful large-scale procurement for any of the individual services or for any joint body. Furthermore, exigencies of combat that drove innovation and cooperation in the joint warfighting art of close air support during the COIN wars (chapter 5) are not yet present in the JADC2 development. Therefore, JADC2 has not been tested in the crucible of combat. Yet we persist, because we think that the need for jointness in command and control is not going to go away or be solved in the next ten years. An analysis of the institutional response, using our pretheory framework, is useful and enlightening for both understanding JADC2 and testing the validity of the theoretical work presented here.

Given the novelty of this topic, we begin with an attempt—albeit necessarily preliminary and thus incomplete—to describe in more detail exactly what it is. In a military context, the first purpose of the so-called internet of things centers on identifying, locating with precision, and tracking many of an adversary's useful military components, networks, sensors, and the like, all of which would be called "targets" in military parlance. The first purpose gives a military force the ability to make use of the second purpose, which is the ability to influence those targets in some way. These options for interference include fooling an adversary, jamming them, interfering with systemic functions via an electromagnetic signal overload (all of the preceding options would be called "nonkinetic" in military jargon), all the way up to physically destroying them with a warhead of high explosives delivered via a missile, rocket, or bomb (the so-called "kinetic" option). After the intelligent network (JADC2) helps with locating and tracking, it is able to assist in engaging and assessing the effects of the engagement via a much quicker cycle than happens in current warfare. At this stage of development, descriptions of what might be possible are quite aspirational, with many writers stressing that artificial intelligence (AI) will change the pace—by orders of magnitude—at which battlefield decisions occur today, culminating in lethal engagement of the enemy's fielded force (Osborn 2020).[1] The ability to interfere with targets, to the degree the adversary comprehends our ability to do so, can be a useful deterrent to avoid warfare or press for favorable peace terms if open conflict has broken out. Working as described here, JADC2 would be intuitively useful for winning a military campaign in ongoing armed conflict.

It seems certain—we can only speculate since it does not exist—that such an interconnected system of sensors, platforms, and weapons would be handy for the US military to have at its disposal. However, the challenges of building it and putting it into use are much greater than those that define other integrated networks we have become accustomed to in civilian life. To understand why this is the case, we first need to consider access to the electromagnetic spectrum. While there are hundreds of signals traversing the airwaves in any first-world country today, various schemes of spectrum management assure that the originators and targets of those signals operate relatively uninterrupted. The second issue that merits consideration is access to data. While civilians may resent and sometimes fear how much access an Amazon, Alibaba, Alphabet, or other corporate entity has to their personal information, collectively consumers tend to deem the risk worth the reward of convenience and continue to share facts about themselves.

The third factor that makes military integrated networks more challenging has to do with machine learning and artificial intelligence. Although it is still a nascent field in relative terms, machine learning and computer processing have matched or exceeded human faculties in many arenas, including complex strategy board games like chess and go. A fourth aspect we need to analyze is the connections we allow the virtual network to have with the physical world. Examples of these connections in civilian use include everything from radio-frequency identification (RFID) tags on a FedEx jet to onetime delivery-person access codes that allow a package to be placed inside one's front door. Since these virtual network connections to physical items and places are, on balance, seen as beneficial by the society they serve, the access between the virtual and the physical is there most of the time. A fifth and final element is the "control" part of command and control or the "human in the loop" problem. We may be happy for a computer to make decisions about when a water valve or control rod should function based on an algorithm; we have not yet made the leap to wide acceptance that a computer's judgment alone should be the deciding factor in when to use lethal force against a human target.

In the context of military competition, every aspect of the arrangement just described is turned on its head. Spectra are noisy; jammers try to deny access to both sides and friendly cell-phone towers are not available to carry signals across long distances. Data from one side are not shared with the other; information is encrypted, things are made to seem what they are not, and spurious inputs are provided as ruses. The clean and efficient (or at least generally uncluttered) circumstances in which algorithmic processing bests human rivals are not replicated during war. Should the first three factors somehow align so that one side could use this integrated network, the "last mile" to delivering effect would be as literally and violently contested as any battlefield in history.

A conundrum therefore arises. An Internet of things is easy to imagine (because it exists today) and easy to envision weaponizing (or "creating effects") because most people in the defense professions and industries can imagine high explosives becoming a part of any network. It seems in the JADC2 case that the joint venture risks putting the cart before the horse with sweeping statements about the promises of surefire victory or deterrence based on a secure, pervasive military network, its ability to identify, process, and prosecute enemy targets faster than the human mind and enemy defenses can react. The shiny appeal of weapons and platforms that operate in concert with this network and the machine learning that drives its lethality has captured the attention and imagination of every service and several defense contractors. For a few years now, any defense contractor showing promise on any thread of this tapestry has had the services knocking at the door, cash in hand. The conundrum lies in how difficult it is to actually build and use this system that everyone is now imagining. The scale of the standalone network that is needed, the security it requires, and the active sabotaging and blocking of this network's effect delivery by advanced adversaries have made the dream of JADC2 elusive in practice. It is fairly easy to assert that DoD finds JADC2 conceptually difficult because of the number of times it has released a "new strategy" (2020, 2021, and 2022 so far) for the concept. Similarly, the number of senior defense and service officials who have been appointed to the role of describing and pursuing the concept has reached double digits (ostensibly whoever was doing it previously to the new appointees did not succeed).

The nascency of the effort puts us in an ideal spot to reflect on and conjecture about what joint cooperation on the topic of JADC2 may look like. The term implies command and control over all domains (land, sea, air, space, and cyberspace) in a joint way (together, supposedly). The way the US military services are organized with the capabilities and assets they control separately would indeed suggest that JADC2 would need to be a joint solution. With the fact in mind that the national-defense enterprise wants and expects a new revolution in military affairs that makes use of the current state-of-the-art technology and integration of capabilities and information networks, how likely is it that a joint venture will be the solution? Or will one service make some notable progress utilizing some aspect of the new technology, as the Navy did in developing fleet-launched ballistic missiles, even making it a key part of its doctrine and organizational construct? Might the topic be one of enduring promise but never really approach the outcomes predicted by Pentagon briefings or contractor brochures? Other sought-after technical solutions and innovations, like the Star Wars missile defense systems of yesteryear or today's hype for a large, high-speed vertical takeoff and landing aircraft, continue to excite but have yet to make it further than lab demonstrations of enabling subsystems (Barajas 2022). However, it is also true that even at its peak of adoption and influence, AirLand Battle as a concept eluded precise definition but was very successful as a joint venture, so a lack of exactitude is not necessarily a death knell for JADC2.

Given the technical difficulty of realizing meaningful progress in the arena of JADC2, it may not seem reasonable to hope for useful predictions about how cooperation around it might proceed. Our optimism is partially grounded in the fact that the pursuit of JADC2 shares many of the exact same institutional, technological, and grand-strategy contextual factors that were impactful in a positive way for AirLand Battle. Though the critical advancements in military technology and organizational behavior that culminated to make AirLand Battle truly joint and successful have not yet happened for JADC2, many of the initial conditions are there. Consequently, whether JADC2 ends up in the historical records as a successful joint venture solution or not may come down to necessary and sufficient factors.

Thus far, we have presented ideas from an extensive body of literature to say that JADC2 is a way to link sensors, systems that deliver weapons, and the weapons themselves via a networking solution that makes extensive use of AI, machine learning, and other computational advantages. However, this description may quickly outlive its use. Since JADC2 is still very much a conceptual capability, the intellectual framing of what will eventually exist is still in flux. Some authors, for example, have referred to it as "layers" that include sensors, communication, processing, decision, and effects (Hadley 2021b). Recently (in early 2022), the office charged with developing JADC2 for the Air Force in the form of something called the Advanced Battle Management System (ABMS) was working hard to simplify its definition rather than putting more things into it, but by late 2022 had been pushed by the service's civilian leader to change focus from this clarifying effort, and an additional leader of the Air Force strategy for ABMS was announced to great fanfare—the third to be installed since 2019 (Insinna 2022). (More on this later when we tackle the relevant forces at play.) The Army seems to be a bit more zealous in making the concept have widespread doctrinal impact. (In that way, JADC2 already mimics the pattern of AirLand Battle development as it relates to the Air Force and Army.) Unlike ALB, the Navy is fully engaged and is a thought leader in many aspects of JADC2; its concepts favor giving JADC2 teeth at the so-called tactical edge, where operational warfighting concepts meet the enemy on the battlefield.

Regardless of the evolutionary pattern by which its technologies and capabilities develop, JADC2 as a warfighting concept still needs to find its way in the big, wide world of military doctrine. Machine learning (as well as to some degree AI) plays a role in most descriptions of JADC2. This sounds compelling at a level of abstraction where the United States "seeks to create simultaneous dilemmas for adversary forces, overwhelming them with too many challenges to counter successfully" ("America's Approach to Command and Control" 2021). It rapidly unravels if someone points out a realistic outcome of this situation: to take advantage of those so-called dilemmas at the pace and scale the network would offer would mean that we would have to allow automated kill-chain closure. In plain language, we mean that a robot or computer would be allowed to use lethal force against a human enemy without any friendly human checks and balances on that force. That is a decisive and irrevocable step the US military has not taken before in warfare or when developing a warfighting concept, and the ramifications are serious.

We have to add a final word here about our JADC2 analysis that did not apply to previous case studies, and that is the matter of classification. New weapons programs are covered under the highest levels of classified protection. The need to shield both progress and setbacks from rivals makes this an obvious approach. To publish our ideas in this book, though, we had to rely solely on open-source articles and unclassified interviews. While there are undoubtedly some dynamics influencing JADC2 that happen behind closed vault doors, most of the general ideas about the concept itself and the bureaucratic influences that impact jointness should carry through without having this part of the perspective. Even if the secret sauce of JADC2 gets developed in a top-secret weapons lab, it will still have to stand on its own in the very open environment of interservice budget competition and congressional politics, among other challenges.

Comparing Joint All-Domain Command and Control and AirLand Battle

Though it is immediately evident that there are surface-level similarities and differences between AirLand Battle and JADC2, it is worth a bit of time enumerating these. We will start first with the concepts. For ALB, the concept evolved over time. It started out with an amorphous understanding and over years of iteration took on a well-defined and well-documented form, the exemplar of which became the 31 Initiatives.

The description "amorphous" fits JADC2 in its current instantiation very well. Today its status is best summed as "all things to everyone"; despite the gallons of ink spilled on what it is generally, very few people on the planet can give you a specific operational vignette on how JADC2 would be used to improve battlefield performance against an adversary. (To be fair, the services dutifully hold at least a few joint exercises each year where they fabricate and test a scenario that offers such a vignette—the rigor, complexity, and potential to scale these are all in doubt, however.) However, one can intuit at least some of the command-and-control system requirements that would be required for a concept of war against China or Russia wherein first-line forces "can smash masses of Russian tanks and Chinese ships in the early hours of a major attack, buying time for the United States and NATO to mobilize" (Axe 2021). In a huge operational area like the Pacific, successfully fighting a potential rival as advanced and formidable as China would not only require advanced platforms and weapons. It would also require an advanced sensing grid, lots of dispersed computational power to make sense of the information absorbed by the sensing grid, and a command-and-control network that could be used to generate effects against multiple targets quickly before adversaries could in turn target US platforms and military personnel.

A raft of articles that the services have generated or inspired on the topic gives the impression that both JADC2 and the ABMS (which is the Air Force's name for its attempt at JADC2; the Army calls theirs Project Convergence and the Navy has named theirs Project Overmatch—more on this later) are concepts that (to use a dismissive bit of Pentagon parlance) "brief well" but are much harder to create in reality. A case in point is contrasting an early description of ABMS that the Air Force released, wherein it described five elements, or "product categories": (1) sensor integration, including satellites, aircraft, ground-based radar, and the like, from both military and commercial networks; (2) data generated by the sensors; (3) secure processing, the ability to keep secret data separate from nonsecret data and yet allow everyone access to data products; (4) connectivity, including machine-to-machine links that are not currently possible between many weapons platforms; (5) applications, for example, a map that shows where all US forces are on the battlefield; along with (6) effects (Hitchens 2020). Some of that clarity has departed with the Air Force leadership who gave us that rubric—it has been notoriously difficult of late to find a general officer who leads the ABMS office for more than a couple of years at most, and many rotate out more frequently than that—witness Brig. Gen. Luke Cropsey's installation in September 2022. Under a former Air Force lead for ABMS, a seemingly less ambitious (albeit temporary) definition emerged focused on interplatform connectivity, or "all about connecting dots" (Eddins 2021). Even that attempt at clarity and simplification, however, was short lived. Right now, anyone looking to identify and describe the joint approach to JADC2 is bound to be confused, if not vexed. As Gen. David Allvin (2022) said of Secretary of the Air Force Frank Kendall, he "is not alone in a frustration about really understanding what ABMS is and isn't, what it should be. . . . Everyone comes at it from their own perspective." And he has acted during his tenure to advance understanding, not least by naming a new lead for the project.

To put a finer point on the confusion, we hearken back to a definition given by a previous proponent of ABMS, former Air Force Assistant Secretary for Acquisitions, Technology, and Logistics Dr. Will Roper. He called ABMS's "true form" "simply a system to make data produced anywhere discoverable anywhere" (Roper 2020).[2] Dr. Roper was a uniquely transformative leader within the Air Force and he offered a compelling vision of improving military acquisitions by disrupting the process with digital engineering. His energy and dedication made him unique in being a mid-level Pentagon official to leave a substantial, enduring legacy, as shown by the admiring remarks of the leaders who have attempted to replace him (Magnuson 2021). His extensive joint experience (Army, Office of the Secretary of Defense, and Air Force, in that order) leading high-visibility acquisition programs with disruptive technologies as a unifying theme means that he is as difficult to replace as he was effective in his various positions (Roper 2019). However, because the current secretary of the Air Force, Dr. Frank Kendall, seems every bit as interested in maximizing the value that ABMS delivers, it is premature to think that the impact of a visionary civilian leader will not be a catalyzing force behind Air Force development of ABMS. And since Kendall was a defense acquisition official who served during the development of some ALB systems, his effort may also impact joint cooperation. Kendall's calendar (through mid-2023) reflected JADC2 meetings with multiservice, Joint Staff, and Office of the Secretary of Defense (OSD) representation, suggesting he understands the personal investment someone of his stature needs to make to see vision become reality.

Our previous case study on AirLand Battle also made much of unique service cultures' lenses of military doctrine. As we mentioned, the Army and (somewhat less so) the Navy are generally better than the Air Force at defining new doctrinal concepts and then using them to drive organizational and operational change. Consequently, it is less surprising that the Army and Navy are making progress in defining what they mean for their respective instantiations of JADC2, while the Air Force is still struggling. In attempting to explain the disparity, we can hypothesize that the Air Force has not yet developed a sense of urgency such that the need to develop JADC2 has attained crisis status. If and when it does, organizational cooperation theory might suggest they would move to "success-based helping" as a strategy, rather than attempting to go it alone. The Air Force has decided to use some of the Navy's data-processing capability as a central ABMS architecture. From this, we may surmise that an appreciation of the need to pursue success-based helping is at least partially underway. In our discussions with senior Air Force leaders in 2022, the Navy's approach to everything from JADC2 to how it messages its force-structure requirement is admired. A simple matter of the grass being greener in adjacent pastures may fuel a more joint effort.

Senior Air Force leaders have yet to agree on the central meaning of ABMS, though, and progress on meaning-making shifts from year to year as leaders depart and new ones take their place. At the end of 2021, the Air Force's ABMS office structure consisted of an Air Force brigadier general who was the nominal lead for ABMS. He worked directly for a lieutenant general who has some interest in a working future ABMS construct, but was not intimately involved in defining it as part of his daily responsibilities. By contrast, Secretary of the Air Force Frank Kendall had a strong desire to see the promise offered by JADC2 but entered office with grave misgivings that ABMS was on the right track to become the Air Force's realization of that vision, let alone lead the joint force, which has been the Air Force's stated organizational goal. He subsequently started a new program-integration office for command, control, communications, and battle management, putting at its head another general officer with an acquisitions background and putting ABMS into the portfolio.

When it is viewed as a totality, the lack of consistent leadership support and excess of conceptual confusion that the Air Force, as the self-proclaimed and Joint-Staff designated lead on JADC2, is displaying makes it less surprising that each service has its own dedicated and independent major effort to develop the JADC2 concept. What is more surprising and foreboding for JADC2 as a joint venture is that there has been relatively little concrete exchange of ideas, resources, or approaches among these efforts. Again, though, JADC2 is very much a concept that is going through rapid development and organizational flux. We have seen firsthand how much senior-leader attention and effort are going in to the Air Force's instantiation of ABMS. As several shouting matches about the topic in high-level meetings have revealed, an adequate amount of passion is extant in the bureaucracy to drive this *somewhere* meaningful. Ostensibly, similar weights of effort are also at work in Army and Navy conference rooms.

Historical Similarities

One of the most obvious similarities about AirLand Battle and JADC2 is that efforts to build both emerged during times of significant change in how the US military perceives the security climate. The context for ALB was a burgeoning Soviet force structure perceived as an imminent threat to Eastern Europe. Through the second decade of the twenty-first century, there has been a similar sea change in how the US military sees China as a likely rival in major, peer-to-peer conflict. To put a finer point on it, the Soviet forces were ten feet tall and bulletproof to the United States in the 1970s and 1980s, and the same US establishment views China in a similar way today. ALB never faced the Soviets, and the test of ALB was a much smaller-scale use of the systems developed against an overmatched Iraq in Desert Storm, but it is impossible to overstate the sense of urgency the Soviet Union provided in its day. Does China, the primary adversary in the current US defense-establishment narrative, offer the same type of prod to the US defense establishment in the current era? Is the impetus China provides felt differently by the Air Force, who knows its long-range strike capabilities will be relevant for the imagined style of fighting? The Navy confronts China weekly in the South China Sea and along key trading routes in the area, perhaps giving it more immediacy of experience and a more acute sense of threat. The Air Force and the Navy both see their assets in the Pacific as being under threat from the advanced weapons systems China is developing. However, within the bureaucratic halls of the Pentagon, the Army may feel the greatest threat, because it would struggle the most to find warfighting relevance in this theater of war without leaps forward in sensing and connecting that JADC2 promises.

A second historical similarity is that along with a change or perception about the security environment, there is a growing sense of urgency that the United States needs to stand up to its challengers. Former Secretary of Defense James Mattis shaped and released the 2018 National Defense Strategy. It pivoted the United States away from combating terrorism as a central focus and instead focused competitive energy on US rivals China and Russia. The degree to which China was at the center of that focus was remarkable. Equally noteworthy was the fact that this document had staying power beyond what one might expect. It outlived the tenure of the leader that pushed it, and the defense establishment became focused on its central tenet (that armed conflict with China is *the* military planning scenario worthy of attention) to a remarkable degree. While there are lots of great arguments as to why we should be planning for other theaters (there are throwaway nods given to a Russia confrontation in the Pentagon that became more poignant after Russia's brutal invasion of Ukraine in 2022), we see otherwise clearheaded leaders focus on China alone.

This singular focus leads to some amount of mental gymnastics in the Army, which—until the 2022 Russian invasion of Ukraine at least—was fighting earnestly to articulate its role and relevance in a potential Pacific Theater in a fight with China over Taiwan. Secretary of the Army Christine Wormuth, at her first address to the Association of the US Army,[3] said that the "Army has a crucial role to play as part of a Joint Force that can deter China, Russia and any other foe while defending the homeland" without saying exactly what that role would be in the imagined fight with China that has absorbed most of DoD's planning bandwidth since 2017 (Wormuth 2021). While addressing the views of "some [who] argue that INDOPACOM should be left to the Air Force and Navy or see the Army as primarily a bill payer to make room for more space, cyber and other high-tech capabilities" (Wormuth 2021), she remained vocal in her defense of the Army's role—and by extension its budgetary share—in all imaginable conflicts. In her preconfirmation testimony, Secretary Wormuth took a similar stance in defending the Army budget, saying, "I don't think anyone would be well-served by looking at the Army as just a bill-payer" (Losey 2021). Anyone who takes General Bradley's dictum about military logistics seriously has to concede Secretary Wormuth's point: whatever advanced equipment and warfighting concepts the services may develop for a fight in the Pacific, providing a sustained supply of fuel, ammunition, and other essentials—as the Army has been charged to do in the Joint Warfighting Concept—will be one of the most difficult aspects of the campaign.[4] Whether the warfighting strategy implied by this line of thinking is sound or not, the sentiment underlines the Pentagon bureaucratic budgetary standard of equal shares for the military departments under the Department of Defense's budget.

As chapter 3 discussed, ALB had a doctrinal concept at its core that grew in influence and momentum along with the approach it drove. The conceptual approach in the current era is somewhat murkier. Is JADC2 the driving concept or is the Joint Warfighting Concept the driving idea? The Joint Warfighting Concept is a construct signed by the Joint Chiefs of Staff and secretary of defense that has focused on getting the services to define "all-domain operations" and achieve effects under that rubric (Hitchens 2021). JADC2 seems to get more press and attention, but neither concept has a moniker that rolls easily off the tongue. In the world of military slogans, both "Active Defense" and "AirLand Battle" were both relatively sleek and stuck in people's minds. JADC2 has interconnectedness as a general theme, but it is still hunting for doctrinal and operational specifics to attach to that theme. This creates a conceptual and terminological gap that stands in the way of wider adoption, but senior-leader skepticism is a much larger and more undermining factor in JADC2's development and adoption. For example, Kendall balked, in late 2021, at the idea mere network connectivity of platforms and weapons would be a panacea that resulted in effective command and control. Because he shepherded the Joint Surveillance Target Attack Radar System (JSTARS)[5] into existence during the ALB era, some Air Force military leaders felt that Kendall's perception of ABMS's potential was initially constrained by what the JSTARS aircraft could do, instead of focusing on "more resilient connectivity, the ability to pass information, and generate better decisions" (Allvin 2022). Accordingly, the Air Force set out to better articulate how ABMS will come together as a system, but much remains to be done.

While the Air Force is still working to overcome skepticism among its senior leaders, it shows further challenges in gaining the interservice trust that AirLand Battle enjoyed when four-star commanders were speaking of the same concepts and when the services were coordinating budget actions to ensure all key elements of those concepts were purchased. However, in the same way that four-star agreement was a boon to accelerating ALB, we can see a similar path to acceptance for JADC2, specifically as the Air Force has imagined it. The current US Air Force chief of staff, Gen. Charles Q. Brown, signed a memorandum of intent to cooperate with the Army on JADC2 (Eversden 2020), and hosted an all-service-chiefs' discussion on the topic in June 2022. These efforts might be symbolic gestures, but Brown (pending US Senate confirmation at this writing) will be the next chairman of the Joint Chiefs of Staff. Thus, if General Brown believes in JADC2 as the Air Force has conceptualized it and the Air Force has managed to overcome its own civilian leadership's skepticism in 2023, JADC2 stands a chance of becoming a more meaningful joint concept.

For now, though, until there is a coherent operational system around which to coalesce and until there is some form of budgetary cooperation in the form of codependent budgetary submissions that marked the unique era of ALB, we would characterize memo-signing ceremonies as symbolic actions of joint cooperation that pervades public discourse among all aspiring leaders but that has not stood the test of actual Pentagon bureau-political battles. In the parlance of crisis-cooperation theory, we would bin this "talking" as the crisis-response activity (Svedin 2009). Navy Rear Admiral Bill Chase summed up the difficulty in taking JADC2 from a white-board concept to battlefield relevance with a pithy observation: "When we all talk about the mission, we're seldom disagreeing. It's when we start talking about who's in charge that we start arguing, like siblings" (Hitchens 2019).

Admiral Chase's observation is current, but not new. AirLand Battle gained traction by incorporating both senior-leader energy and a more grassroots effort to invent new doctrinal concepts. It came under fire when Army officers perceived the authoritarian imposition of an underlying doctrine they did not like, one that made them too dependent on another service (the

US Air Force in that case). JADC2 cannot survive as a joint concept without navigating the same treacherous territory and without skilled leaders keeping the tensions in balance.

One way that skepticism toward new concepts can be overcome is through operational implementation or field unit testing of an idea in actual or simulated combat. ALB concepts imbued US and NATO military officers with a certain optimism as they were rolled out throughout the 1980s, with leaders no less influential than the Supreme Allied Commander facile in describing how Air Force surveillance technology would pair with Army weapons to overcome the military threat posed by the Soviet Union (Canan 1984). While there were some reluctant reactions from entrenched communities to the new doctrine that ALB offered as a theoretical cornerstone, JADC2 has not matured or coalesced to the point where warfighting commands react to it—either positively or negatively. The common reaction to the general concepts is agreement that such an advance in warfighting would bring great capability to the joint force. Then, after they start trying to envision how it would be implemented in practice, they have a hard time getting to specifics or seeing how current warfighting principles would get in the way. Hence, the early skepticism expressed by Secretary Kendall is a common second-order reaction to JADC2 at its current development phase. A less-common reaction, but one that serves as a counterargument to JADC2, is alarm that the United States would go to such great lengths to create a centralized, vulnerable system of data collection and dissemination that could fail in the highly contested electromagnetic environment that would accompany conflict with China (Seip 2020).

There are a number of ways that the development of the imagined capability of JADC2 could become bogged down as military thinkers try to imagine its use in a warfighting operational concept. A useful example is autonomous targeting, a capability frequently mentioned in connection to JADC2 concepts. Envision a system capable of identifying dozens, hundreds, or even thousands of targets on its own and generating a weapons-quality targeting cue for a weapon to hit or in some other way affect the target. Now envision thousands of weapons on autonomous platforms in a battlespace helping keep pressure on the enemy and our own humans out of harm's way. Descriptions of the "Assault Breaker Two" make the tie between ALB concepts and the new promise of extensive networking, artificial intelligence, and machine learning explicit (Axe 2021). This sounds like a great combination full of tactical promise until one considers the traditional US requirement to have a human in the loop checking and deciding before any ordnance is released. With regard to aerial attack, lately real-time consideration of rigorous rules of engagement have slowed down target engagements to a comparative crawl—a remnant from a few decades of counterinsurgency warfare

(Pede 2020). This would all likely eventually become moot in a true near-peer fight against the likes of China, but asking a military shaped by two decades of counterinsurgency fighting to imagine it is challenging. Bluntly, rules of engagement (ROE) that developed in Iraq and Afghanistan "may get you killed" (Pede 2020) in a China fight.

More compelling counterarguments, i.e., arguments that do not go away even if we overcome overly restrictive ROE in the reality and existential threat of a near-peer fight, are those that point out almost certain escalation from China if the United States or a coalition begins to attack and blind the military systems China uses to detect attacks against its homeland. China favors a total-war strategy. Furthermore, although it has a purported no-first-use policy when it comes to using nuclear weapons, "Chinese officials have privately stated that China might respond with nuclear weapons if its nuclear forces were attacked with conventional weapons" (Kristensen and Korda 2020). Since part of China's nuclear early-warning systems is integrated into its overall military command-and-control structure in a way that makes them indistinguishable to an attacking nation from conventional systems, the stakes are raised very quickly and the case against building too effective of a command-and-control system that could attack China becomes a reality.

Parallels to the previous ALB case are in a sense impossible at this point: JADC2 has not become a functioning concept or system yet, at least not to the degree that one could identify a tangible center of joint cooperation or any other kind of inflection point suggesting a decline in collaboration to a lower, nonjoint state. JADC2 remains an arena of potential joint cooperation. What we can say is that there are specific doctrinal, operational, strategic, moral, and political challenges to JADC2 that will have to be overcome for JADC2 to become a meaningful joint concept. The Air Force needs a more compelling operational concept that explains how JADC2 promises competitive advantage to this service before it can hope to overcome resistance to joint collaboration and get other services to give up—or even comingle—their JADC2 efforts. As previously mentioned, the Air Force has pursued a Navy-developed concept to provide combat platforms with weapons-quality target information. The Navy has come further than the Air Force and has the only currently viable solution for that facet of JADC2. It is possible that the Navy will lead the Air Force into finding what it needs, the way the Air Force led the Navy into the JPATS trainer. In addition to these steps, specific tactics, techniques, and procedures regarding how JADC2 will be used in battle will have to follow in short order before widespread adoption can occur. To the degree that these steps involve ROE (the laws of war) or intersect with wider moral dilemmas, they will require a coherent address by the Air Force, Space Force, Army, Navy, Marines, and then the joint force.

In the previous case studies, we looked at successful joint ventures in a wide range of political and organization contexts. At the analytical level that deals with external relations, congressional influence—which frequently emerges as a source of pressure for joint cooperation—on JADC2 is not in clear view yet. What we mean by this is that JADC2 concepts have not yet solidified enough within the Department of Defense to get a satisfactory airing on Capitol Hill. By way of proving this, in its fiscal year 2022 budget submission, the Air Force had earmarked a few hundred million dollars for ABMS. Because Air Force officials could not explain what exactly this money would buy to OSD, the money was removed from the Air Force's budgeting authority and reallocated to other defense priorities. This is one of the most damning rebukes OSD can give to a service program, and is likely in part a source of the skepticism that tempered Secretary Kendall's initial interaction with ABMS. At any rate, until there is a coherent direction from OSD about what is to happen with JADC2 (or another service or defense contractor approaches Congress clandestinely because it is frustrated with the Air Force's lack of progress on the matter), legislators, as the principals in this principal-agent relation, are unlikely to act because they do not have sufficient knowledge or the information they need about the matter.

In other cases, the pressure that Congress has put on the services has supported the success of joint ventures. Congress as a powerful external stakeholder would theoretically be able to push for the services to develop a coherent JADC2 system. However, Congress's incentive to act in this direction as principal, to push for successful JADC2, runs up against the chaotic defense acquisitions process, which lends individual legislators considerable power. Thus, Congress's potential as external pressure forging successful joint collaboration remains unclear for the time being. This arena is not without hope, though. Christian Brose (2020), both in his current defense-industry strategist's role and as a former Senate Arms Services Committee staff director, retains influence within the National Capital Region, has adequately grasped the operational imperative of JADC2, and can articulate to a legislative audience why their interest might be useful to force the convergence of concepts and materiel that is lacking today.

In a similar vein, while scholarly debate and research can influence joint programs and joint cooperation, the exact nature of the JADC2 challenge is not well enough defined to elicit a meaningful response from this external actor. There is certainly a great deal of academic and popular press writing on the topic, but the discrepancy between the unclassified sources on the development of military technology available to civilians and the classified information available to military leaders makes academic discourse on this topic less relevant. As a result, novel approaches, such as DARPA's System of Systems Integration Technology and Experimentation (SoSITE) and various defense-contractor approaches in the JADC2 realm, remain veiled to a general audience. Consequently, the solutions offered in open press seldom have practical relevance.

If congressional and academic actors' impact on JADC2 as a joint venture is unclear, the role and impact of the defense budget are not. Across the Department of Defense and within the US military services, the perception of a diminishing defense budget over the coming years is a definite factor. This perception is important. In inflation-adjusted terms, the current DoD budget is greater today than it was during the peak of the Reagan buildup in the 1980s. However, the purchasing power of that budget is smaller today. Three factors drive a decreased purchasing power: (1) growth in personnel expenses, (2) increasing defense bureaucracy overhead costs, and (3) escalating acquisition program costs (Serbu 2021). In this sense, JADC2 could not be in a more dissimilar context than the one that generated AirLand Battle progress. The mid-1980s witnessed defense budgets growing by leaps and bounds, and the services were willing to trust the tide that raised all boats. Today the services feel so pinched or constrained that they perceive that they cannot even meet their core mission sets. The budgetary largesse that caused jointness to flourish in the 1980s is not in sight today. Although descriptions of service efforts to "get serious" about JADC2 abound (Freedberg and Hitchens 2020, e.g.), the current uncertainty about the nature of JADC2, along with severe budget austerity within all the services, makes these claims worthy of skepticism.[6] In spite of a much-touted US defense budget bump that came in 2022 while Russian aggression toward Ukraine lit the world stage, most of this money came with predefined congressional spending direction; very little would allow for additional JADC2 research, experimentation, or procurement.

Next, a doctrinal subtext impelled ALB forward, but that subtext is missing for JADC2. As we have already discussed, JADC2 lacks doctrinal maturity but it also lacks a major warfighting command that wants to adopt its capabilities. Since it is now more of a systemic concept than an operational or doctrinal concept, the latter will have to come into better focus before the output can be specified and warfighting communities can start to see their advantage and start to advocate for them. To use a favorite Pentagon cliché, there is just not enough "meat on the bone" of JADC2 for most of these things to come into existence yet. If JADC2 does not get past the critical hurdles of actual application and, with that, buy-in, the concept will not mature into an actual joint collaboration venture. However, simply walking away from developing JADC2 capabilities may not be an option for the US military. There may be some inevitability in technological progress and an interlinked battlefield may well be the next step in technological advances in the same way that airpower became part of modern warfighting. Consequently, the United States and its adversaries will need to pursue an effective operational concept and technical solution—even if "our jointness is all askew" (Chiabotti, 2014).

Testing JADC2 as a Joint Concept

We aspired for economy of expression at the beginning of this chapter. Therefore, as we set out to describe how JADC2 has been tested as a joint concept in the same way other concepts in this book have been examined, we could simply write, "It hasn't been," and move on. There is, however, a little bit more to say that may be useful in our predictive and framing theoretical efforts.

With a bit more charity, we can say that certain JADC2 concepts have been tested in limited ways, and these tests frequently have had a context of interservice cooperation. As the services labor to build JADC2 concepts for employment in warfare, a common observation and critique has to be that these are "laboratory-type" experiments that do not capture the grit and static of the battlefield. The Army and Air Force do the lab-style experiments in pristine conditions, testing concepts in exercises far from any theater or battlefield. Even in those cases, there has been a notable lack of jointness. The three-star general charged with design and implementation of Air Force future concepts, Lt. Gen. Clinton Hinote, took a wistful, maybe-next-year tone in his description of Project Convergence "Live" 2021, the Army's capstone JADC2 experimentation at the Yuma Proving Grounds, lamenting that "Our play was pretty minimal this year, and it showed. I get the sense that the Army leadership is both frustrated and hopeful that they can get us to participate in a much larger way" (Hinote 2021). This lackluster demonstration of jointness occurred despite the earlier-described and well-publicized memorandum of agreement (Lacdan 2020) signed by the Army and Air Force chiefs of staff in September 2021. Admittedly, this agreement focused on data-sharing standards more than exercise schedules, but in the past Air Force participation in the Army's JADC2 experiments had been more prominent. The experiments focused on connecting geographically dispersed Army and Air Force nodes on a common network, sharing information via battle-management and command-and-control systems, and displaying information in a format that could enable decisions by field commanders (Kester 2020).[7]

Rather, this failure-to-launch theme brings up a larger question: Why is the JADC2 concept not compelling enough to have mustered a common joint vision after being around in the defense establishment for five years? The GAO has released a report criticizing the Air Force for its management of the program, specifically critiquing the fact that the Air Force is not yet managing ABMS as it would a formal acquisitions program for other major weapons systems (US Government Accountability Office 2020). Ironically, everything in the report implies that the Air Force has had unbounded exuberance for ABMS—so much so that it abandoned its usual bureaucratic strictures and discipline in pursuing it. The GAO report says,

> The Air Force started ABMS development without key elements of a business case, including: a) firm requirements to inform the technological, software, engineering, and production capabilities needed; b) a plan to attain mature technologies when needed to track development and ensure that technologies work as intended; c) a cost estimate to inform budget requests and determine whether development efforts are cost effective; and d) an affordability analysis to ensure sufficient funding is available." (US Government Accountability Office 2020, ii)

The enthusiasm the services show for JADC2, the Air Force in particular, stands in marked contrast to their ability to define precise parameters or processes for acquiring such a thing.

By contrast, the Navy conducts its JADC2 experimentation with real platforms in actual geographic areas where it thinks armed conflict might occur and validates (to the point one can without releasing ordnance) that the concepts would be viable for targeting. Hence, it seemed at the end of 2021 that the Navy was a bit more serious, more committed, and further along in its pursuit of successful concepts. The "Dumb, Devious, Defiant" trope we have invoked before may be helpful here: the Navy's rugged individualism and track record of introducing new doctrines that become new platforms and then new professions with promotion career tracks may be JADC2's best hope. Single-service pursuit to successful ends may be exactly what is needed to make JADC2 a real entity, because the other services will then have a strong incentive to join with the successful approach. Several knowledgeable Air Force insiders have expressed admiration for the Navy's apparent no-nonsense approach in pursuing JADC2, noting that they wish the US Air Force could be as articulate and practical in its instantiation of ABMS as the Navy has been in Project Overmatch. The Navy, in its "Defiant" approach to go it alone on JADC2, at least at first, may turn out to be an ideal organization incubator for meeting a challenge no one yet knows how to master.

Theoretical Explanations of Jointness

As in previous chapters, we surveyed the general and military-specific social sciences that might explain (or predict) behavior with regard to JADC2. Since much of the concept is still indeterminate any application of theory is likely to be indeterminate as well, but we can offer a few observations that might be helpful.

General Theories

PUBLIC GOODS

JADC2's intuitive utility in prevailing against the United States' most threatening rivals qualifies it as a public good. The services alone have the expertise and budgetary authorities to make this kind of concept militarily useful. To put a finer point on it for our study, it is a public good that the services can provide and they would provide it to the fullest degree by working together. Since there are only a few services and they labor on the problem in plain view of one another, it is a small-population public-goods issue according to Mancur Olson, making the free-riding less of a problem. The real question is whether social pressure will build to a point where the services take compelling action. If sufficient pressure does build, we can think about whether this social pressure is likely to come internally from within the services or externally. The question of pressure is raised by other theoretical lenses we have applied as well, such as crisis cooperation, organizations, and other external factors that influence military priorities.

ORGANIZATIONS

The organizational-politics lens reflects the same possibility of bifurcated behavior. If threats to US security are perceived as serious *and* JADC2 is a viable solution to meet them, Type I, straightforward behavior to develop it is the expected outcome, nominally supported using the services' Type II bureaucratic processes to get through the heavy lifting of military acquisition. If the sense or urgency is not as great and bureaucratic politics becomes the operative dynamic, however, we can expect a languishing Type III response that delivers no capability or one that is very watered down compared to its imagined potential.

CRISIS COOPERATION

On this theme, the emergence of a threat-induced sense of urgency is the deciding factor in what type of behavior emerges from a crisis-cooperation lens. Thinking about a hypothetical loss caused by a JADC2 gap allows us to frame why organization theory and crisis-cooperation theories might predict continued stasis for a few years. Absent a clear, first-order threat to the United States,[8] the need for JADC2 is unlikely to drive anything but continued bureaucratic politics, that is to say, muddling through in the direction of the current status quo. Likewise, crisis-cooperation theory would predict that the services will continue to engage in talking with one another, but will not fight with one another in a way that will bring about meaningful collaboration until there is more of an immediate threat.

PROFESSIONS

As we have highlighted a few times now, subgroups' roles in military doctrine are important, and these might drive control-based innovation. A complicated, would-be relationship emerges between one of the subgroups we looked at in the chapter on CAS and JADC2. One of the groups within the Air Force most eager to embrace and test out JADC2 (including ABMS) concepts has been the Tactical Air Control Party (TACP) airmen that we met in the preceding case study. In a growing body of literature, these professionals observe the practical problems of JADC2: for example, "Air Force systems are technically capable of connecting to any military service's network. However, they only actually have the authority to connect to the Air Force's network" (Leland et al. 2021). They offer grand solutions to these problems: "both dimensions of authorities—the authority to connect to networks and the authority to direct assets—need to extend across all the services and be pushed down to the tactical level" (Cowie et al. 2021).[9] However, as we discussed earlier, TACP themselves represent a form of interaction with the Army that is seen in the larger Air Force cultural construct as a burden, hence falling outside of the mainstream focus of how the Air Force views how airpower should be used to win wars. The prominence of TACP and the systems, style of fighting, and joint battlefield urgency that drove it during the campaigns in Afghanistan and Iraq evaporated in planning for a fight with China. In fact, TACP have doubled down in tying their continued relevance to being the vanguard of ABMS. The head of the operational portion of the ABMS program in the Air Force from 2020 through this writing, a former commander of a TACP wing, has urged them in articles, emails, and social-media posts to do as much in the wake of a significant cut to TACP force structure in the Air Force's proposed budgetary submission in late 2021.

The irony of this situation is that the proposed approach is strategically and tactically sound. In terms of having the knowledge base, technical capability, and sense of urgency required to further ABMS and JADC2, TACP are well qualified. However, the situation is just as politically unsound: TACP within the Air Force are viewed as an "outsider" career field, airmen in name but soldiers in alignment and view of the world—hardly the community the Air Force is likely to entrust with leading its revolutionary approach to warfare, joint or otherwise. Recent proposed budget cuts that will take the TACP enterprise to approximately one-quarter of its size in 2020 over the next decade prove the teeth of the Air Force's organizational approach to this subgroup. The subgroup's strong identification with CAS also prevents it from being willing to be trained to do other tasks that might be useful for

JADC2 (such as air-battle management) and leaves it relegated to driving someone else's C2 node to the forward edge of the battlefield. It's a noble gesture, but largely hollow. Our assessment on whether this will become a driving dynamic is, "probably not."

AGENCY

As is the case with others who have attempted to explain the organizational behavior of the military as it pertains to the production (or not) of public defense goods, principal-agent theory is a very useful lens. From the perspective of the services (analysis level a), OSD and the Joint Staff (analysis level b), *and* (at least ostensibly) any external audience (analysis level c), the existence of JADC2 is a useful public defense good. However, it is only from the among services' coordinating organizational level (Joint Staff and OSD) and from external stakeholders' perspectives that the *joint* pursuit of JADC2 is intuitive. For these principals, joint cooperation is highly desired or will be in short order. The services are the agents providing the good, though. From their perspectives, distinct incentives to go it alone and to build the system that will unite everyone are at play. Therefore, if the services do not shift their view of the situation vis-à-vis the United States' peer rivals, agency theory predicts a glum outcome for joint collaboration absent strong external pressure, a possibility we will touch on in the next section.

Military-Organization Theories

MILITARY INNOVATION

At this stage of JADC2 development, with a relatively uninformed external audience, Rosen's military-centric mechanism for explaining military innovation seems dominant. Most of the thinking about JADC2 seems to have come from service-internal sources, though as noted OSD has expressed its skepticism by removing some amount of JADC2 spending that it thought poorly targeted. The military subgroups who will be operating the systems are ostensibly those who are already in command-and-control career fields; no internal jostling or asserted need for a new career path to foster innovation has yet taken shape, TACP interest in ABMS aside. It is also not difficult to see that if the United States had a significant military loss attributable to a lack of an effective command-and-control system, congressional and DoD interest in JADC2 would increase quickly. What we can say about the organizational level that coordinates among the services and its impact on JADC2 (analysis level b) is that OSD's skepticism about what the Air Force planned to buy forced the service to go back to the drawing board and be more specific about its concept of operations and systems and other essential materiel and doctrinal concepts. While not fostering innovation per se, the DoD structure represented by the aforementioned spending rebuke imposed some rigor that prevented the Air Force from being fiscally undisciplined in its pursuits.

CIVIL-MILITARY RELATIONS

For the same reasons as there are possible forks in the road of service behavior when viewed through other theoretical lenses, civil-military relations suggest that control of JADC2 will remain firmly entrenched with the services, and that the general population will have to just trust that the military will do the right thing through objective control as Huntington proposed. While it is conceivable that a visible failure to develop effective JADC2 might create a public outcry and tip the balance toward more subjective control, it is more likely that an interested external stakeholder like Congress will intervene before this topic becomes a civil-military trust issue.

SERVICE CULTURES

Service-culture tension is in full view here, starting with the Army-Air Force clashes we have noted in previous chapters. The Air Force loves to have domain awareness in the air, and it has allowed an entire portion of its fighting force (RPAs and IRS) to develop to that end. As noted in the case study on CAS though, there is a tendency to cut back on the enablers that allow close cooperation between the Army and the Air Force, unless there is a pressing battlefield need for those enablers. JADC2 inhabits a middle ground between domain awareness and cooperation enabler. Since the campaign in Afghanistan has ended and the campaign against Al-Qaeda and ISIS have become less kinetic, the Air Force has followed the historical pattern and has cut TACP as an enabler, proposing a 50-percent cut in the current program objective memorandum that defines the defense budget (Willis 2021). Therefore, leaders in the TACP career field perceive it to be in crisis and that becoming leaders within ABMS and JADC2 could be their reprieve. Because TACP inhabit a central role at the joint intersection between the Army and the Air Force, some military strategists have advocated specifically for this arrangement (Birch, Reeves, and DeWees 2020). In the COIN case study of chapter 5, we noted that senior-leader energy that focused on TACP development seemed to follow the immediate demand signal from the battlefield. Regardless, this senior-leader involvement was followed up in tangible ways, including revising formal agreements. With that observation it is reasonable to conjecture that the recent MOAs signed regarding JADC2 cooperation may signify a path to real cooperation after all.

Another examination of service cultures offered earlier in this book may explain why the Navy is further along than both the Air Force and the Army at fielding operationally useful JADC2 systems. Recall that the Navy's feudal service structure means that it has to address the concerns of all of its component branches; effective JADC2 is a way to make all parts of the Navy successful in combat. In the Army's and the Air Force's monarchic structure, it is more likely that a leader from an important internal service tribe would take a longer wait-and-see approach, thinking that their specific service can still be effective in combat without a revolution in sensing or targeting that JADC2 would provide. As an example, Air Force fighter pilots signify an important service tribe and they may be least likely to view the gap in JADC2 capabilities as a crisis.

DEFENSE DEPARTMENT AND JOINT STAFF

Here a theme arises that we have broached in the other case studies: the joint imperative from the Joint Staff or OSD is relatively weak. We have mentioned that OSD previously cut money from the Air Force's budget for ABMS, complaining that the service's explanation for what it intended to do with the money was not detailed enough. This is a significant rebuke and influence of a kind to be sure, but perhaps not one that forces constructive cooperation.

Additional evidence points to the Pentagon's overall ineffectiveness in driving joint cooperation. The Joint Staff's Joint Requirements Oversight Council (JROC) in the summer of 2019 gave the Air Force the lead for JADC2-related technology testing out of Nellis (Hitchens 2021). Our interviews revealed that senior leaders in charge of or heavily involved with ABMS in 2021 and 2022 were unaware of or only passingly familiar with that notion. While an advocate of jointness might hope that Joint-Staff directives about JADC2 experimentation might have more teeth or a half-life longer than one year, the fact that they do not is not very surprising. All of the services' requirements portfolios are vast. For example, three-star flag officers charged with force design and requirements definition for the services need to have familiarity with hundreds of programs across dozens of portfolios, and the details of both the requirements and the defense spending directed at them fluctuate continually. Furthermore, turnover in the relevant offices is rapid, particularly among uniformed personnel. Finally, the COVID-19 pandemic devastated the kind of person-to-person handover and knowledge transfer that historically has maintained continuity in pursuit of complex interservice projects like JADC2. We infer from observation as well as from our previous case studies that without either internal or external motivation for interservice cooperation, the mid-level of the US defense bureaucracy will not fill the void.

OTHER EXTERNAL FACTORS

From our earlier case studies, we can infer that the bureaucratic structure of the services disincentivizes achieving a public good for the same reasons that societies tend to vandalize their commons, but the services will sorely miss a

JADC2 capability if they arrive on a highly contested battlefield without it. It is too soon to guess what will happen, but we can think about what might close the observed gap. Will an informed, external principal like Congress step in with time to spare? Or will it take the emergence of a new battlefield to drive the proper sense of urgency? In the immature case of JADC2, beginning with the notion that the concept itself could be a public defense good makes it easier to apply other theories in our efforts to predict whether meaningful joint cooperation will emerge, and is useful in terms of our final goal in this book of offering a framing theory of what drives joint cooperation.

Concluding Thoughts on JADC2

Perhaps the most insightful understanding about JADC2 and joint cooperation comes from previous work about bureaucratic cooperation. The Army and the Air Force have talked extensively—generating hundreds of stories in 2020 through 2022—about their visions for JADC2. The Navy has spoken relatively less about its vision and about the concept overall—according to Vice Admiral Jeffrey Trussler, the deputy chief of naval operations for Information Warfare and director of naval intelligence starting in June 2020, "We just don't talk about it that way [that the Army and Air Force do]" (McLeary 2020)—except for some prominent successes and organizational restructuring to go after the problem in depth.

We mentioned the 2021 Army JADC2 experiments earlier. A small detail but a telling element of Lieutenant General Hinote's e-mail to the wide Air Force audience was a question about whether the Army's Joint Systems Integration Laboratory at Aberdeen Proving Ground could connect to the Air Force's Shadow Operations Center, an experimental and training command-and-control organization at Nellis Air Force Base (Hinote 2021). It can and does, but the vignette reveals a simple point. The sheer number of cooperation mechanisms available, the low-level activities continually underway, and the wall around what can be discussed publicly and what happens behind classified vault doors mean that this a very complex ecosystem in which to maintain one's sense of orientation and situational awareness. That the Navy's pursuit of Project Overmatch seems so far to have resulted in both conceptual and materiel gains that outstrip that of either the Air Force or Army, despite the well-touted cooperation of the latter, is likely just because we have sampled the stream of history before it flows to a logical conclusion. If we point back to the tight wrap of security around classified programs, there is no doubt that Air Force-Navy efforts were underway in that context; they just have not made it into open sources yet, and therefore are not included in our preliminary analysis.

Consequently, rather than letting antijoint pessimism get the best of us, it is more reasonable to note that some of the progressive steps that happened to enable ALB simply have not happened yet in the context of JADC2. We would also do well to remember that many of the aspects of ALB depended on some very leader-specific behavior. Furthermore, some of these were fairly extreme, such as an empowered Army Training and Doctrine Command general becoming empowered when his previously smothering four-star boss got terminally ill. Because of the unique journey that previous joint endeavors have taken, we cannot credibly predict how ABMS, Project Convergence, or Project Overmatch would culminate in a successful doctrinal concept or viable systems of systems for the battlefield, let alone if they would lead to genuine joint cooperation. There is very real concern that the Air Force should be the leader and integrator of the overall communications architecture, but that they are not (for a variety of reasons) building a system that can scale to accommodate all players on the battlefield (Hadley 2021b). A recent Center for Strategic and International Studies report underlined the same concern, and pointed out that the Air Force's two current fifth-generation fighters—built by the same company—cannot communicate directly with each other without a third-party communications node (Harrison 2021).[10]

We are left for the time being with an indeterminate view of JADC2 development. While the conceptual bug seems to have bitten the three services with the same amount of fervor, development paths have not followed the hope that the "J" in JADC2 would imply. However, this is endemic to how the Pentagon operates. Service behavior at that level normally follows single-entity interests in pursuit of budget share. There are certainly structures in place that can drive JADC2 to greater clarity of definition and commonality among the services. The main problem seems to be that a crisis or sense of urgency of sufficient magnitude has not arisen to kick these structures and the force they can bring to bear to life. Yet there is still plenty of time for senior-leader involvement, operational imperative, or outside influence to make that "J" ring true. We can root for our favorite, but the final account of development in the rough-and-tumble joint arena has yet to take place. We wait with bated breath to see how that turns out.

Chapter Seven

An Empirically Informed Theory of Jointness

*If we then ask what sort of mind is likeliest to display the
qualities of military genius, experience and observation
will both tell us that it is the inquiring rather than the
creative mind, the comprehensive rather than the
specialized approach, the calm rather than excitable head
to which in war we would choose to entrust the fate of our
brothers and children, and the safety and honor of our country.*
—Carl von Clausewitz, *On War*

*But what is chance? What is genius? The words chance and
genius do not denote any really existing thing and
therefore cannot be defined. Those words only denote a
certain stage of understanding of phenomena.*

*I do not know why a certain event occurs; I think that I cannot
know it; so I do not try to know it and I talk about chance. I
see a force producing effects beyond the scope of ordinary human
agencies; I do not understand why this occurs and I talk of genius.*
—Lev Nikolayevich (Leo) Tolstoy, *War and Peace*

Theoretically Reframing and Reconceptualizing Jointness

This chapter is primarily geared toward our academic audience and those
practitioners who are interested in conceptualizing joint military collaboration theoretically. The bulk of the practical conclusions and implications of
our study, which may be of greatest interest to practitioners and leaders in
the field, are discussed in chapter 8.

Like Martel's (2007) work to develop a framing theory of "victory," our
study of the realities of "jointness" presents a conundrum of definitional
ambiguity and political complexity.[1] In this chapter we sketch out a refined
framing theory of "joint military collaboration" from the one scraped together in chapter 2 and a revised and reconceptualized definition of "jointness" in light of the successful cases of jointness we examined (chapters

203

3–6). The case study method we chose to help us map and explain jointness leaves open the possibility of equifinality. In other words, in any given scenario there may be multiple paths to achieve what we describe as joint action as well as additional ways for which we have not accounted. Consequently, further study is required to move forward from the descriptive effort in this study and toward a more fully formed theory of jointness with greater predictive power. However, the facets we did examine and discuss have utility on their own.

Our framing theory has four components that, in broad strokes, identify and connect the why, what, when, and how of joint military collaboration. These components boil down and combine many of the theoretically derived factors from prior research. Our discussion of the four components brings the theories discussed in chapter 2 into better focus by summarizing their multiple appearances in our case studies. Thus, we narrow down the wide range of academic theories that have bearing on jointness to the most useful dimensions to assess jointness. As we walk through each component of our refined framing theory of institutional jointness, we highlight how each component contains factors that support or undermine jointness and may help us predict whether specific interservice joint ventures will be successful. We conclude the chapter by distilling a revised working definition of "jointness."[2] Building on prior theoretical work and informed by the case study observations, we hope that this tighter definition will provide clarity that augments previous discussions and makes future work on jointness more fruitful.

The first task is a first-iteration explication of the trends that emerged from this study's small-n examination of joint institutional cooperation.

The Components of Jointness

This study identified four components that proved useful for characterizing, comparing, and analyzing the cooperative examples it examined. It is helpful to envision these variables as existing across four corresponding spectra; they are multivariable characteristics that can change over time in a given example. Using four separate, time-dependent variables to describe any phenomenon suggests a high degree of complexity, and this study finds that the structural, political, and technical variables of jointness are worthy of such characterization. It comports with Svedin's (2009) observation that cooperation on a national security dilemma "demonstrates the complexity of organizational interactions in crises" (126).[3] The dimensions described here attempt to gather up many interlinked elements that influence one another in a way akin to an improvisational jazz ensemble. While there is a discernible pattern and rhythm within the whole, the dominant player changes over time, and the theme with which the piece started may or may not be intact at the close.

Component 1: The Motivation for Jointness

This study found that no joint cooperation began without some type of outside influence, whether it came in the form of world events (war or the threat thereof) or involvement from the higher strata of the defense establishment.[4] This is unsurprising, since on balance the general organizational theories applicable to military bureaucracies illustrate that joint institutional cooperation would *never* happen on its own. Prevailing theories about collective goods, bureaucratic competition, professions, and the principal-agent dilemma all suggest that the trajectories of the self-sufficient bureaucracies composing the military would never elect to play in unison without an outside conductor. Crisis-cooperation theory and its presumption of collective action are an exception, but its very title reveals the presence of an external influence—there is no "crisis cooperation" without a crisis.[5] Despite the leanings of the case-study analyses toward viewpoints espoused by Barry Posen, the framing-theory component proposed here nevertheless includes the possibility that internal service or among service coordinating level initiatives could in some cases be the root cause of joint cooperation.[6] Allowing this end of the spectrum a possibility of existence transmutes Stephen Rosen's idea about the roots of doctrinal innovation to the realm of jointness.[7]

For AirLand Battle, a visible external influence was growing US concern over the Soviet Union and its Warsaw Pact–undergirded military capability. In the case of the JPATS, the steely gaze of Congress on failed acquisitions programs provided the Air Force palpable motivation to take charge of its own training fleet. In the case of contemporary COIN air power, the urgency and lethality of war forced the services to consider both command-and-control and technical issues that they had been able to skirt as a matter of peacetime habit. For all three cases, some amount of fear was operative in creating either the external impetus (e.g., there was trepidation in US national security circles writ large over a military defeat in Central Europe) or the cooperative military response (e.g., the Air Force worried about losing its autonomy in selecting trainer aircraft after the T-46 debacle, the Army and Air Force both feared the wrath of Congress if they botched anymore air-ground coordination in Afghanistan or Iraq).

Characterizing external motivation for institutional jointness is a qualitative *and* quantitative exercise. War is qualitatively different from a congressional hearing, despite the analogies to combat that beleaguered flag officers might share after a particularly hostile session on Capitol Hill. Both war and domestic external stakeholder interests can vary in intensity though, even within the same long-term chain of events. Irate senators in the mid-1980s drove the Air Force to seek common trainer-aircraft positions with the Navy, though as the intensity of that displeasure died down and other concerns came to the forefront of debate, the fervor to demonstrate visible institutional jointness on that front also died down. Major combat operations in 2001 and 2002 seemed to focus joint cooperation with an intensity that steady-state COIN did not muster later.

With regard to AirLand Battle, the quantitative and qualitative aspects are even more nuanced. The Soviet threat was never quantified with great accuracy, yet it captured the attention of the entire national security establishment, including the executive and legislative branches of the US government. While neither the president nor Congress explicitly directed the creation of AirLand Battle or a concept like it, Army leaders found in the Soviet Union a problem of sufficient severity and magnitude that their proposed solution became an answer that gave structure and direction to national-defense planning (and spending) for several years. The development also offered elegant redemption following the ignominy of Vietnam.

The case studies allowed for some preliminary conclusions about the effectiveness and staying power of external influences. A perceived existential threat embodied in conventional Soviet military power provided Army and, later, Air Force officials with the potential energy to overcome normal anticooperative inertia. This crisis did not dictate a specific type of response, demonstrating the variable's high interactivity with capable, constructive leadership (which we describe as the fourth component of joint cooperation). The power of an existential threat to keep joint cooperation alive—it never really subsided until the Soviet Union fell and the Cold War ended—seems to have had remarkable longevity.[8]

The motivation caused by Congress's anger in the JPATS example drove a specific response: cooperation with respect to trainer aircraft. Here again, though, even when Congress led with a specific suggestion to acquire the T-45, military leaders operated creatively within the problem-set context to find a different solution—one more amenable to the Air Force's preferences and closely enough aligned with Navy priorities to achieve sister-service buy-in. Here, there is ambiguity about the duration of influence. On one hand, the outcome was a purchase of common military equipment that both services will have used for over three decades. On the other, acquisition programs are by nature long-term sagas; once full-rate production begins, they almost always reach their scheduled culmination, though Congress has remained steadfast in holding the services to their purchase commitments decades after the first spark that began the program. Perhaps a more reasonable quantifier of the duration of the joint commitment was the interservice training program it spawned, which lasted just over a decade. This is in itself considerable joint longevity. We determined that the influence that drove the common training programs—a DoD "good idea" rather than explicit congressional direction—differed from the conditions driving shared acquisition. The strength of the initiating force explains some of the observed difference in the staying power of external influences on nonacquisition matters. Returning to the jazz analogy, Congress tends to be a virtuosic best player in matters of peacetime facilitation, but it tends to sit out the set when the tune is that of an ongoing war.

Expert observers, including members of Congress themselves, have noted that the legislative branch is out of balance with respect to appropriations, micromanaging acquisitions but ignoring glaring omissions of national security strategy. Senator Sam Nunn decried Congress's use of its purse strings to parochial ends while it failed to shape a collective national strategy: "Annual authorizations provide a strong lever to influence defense policy and provide broad oversight. Unfortunately, we have come to abuse that lever; as the old saying goes: 'We have found the enemy and it is us' " (1985, 253–250).[9]

For those infrequent times when the legislature does get involved in the strategic or command-and-control aspects of national defense, it tends to be a shakeup of major proportions. The most recent, and arguably most influential, example in US defense history is the passage of the Goldwater-Nichols Act. The legislation resulted in profound changes, including the favorable outcome of populating Joint Staff billets with a better caliber of officers, but Sapolsky, Gholz and Talmadge (2009, 53–55) rued its continuation of a relentless trend toward defense centralization. They believe the legislation and other changes like it have had a chilling effect on the competition among diverse military options. For this reason, perhaps it is best that Congress's forays into this aspect of defense policy are generational rather than routine (Locher 2002, 29).[10]

As for the unique influence of war, its power is clearly in view in the response to Operation Anaconda and preparation for major combat operations in Iraq. As in de Tocqueville (1840), this study "finds no ill of war," at least as far as its ability to provide the sense of urgency prerequisite for jointness.[11] Despite, and perhaps because of, open interservice fighting, the conflict witnessed a concerted effort to cooperate on the topic of air-ground command-and-control structures, historically one of the thorniest interservice problems. While fear of losing—or, more precisely, appearing to perform poorly—in battles after Anaconda drove much of the response, there was also a visible return to dysfunctional form when the intensity of conflict dwindled to that of steady-state COIN. The middle years of the Afghan and Iraqi conflicts witnessed a return to interservice squabbles based on operational preference and competing theories of victory. Band-of-brothers camaraderie yielded to fractious infighting, suggesting that the intensity and staying power of war to effect institutional cooperation analogize best as a flash in the pan, not a pressure cooker.[12]

Component 2: The Potential of Joint Collaboration

The second component speaks to the type and potency of jointness available. Joint projects that fit into preexisting institutional structures are more likely to succeed in producing helpful cooperative outcomes. There may be a great deal of external influence available to spark movement against bureaucratic inertia, but if no visible joint solution to a given problem emerges, the opportunity may go unexploited. If a joint solution drives too much against the grain of one or more of the involved services' preferences, there may be some cooperation, but it might be stunted, uneven, or short-lived. Finally, if a joint solution is readily available that can absorb energy imparted by outside influences without upsetting too many parochial interests or trespassing on bureaucratic turf, the chances of a joint institutional cooperation success may be high. This work observed that there are different levels of difficulty involved in facilitating jointness; they are in part dependent on the type of joint solution available.

The type of cooperation that presents itself is an important distinction. Although defense professionals have not reached unanimous agreement on what the essence of jointness really is, their opinions in the aggregate reveal that it has distinct materiel as well as organizational facets.[13] All three case studies in this investigation revealed aspects of both. In AirLand Battle, command-and-control compromises affecting doctrine and tactics (which are difficult to work out because the services are ideologically stingy) accompanied acquisition plans for weapons systems to make them possible (which are by comparison easier to program into a budget, especially an ample one). In the JPATS case, planning for the acquisition of an airplane and its associated support (a comparatively easy budgetary effort) preceded the organizational adjustment that facilitated joint training (a more difficult meeting of the minds about fundamental training and the pedagogy of flying). In Afghanistan and Iraq, the services addressed their visible shortcomings in the coordination of CAS and ISR with some organizational changes—recall Monte Cannon's description of "pasting over seams with still more liaisons . . . to engender battlefield integration" (2012, 277)—but the bulk of the response was materiel in nature: newer and better weapons, communication devices, and situational-awareness systems used brute technological capability to overcome the less-tractable institutional problems.

The cooperative-potential facet of institutional jointness interacts with the motivational causes described earlier. An observation that "Congress finds it extremely difficult to take responsibility for ongoing military operations" in part explains why its outside influence so often goes to materiel improvement rather than participating in the "unequal dialogue" that would ostensibly lead to improved command-and-control measures.[14] It is easy to dismiss Congress's responsibility in this area and point to the executive's commander-in-chief prerogative, though simple electoral politics and the political liability of critiquing actively fighting forces probably drive closer to the heart of the matter. The threat of congressional inquiry does offer quite a bit of leverage over the military, but experience shows that it ultimately affects acquisition more than doctrine. When doctrinal topics are broached, though, the services listen carefully for the lead player's theme. Part of the reason the services paid so much attention to improving CAS procedures after Anaconda may have been their recall of the contentious hearings about air-ground coordination Congress conducted after each of the twentieth century's major conflicts.[15]

True to form, Congress was open in its inquiry about CAS as the wars in Afghanistan and Iraq came to a close, although the specific focus was mostly a debate about materiel—rather than command-and-control issues, it centered on the efficacy of the A-10 and the Air Force's desire to retire what the Army sees as its "best" battlefield air support.[16]

This represents a squandered opportunity. Congress could maximize its leverage over command-and-control matters if, instead of waiting to critique in the aftermath of failure, it used its influence, the power of the purse, and the leverage of inquiry to force joint action in those mission areas that the services have characteristically demonstrated unwillingness to practice—the difficult work of fighting together to perform missions that neither view as their most-favored skill sets.[17] Such an outcome relies heavily on the interplay of external and internal leadership.

The competitive roots of the Army–Air Force dispute are clearly available and beg for adjudication by a higher authority. The Army, in its conception of mission orders, wants clearly delineated responsibilities among subordinate commanders who can pursue an overall commander's intent. At the same time, it seeks to attain increased standoff from the enemy force via improved intelligence systems and longer-range weapons systems. This push for ever-wider influence by lower-echelon commanders puts the Army in conflict with the Air Force, which perceives an infringement on its self-ascribed, theater-wide responsibilities that accrue to a high-echelon leader with central authority. This conflict will only gain frequency and volume as weapons systems' standoff ranges and ISR capabilities grow.[18]

Another standoff looms in the fundamental arena of what is considered combat and what services recognize as such. A recognition that the US military "freely and noisily" recognizes "exceptional suffering and hardship"—and less so tactical prowess or strategic clear thinking—raises the specter that the armed services, out of a skewed sense of "honor," reward actions that put troops in harm's way and close contact with danger over those that advance a winning cause (Lexington 2014b, 33). Displays of emotion over the validity of combat-action awards and charges that operators of advanced weapons and intelligence systems are somehow exhibiting cowardice threaten to negate their value.

Thorny as these issues may seem, they directly affect combined-arms effectiveness and efficiency. Any effort labeled joint, whether it occurs in the realms of bureaucracy, acquisition, or combat, that does not address these competitive issues is either a waste of time or doomed to failure. Meaningful joint efforts will confront those areas of doctrinal preference where the services are most at odds, but they will not attempt to force ideological agreement. They will instead perform the hard work of adjudicating, as facts allow, among the best options the services can offer. With an accurate set of data in hand, external defense leaders can direct the *best* approach—whether it is a unitary or cooperative joint effort—that the services, stuck with fairness as an inevitable outcome of bureaucratic logrolling, would not pursue on their own.[19]

This dimension of jointness drives a tradeoff between institutional and materiel improvement. Materiel solutions tend to be easier to implement, as they are less likely to interfere with any service's strategic preference or theory of victory outright. Yet materiel solutions have a short-lived effect, as they are almost always overcome by enemy ingenuity in the field. Interorganizational improvements, most notably command-and-control schemes, are a type of cooperation that is more difficult to initiate but also more likely to provide lasting results. Somewhere in the middle lies training, which brings extant materiel and organizational habits together in rehearsal, where those with a willingness to look can identify shortfalls and remedies in both areas. Practitioners of jointness should realize that training efforts do not wither solely because of inconvenience and expense. The very act of exercising jointly forces confrontation of service ideologies, a challenge to institutional thinking that normal bureaucratic forces seek to quash.[20] Service members who are interested in improving joint efficacy have a responsibility to advocate for overcoming institutional barriers, and external leaders should intervene on their behalf when they cannot.

Component 3: Aligning Multilevel Interests

This component of jointness acknowledges the presence of structural fixtures and fissures in the defense-establishment bureaucracy that can help or hinder the pursuit of institutional jointness. While the demarcations this work used to describe the organizational levels of analysis (within individual services, groups coordinating among the services, and external relations) are somewhat arbitrary, they proved useful as a qualitative descriptor of how and where opportunities and obstacles for jointness arise. The theme that emerged across the case-study analyses, though, was the need for coalescence of interests as a precondition for joint cooperation. This renders the exact points of division among the organizational levels of analysis somewhat irrelevant; it is enough to know that higher and lower organizational-level interests converge in successful endeavors.

AirLand Battle's roots lay in a perceived security imbalance that created concern at the highest levels of US government, in turn transmitting this concern to the military and other subordinate organizations. Thus, an interest in providing a security solution existed throughout the external defense establishment. That interest penetrated into the Army, representing the internal service perspective, where two influential TRADOC generals saw a link between the external stakeholders' interest and their institution's desire to escape the doldrums of Vietnamese COIN warfare.

Shrewdly, these leaders reached out to a subset of the Air Force—the Tactical Air Command—that happened to be growing in influence and would soon represent the most dominant vector of its parent institution's thinking. As interests coalesced across services, AirLand Battle found acceptance in the groups coordinating among services (both the Joint Chiefs of Staff and the regional combatant commanders), because its constructs could effectively address interests there as well. Having constructed a system that unified interests across all three levels to answer an externally imposed security problem, AirLand Battle was capable of sustaining its own momentum and became a stable system of military thought for several years.

Aspects of the JPATS effort were similar, though impetus arose first not from world events but within the external defense establishment (i.e., Congress), who transmitted its wishes to the Air Force. Just as in the case of AirLand Battle, military leaders in the Air Force's training command proposed a plan that met Congress's requirements and their own preferences—in this case the internal service desire was to structure primary pilot training in accordance with service philosophies. Mimicking the outreach of TRADOC generals fifteen years earlier, Air Training Command (ATC) enthusiastically embraced its Navy counterpart and presented a plan that met nearly all of its sister service's desires for a trainer, including avoidance of many of the risks attendant in leading an acquisition program. After teaming together, both services worked to present their approach simultaneously to groups coordinating among services in briefings to the Joint Staff, even while they built momentum at the external relations level through continued salesmanship on Capitol Hill. Once DoD became involved, directing an increase in the scope of jointness through a mandate for joint training programs, a coalescence of interests was complete. The prospects for joint success were then strong enough to overcome the friction of what proved to be a difficult competitive contracting process.

The COIN air power cooperation, in contrast with the first two examples, represents the weakest coalescence of interests. Like AirLand Battle, its cause was an externally imposed security situation—antiterror conflicts prosecuted in Afghanistan and, later, Iraq. While the fear of losing or performing badly in combat drove an initial push to overcome two of the most chronically neglected aspects of combined-arms cooperation, the mutual effort was patchy at best. Congress never really became involved at the external stakeholder relations level; its perennial unwillingness to comment on ongoing combat operations does not offer a substrate for jointness to take hold. Stepping in to the external breach at times was the DoD. However, drastically different guidance from two successive secretaries of defense, especially on the topic of ISR, led to more confusion and finger pointing about why cooperation was poor and why there were fewer clear mandates to seek joint solutions. The DoD, like Congress, muddled through on the matter of overarching military strategy, never addressed head-on the competing theories of victory, and accepted a series of compromises among the various light-footprint, regime-change schemes and troop-intensive COIN strategies advanced.[21]

Another factor germane to this component of jointness, derived from observing CAS cooperation in Iraq and Afghanistan, is the tremendous preference gap between the services in style of warfare and theory of victory. When organizations are unconstrained in their pursuit of goals, when their aims are aligned with institutional values, and when they are able to use processes with which they are comfortable, they are better able to cooperate with peers. On the contrary, when they are faced with tasks that fall outside preferred or core competencies and are reliant on lateral organizations to complete those tasks, the potential for conflict is high.[22] The Army made a reluctant embrace of COIN warfare, perhaps in a bid to remain relevant in a war that started with an impressive showing of air power combined with small-footprint ground forces. The Air Force failed to embrace this philosophy, and was pushed there forcefully by mid-conflict coercion by Secretary of Defense Robert Gates. The Air Force's initial perspective was not an irrational position, as the large-force idea did not receive support from the DoD until halfway through the conflicts—it then took the forcible removal of senior Air Force leadership to make the cultural change clear. Throughout the conflicts, the Air Force's ideals of centralized control and strategic effect competed with the Army's boots-on-ground, personnel-intensive theory of victory and its demand for support to surface forces—apportioned according to geographic area, not necessarily by strategic importance.

As Builder, Ehrhard, Donnithorne, Cannon, and others have described in detail, the services have preexisting notions about the type of conflicts in which they see their greatest opportunity to contribute. A recurring theme in the contests they visualize is "high-intensity warfare against a sophisticated enemy" (Builder 1989, 132–133).[23] This is a visionary backdrop against which commanders can easily motivate their tacticians. It is easier to sell to the nation, too, which is why advocacy for COIN warfare expertise makes only sporadic appearances in US defense history. The sweeping scale of major combat, along with its promise for joint and combined arms, does not constrain discussion within the bounds of roles-and-missions infighting the same way COIN does. There is plenty of worthy employment to go around; no service has to abandon its primary mode of fighting or the means with which it fights. The preference for this kind of warfare and the urge to be prepared for it is why the Air Force will pursue the F-35 and sixth-generation fighters even during a time of flatline or shrinking defense budgets, for example.

By contrast, when the resources available to originate, train, and equip forces for unconventional conflict begin to disappear under budget constraints, the services' thirst for cooperation quickly gets sated. As budget contraction in 2014 revealed, *intra*service disputes alone are enough challenge for the internal defense organizations; the attention available for pursuing and executing cooperative interservice programs is predictably low (Kostro 2014, 2–3). As an Air Force budget planner remarked (and we have observed), "The system rewards a lot of bad behavior" (Babcock 2014).[24]

Cannon prescribed a common vision of what the joint force was to accomplish as a means to reconcile competing theories of victory and reduce the principal-agent preference gap; he recommended brute-force joint constructivism for each new case. As chapter 7 discusses, he is in good company with many senior flag officers (and other defense professionals) who have an innate trust in the ability of a singular concept of operations to drive jointness by overcoming interservice friction and competition.[25] Cannon (2012) was correct to say that traditional means of agency-theory enforcement— monitoring and sanctioning of wayward agents—by themselves would be ineffective.[26]

He instead advocated the practice of "constructed" joint cooperation, and the case studies examined here reveal that there are several tools available to the would-be constructivist.[27] One means involves causing the coalescence of the interests of one service subgroup with the interests of another (TRADOC with TAC for AirLand Battle, training command pilots in ATC and NATC in the case of JPATS, and Air Force UAV operators with the Army's increased demand for remote ISR). As organizational cooperation literature teaches, effective facilitation of cooperation is the ability to find unused "organizational slack" and then coordinate, communicate, and adapt in kind among organizational goals, preferences, and work procedures; the result is organizations that work on a task for a period of time as an apparent single body, even though in reality the unitary actor consists of groups with many disparate and divergent goals.[28]

Combat, or the impending threat of it, puts the fear of failure at the forefront of military personnel throughout the ranks, and reaches up into the external relations levels of the defense department. As a senior Joint Staff officer wrote, "jointness . . . is the highest it has ever been, but there is still lots of room for improvement"; his chief concern was "that we will lose what we have gained from Iraq and Afghanistan after we draw down in Afghanistan" (Roberson 2014). He also indicated that the services would "fight to maintain that level" of cooperation, but that with budget cuts would come a loss of opportunity for joint exercises—the only way to maintain joint capabilities short of getting actual combat experience (Roberson 2014).

Ironically, though wartime contingencies present the most obvious and pressing need for cooperation, they are also likely to incite visceral and reactionary responses among the services, which instinctively revert to their operative theories of victory and operational preferences. While war may provide an initial impetus for institutional jointness, it in no way guarantees that interests will align over a long time period. The example from COIN air power is instructive as follows: the Army and the Air Force immediately recognized a reason to improve joint integration, but found the effort difficult to sustain as competing theories of victory and operational preferences prevented their interests from coalescing. Organizational slack is at a premium when it comes to services defending their existing habits of warfare.[29] Subgroups comprised of specialists within each service may have a common viewpoint of the battlefield that *could* coalesce into jointness, but subgroup affinity rarely overcomes the stronger influence of service identity.[30] Compounding the problem is the reluctance of the higher external defense establishment—Congress and the president—to take part in the unequal dialogue that could break the resultant logjam. The need for someone willing and able to overcome these daunting challenges speaks directly to the fourth and final dimension of institutional jointness.

Component 4: The Role of Individual Leaders

If outside influences usually drive institutional jointness, it seems equally evident that specific individuals will play an outsized role in any cooperative efforts that come to fruition. This trend is in view in all of the case studies incorporated in this work. A few of the relevant factors at play in this component are the leader's position within the defense establishment, his or her ability to build consensus across the levels of that structure, and the arena in which he or she seeks to pursue a cooperative effort.

For AirLand Battle, all evidence points to Army TRADOC generals as being the most influential thought leaders who gave the concept its final shape. General Starry built on the foundation that General DePuy had laid, but they each responded to the external demand for a security-problem solution.

Both cultivated relationships across service lines and socialized their entire effort widely within the defense establishment. With cooperative momentum thus imparted, the greater effort benefited from the instincts of two joint-minded service chiefs who gave the effort its most visible cooperative entailment, the pursuit of the 31 Initiatives.

The JPATS case study hearkened to the concept of a heterogeneous engineer, one who worked among available nodes of opportunity to cause a stable system to emerge. Using the advantage of congressional pressure and the political appeal of institutional jointness, Air Force executive leaders and action officers, later accompanied by Navy counterparts they had convinced through careful persuasion and accommodation, drove the idea of a joint primary trainer to its status of acceptance throughout all strata.

This investigation found that COIN air power support suffered from a dearth of common interservice interests and thus had to settle for mostly technological innovations to foster joint cooperation. Only a few (relatively underwhelming and short-lived) command-and-control advances appeared. Where progress does appear, though, in both in operational and materiel innovations, the leadership of individuals was pivotal. The example of Rover video data-link development is particularly compelling, as it witnessed individuals from deep within their respective services make an exceptional plea for the new system that enabled novel combined-arms practices. Here, the strength of an idea that prevented *losing* on the battlefield cut through the bureaucratic noise. With their enthusiasm and willingness to build consensus, interest coalesced across all levels of the defense hierarchy, and the nice-to-have system became part and parcel of the joint institutional effort. In many CAS scenarios, it became the prime facilitator.

Heterogeneous engineering can be frustrating when one is working against many dissociative forces and has little bureaucratic power, but the lesson here is that it is nonetheless possible. The person attempting must have an iron will and the ability to persevere in the face of detractors and criticism, because "[a] majority of the interaction between organizations in crises"—and the nature of military organizations makes virtually every cooperative opportunity fit such a description—"both in decision-making situations and over the course of a crisis, are conflictual" (Svedin 2009, 125).[31]

Human beings, while demonstrably capable of cooperation, still fail at it, even when it would be individually or collectively beneficial to them.[32] Exhibiting a sufficient degree of self-assurance in a combative situation such that one remains able to pursue a common good is in tension with the threat of outright aggression, which acts counter to all cooperation, including jointness.[33] Constricting channels of communication more than incompatible radio systems and causing more angst than differing philosophies about what ultimately wins a war, superiority complexes degrade joint relationships faster than unsuitable materiel or enemy action.[34] The acidic effect of unrestrained egoism on joint relationships has in the past appeared as acrimonious exchanges in the press or a refusal to communicate.[35] Our CAS case study suggests that some of the most damaging relations are those continued through a veil of polite behavior in support of institutional jointness that masks a complete disregard for the perspective and experience of another service.

The destructive nature of arrogance—real or perceived—works through at least two distinct mechanisms: a loss of faith and goodwill between peers and increased conflictual behavior by an organization's members, exacerbating a cycle of diminished trust.[36] Verbal salvos in these wars of words appear as offensive or defensive in nature, depending on an observer's perspective. An example is Charles Dunlap's (2012) statements about the Army and Marine Corps's joint effort on a 2006 counterinsurgency manual, which he found to be insufficiently informed with an "air-minded" perspective (63). Dunlap's provocative phrasing ("in reality, American ground force commanders often do not understand how to use air- power effectively and efficiently") creates a tone that suggests a persecution complex ("Airmen believe that U.S. ground forces are the finest in the world. Unfortunately, that feeling evidently is not mutual") and slanted evocation of Operation Anaconda ("although fixed-wing air power eventually rescued the operation from serious difficulties and accounted for most of the terrorists killed in the operation, the Army commander nevertheless denigrated the Air Force's efforts in a subsequent magazine interview"). This style may play well to air power proponents as a defense of essential points that would otherwise go unmade (Dunlap 2012, 68–69). To many, however, it may seem an offensive strike by air power advocates against surface forces, carried out, as Leo Tolstoy described, "with a boldness characteristic of people employed in country not their own" (2009 [1869], 1319).

Given the barriers to joint cooperation, a frustrated military leader might opt for a go-it-alone approach in trying to force what she perceives to be necessary changes. However, for the same reasons that the obstacles are effective, such behavior is likely to doom any reform efforts, and is anathema to any definition of "jointness." One need not be as radical or petulant as Billy Mitchell, either. Simply failing to find another group with shared interests when trying to invigorate an initiative may doom an initiative to obscurity. National bureaucracies offer too many processes to delay promising ideas and too many opposing constituencies to ignore these kinds of forces. Having attained a coherent vision that unites a few, the purveyor of jointness must be creative in finding out how to interest more participation in that vision. Compromise and heresthetic art may be required to keep the vision intact and overcome ubiquitous bureaucratic friction.[37] Luckily, opportunities for logrolling abound as well. Talented military leaders with some degree of political acumen are therefore absolutely essential if institutional jointness is to be a reality.

Even in praising the actions of so many individual military leaders, though, it is necessary to acknowledge the frequent appearance of external forces in their motivation. All of the case studies in this work were contemporary with the development of or followed the passage of Goldwater-Nichols. Is it reasonable to speak of a "joint-minded" military leader outside of this context? Would implicit fears of retribution for poor combined-arms performances be ingrained in the military psyche absent legislation and days of probing testimony? Would the musicians show up if Congress did not print the score? Proof of negative, counterfactual outcomes is impossible, but it seems that quite a bit of credit is due to the external work that made "joint" and "purple" a more earnest part of the Pentagon vernacular. This returns us to an appreciation of the external motivators for cooperative military action, and brings the dimensions of jointness back together in a full circle.

Combination of Factors

The four components we use to describe jointness are individually complex; taken together, the picture they provide of jointness threatens to be intricate in the extreme. However, a simple visualization of the probability of cooperation afforded by any single component allows a first-order combination that may provide an idea of the likelihood, utility, and staying power of the cooperative efforts that arise for a given security issue. If the components are imagined as lenses lined up in front of a light projecting the prospect of jointness, then the more of the lenses that are positively aligned with one another, the brighter and clearer the picture of jointness is likely to be in the specific instance it refers to. The resulting view of the potential for successful joint service collaboration depends to a great extent on how the four factors of why, what, when, and how that shape jointness line up.

Elements of Definition

Foundations of a Better Definition

The quest for a framing theory continues with a search for a more precise definition of "jointness."[38] Chapter 1 established that the term is used in many contexts and with diverse meanings; such flexibility may afford political utility, but is unhelpful for measuring the value of proposed or observed joint efforts. Martel described the definition and theory of "victory" to be undetermined and subsumed in the pursuit and philosophy of strategy. "Jointness," since its pursuit is often assumed to be a necessary component of military strategy, exists in a similar state of subordination to a larger abstract concept.

The search for a refined definition begins with an examination of the word "joint" itself. In ordinary usage, the sense appropriate for this investigation is that used in most dictionaries' first or second adjectival form, denoting the united activity of two or more. Its etymology derives from the biological meaning of "joint," the Latin *junctus*, a participle of *jungere*: "to join." Both are related to the Modern Latin *jugularis*, the all-important joint at the top of the spine through which the lifeblood flows and atop which the decision-making center rests, an apt metaphor for the command-and-control meanings that pervade military usages of "joint." Another definition from normal usage, "shared by or common to two or more," is the origin for the commonality that "jointness" often implies in military usage (Webster's Illustrated 1990, 906). The glossary that purports to define all military terminology provides the specialized definition that motivates this study, connoting "activities, operations, organizations, etc., in which elements of two or more Military Departments participate" (Joint Publication 1–02 2010, 139). To a US military audience, 'joint' is distinct from "combined," which identifies "two or more forces or agencies of two or more allies operating together" (Joint Publication 1–02 2010, 139).[39]

Martel noted that the descriptive utility of "victory" struggles to differentiate magnitude; it can indicate prevailing in everything from a tactical skirmish to the grand-strategic vanquishing of an enemy (Martel 2007, 85–87). "Jointness" suffers from a similar—though qualitative, not quantitative—malady in military jargon, because the bar for its use is also low. It merely takes the participation of "elements of more than one military service" to earn the label, an overbroad criterion that is not helpful for separating the helpful from the showy or deleterious (Merriam-Webster 2013). Yes, "joint" indicates the participation of two or more services, but "activities" and the other nouns in the definition can include almost anything, including unproductive arguments and spiteful sabotage.[40] This meaning allows the Mitchell-Moffett exchanges, the "Revolt of the Admirals," and Army-Air Force arguments over Operation Anaconda to qualify as "joint fights," but certainly not in the helpful, synergistic sense the Joint Staff's publishing enterprise hopes to imply in its glossy pamphlets. It is necessary, therefore, to modify the definition of "joint" with some additional constraints that give it a clearer and more useful meaning. Pragmatic delimitation comes from considering those things that are useful and desirable to be achieved from joint institutional cooperation and negative or harmful consequences that it might help avoid.

Proponents of jointness have an implicit expectation of interservice cooperation and that they will see multiple services working or acting together for mutual purpose or benefit. This may further imply that they share resources, be it equipment or even personnel. And it could go even further to an assumption of shared goals, credit, and glory. To some who think along these lines—including those who drafted early defense reform legislation— *unification* into a single service is a goal of jointness. The purpose of every defense-reform initiative since the turn of the twentieth century has focused on unification through the centralization of power in a defense bureaucracy of growing power and authority. The realities observed in the previous US case studies suggests that this means of encouraging cooperation is a bit too hopeful—on some facets, it probably exacerbates agency issues instead of solving problems, burying them in an impenetrable central bureaucracy. Nonetheless, some degree of *integration* of the services to accomplish coherent action is undoubtedly an expectation throughout the defense establishment, even internally within the services, which at their best realize a need for assistance and the potential for genuine synergy.

Additional positive consequences of jointness include efficacy and efficiency. Monte Cannon argued that the pursuit of each, through improved command-and-control measures, was a worthy goal for the study of jointness (2012, 22). Efficacy is the capacity to produce a desired result; it is potential effectiveness and the ability to do what one wants. Efficiency, the ability to do or produce without waste, can lead to effects with minimal or comparatively fewer resources. An essential part of the definition of jointness should thus be one that implies a pursuit of efficacy and efficiency.

On the battlefield, where efficacy is the paramount measure of jointness, interoperability, or being capable of reciprocal use by other services or countries, is a worthy pursuit. Not only does interoperability ensure that people pursuing common military objectives can work with one another's gear, it may also lead to economies of scale in production and allow more bang for the defense budget.

The studies reviewed in this investigation suggest that commonality, or sharing features or characteristics, may promise efficiency when missions are very closely aligned (as in primary pilot training), but that often the dissimilar fighting styles of the services may make them less efficient than independent purchases. The services, even if they can be persuaded or forced to purchase a common weapons system, see a threat in complete commonality, and will do everything in their power to make their version of the system distinct. This is showcased in the T-6 program as much as it has been with the F-35.[41]

Another pillar upon which support for jointness rests is a trust in its ability to reduce the waste produced by the hulking, redundant bureaucracies that backstop national security. To explore this angle, one must examine structural threats to efficacy and efficiency. The waste observed when the services act together consists first of the overlapping resources they expend because none is fully interdependent—think of the "air force" that each branch of the military maintains, for example, or the overlap between ground- and air-launched "artillery." The second root of waste and inefficiency is shirking—evading work, duty, and responsibility—that components exhibit because their core ideologies do not align with the theory of victory that underpins a given operational strategy.[42] An illustrative example that includes a temptation to shirk was the Army's boots-on-the-ground model of victory in Afghanistan and Iraq, which competed with a regime-change-and-leave model espoused by members of the Air Force and, notably, DoD leadership early in the conflicts. While an attempt to ascertain the *best* model does not seem to have taken place, there is little doubt that the Army, and then the Air Force, tacitly or explicitly disagreed with the theories being used.

Other negative entailments of bureaucratic behavior observed in the case studies include segregation—excluding or isolating others from the main body or group. Aside from the loss of communication that inheres in the shadow of segregation, it may also lead to a hoarding of resources and quarreling. When allied forces are segregated from each other, they tend to put in place strict control measures to prevent fratricide and other negative externalities when a unit crosses another's boundaries. In an information-intensive war, this quickly creates seams where insurgents can hide or data can get lost with the effect of diminishing overall potency in warfighting.

Part of the difficulty in defining jointness is the lack of a framework to relate it to victory. The aims of jointness, no matter what specific entailments one believes it ought to include, are all ostensibly to enable the military to attain victory more easily. Joint aims might be focused on the relatively short-term goals of a battle or a campaign, or they may serve to ensure victory over a longer term, as by improved training methods. However, as Martel pointed out—and the preceding case studies underlined—there is no common theory of victory (Martel cited in Cannon 2012, 296) in jointness.[43] To the contrary, there are three or four competing theories of victory, and the services that brandish them show no signs of an ideological surrender. Cannon determined that, for joint operations, a well-articulated mission statement toward a common goal was most likely to overcome "the services' disparate views of war and their attendant visions of victory" (2012, 298) that are the core of the joint problem.[44]

Command, in the sense of the word applicable to joint operations, is a military leader possessing or exercising controlling authority. Cannon, following a Clausewitzian maxim, specified that joint commanders should "formulate strategy that links military means with the desired political ends" (Cannon 2012, 114). Control in the way it is most commonly used in a military setting means the act or power of directing or regulating actions; in this sense, it indicates the degree to which a commander maintains the "coherence of decentralized operations within and between warfighting domains" (Cannon 2012, 16).

Given the disconnects between the cultures and between the theories of victory that this study and others (Cannon 2012, Donnithorne 2013, and Ehrhard 2000) have exposited, however, there is no single command-and-control method, system, or even general principle that we could expect to emerge. Unification is elusive and will remain so, even when externally ordered, as long as the services have individual authority to organize, train, and equip their forces. Even if Congress were to change these structural arrangements of force development—a move so politically risky as to preclude its attempt—fundamental differences about how military forces achieve victory will persist. There enduring disagreements about the best way to reach efficacy and efficiency are likely to persist, and honest conversations will acknowledge that this area of contention needs to be addressed. Unless these fundamental disagreements are both voiced and resolved each time they arise, we can expect service cultural preferences and competing ideologies of victory to remain the most corrosive elements threatening the positive objectives of jointness.

An improved definition of jointness, therefore, recognizes the inevitable competition that appears in any effort to make independent organizations cooperate. It carries with it the implication that some entity or decision-making process must settle fundamental differences in understandings if efficacy and efficiency are to be the result. Joint publications that pretend all the services agree—as many of the Joint Staff's publications naively seem to do—may be regarded as masks on reality at best and complete delusion at their worst. This is Cannon's fundamental point: authority to resolve these kinds of disputes nominally lies at the organizational level that coordinates among services with a joint force commander, but the joint commander at best functions to broker weak compromises between components (proxies of their sponsoring services) striving to fight their own battles. This study has observed that meaningful movement toward jointness comes primarily from external influences: threat of failure in wartime and the authority of Congress (with cameo appearances by the DoD) at other times. Once we admit that the services internally and the groups coordinating among services are even less apt to choose a path of joint cooperation than they are to innovate new warfighting methods and doctrines, we will reduce the naiveté the US defense establishment has demonstrated about jointness.

Such realism would also address a final aspect of the inherent tension among the services that belongs in our definition: jointness for jointness's sake is an unwise proposition. If a mission or task can be performed most efficiently by a single service, then that service should be allowed to do it. If there is debate about a service's ability, a shortfall in overall capacity, or a synergy to be exploited from joint action or acquisition, then jointness should be adjudicated and pursued. Absent compelling evidence that it is worthwhile, though, jointness can become a Sisyphean exercise in pushing together two opposing forces. But were one service allowed to win the right to carry out the mission solo, it would spur the others who did not on to develop more options for the next round of competition.

Our Definition of Jointness

To narrow the definition of "jointness" to a useful degree of precision without ignoring essential facets of its usage, this work proposes the following meaning: *the directed efforts of two or more services to achieve battlefield efficacy or peacetime efficiency, pursued when required effects or efficiency are unavailable through competitive, single-service means.* This definition is first of all normative; it recommends that jointness prescribe certain actions rather than being simply a descriptive term of art. This working definition advances a hypothesis that "jointness," in addition to its multiservice connotation, should demand utilitarian adjudication among other available means.[45] The US military should only pursue joint action when it is assessed in good faith against other available alternatives, and it should strive, as far as politics allows, to set aside the ubiquitous notions of "fairness" that plague contemporary philosophical discussions.

Jointness is desirable to the point that it allows military services to achieve an effect on the battlefield or efficiency in performing necessary tasks that they would not be able to attain on their own (Johnson 2014).[46] Equal service representation, uniform resource allocations, or matching media face time are of little value in this formulation.

Jointness as a contemporary term has an agreed-upon descriptive meaning, but flounders in providing prescriptive structure to military-strategic problems because the meaning is so wide. Labeling something "joint" does not currently acknowledge the implicit clash of service cultures and theories of victory that happen anytime a difficult decision confronts the US military. A better usage of "joint" would be one that, paradoxically, emphasizes competition over cooperation. It would align with the recommendations of Sapolsky, Gholz, and Talmadge (2009), who found value in moving toward greater decentralization and competition among the services.

When one service can present an idea that offers it a "chance of disproportionate gains or lower losses" (Sapolsky, Gholz, and Talmadge 2009, 164), ensuing competition gives the external defense establishment real options to choose from; it need not settle for whatever backroom compromise seems least objectionable to the Joint Chiefs.[47] The services should compete against one another in the realm of ideas, and their best offerings should face off against one another. Most often, an interservice amalgamation of approaches will still result, but it will benefit from the underlying competition.

Such a prescription is admittedly difficult for two reasons. First, it takes a whole military career—and a host of special security clearances—merely to be acquainted with the universe of available service capabilities. The people who attain these positions have certainly been steeped in the ideologies of their sponsoring service, though, and have likely attained their rank by demonstrating unwavering fidelity to a few parochial dogmas. Unfortunately, their upbringing makes them less likely to honestly engage competing ideas if it appears that the clash might result in discrediting a premise foundational to their services. The second reason is that all parts of the defense establishment gain in the short term from an amorphous definition of "jointness."[48] Programs are more likely to garner support, whether in the Pentagon or in Congress, if they carry a "joint" label. Up to this point, "jointness" has not lost the luster that condemned "commonality" to the dust heap of Pentagon usage—broad enough swaths of the defense audience assume it means "cooperative," "efficient," or "inexpensive," depending on their perspective, that programmers do not have to explain why their project is necessarily any of these things or why it is superior to a nonjoint option.

This study recommends a *via media* with respect to two facets of the problem of jointness. The first is to realize that the term "joint" has different contextual meanings. The second demands that external authorities weigh the options offered by the individual services, but they must always appreciate that these courses of action are tainted by bureaucratic self-interest. This requires much of the US civil-military establishment, as any helpful change would. Because of the tendency toward bureaucratic stasis in the absence of a crisis, jointness precludes a state of ideal existence in a way that students of strategy come to appreciate the elusiveness of an endgame.

Strategy and jointness are arenas in which mankind must continually strive. Empowering and holding senior military leaders accountable for meaningful strategies is out of vogue, at least in the estimation of observers like Eliot Cohen and Tom Ricks. It will require larger amounts of trust, communication, and feedback than currently exist in the defense establishment, and channels of communication must be forged or cleared anew for such a vision to succeed.

Though the same barriers to a clear definition of "jointness" exist as do those with regard to a useful definition of "victory," striving for greater clarity in both is an objective worth pursuing. The quest for a normative, competitive, and nonideological definition of jointness is at once the best way to focus on a nearly intractable problem even as it highlights the unlikelihood that military organizations will ever tackle such a problem with objectivity. It is a jumping-off point, suitable for experimentation and critique.

Ramifications of Our Definition

The definitional elements described earlier hint at the beginnings of a theory of jointness, inasmuch as they relate a few observable and quantifiable variables (multiservice participation alongside efficacy or efficiency) through a singular statement (Easton 1953, 56–57). Furthermore, in line with Easton's efforts, this statement of jointness edges toward an empirically testable general theory. Stated simply, joint options should be formulated and tested against unilateral options, adopted when they provide a greater potential to advance national ends, and rejected when they fall short of this test.[49] This approach is simplistic on its face, but in reality, questions about jointness rarely receive this much rigor. Instead, they are tested against even more amorphous concepts such as fairness and service-specific doctrines that hold sway in debates. Theory, according to Crick's formulation, explicitly *excludes* doctrine or ideology, and this proposed definition of jointness argues against the normative prescriptions of specific service doctrines (1973, 12–14). Indeed, this investigation has found allegiance to service doctrine to be a frequent enemy of beneficial jointness, hindering the efficacy and efficiency it can otherwise help attain, even while doctrinal practices remain a key enabler of military effectiveness.[50] It is a deep quandary indeed.

Without stipulating the exclusion of doctrinal measures, the meaning of and utility of jointness can vary depending on one's philosophy of war and ideas about strategy. At an extreme, component or service advocates might argue that any capability in their most important domain that gets sacrificed in the pursuit of jointness is too high a price. The other extreme approach is one that blurs "the military" into a homogenous organization without domain-specific boundaries, an approach that relies more on an inchoate joint doctrine than any objective, comparative adjudication.[51] Excising high-level doctrine from decisions of jointness is the only likely way to avoid emotive responses and exaggerated claims that do nothing to advance sober military decision-making. (It is also not terribly likely to happen as long as senior military figures are the ones making the decisions.)

Such an approach to the definition of "jointness" precludes some of the most troubling artifacts that come to mind when the term is invoked. The promulgation of "least common denominator" joint doctrine is a primary example (Johnson 2014).[52] Joint publications should not strive to include only those noncontroversial statements to which all services could agree (or to which they exhibit apathy)—this approach fails the definitional approach because it precludes maximum efficacy. Such an approach *minimizes* the efficiency that doctrine might be able to provide. Don Snider (1996, 25) surveyed three "inputs" to military activities in which he found increasing evidence of jointness: (1) common warfighting strategy, (2) increasing number of joint doctrines, and (3) increasing opportunities for joint training evaluations. This investigation finds little meaningful evidence of progress in the first two areas. Warfighting strategies remain firmly rooted in competing service theories of victory. Joint doctrines at a high level of abstraction are, unfortunately, simply the pablum that gets past the services' ideological screens for publication. What would be better is a sustained and spirited debate about the most compelling theory of victory as it applies to a specific security problem or an ongoing conflict.

At the opposite end of the spectrum of abstraction, there are several useful, lower-level joint publications that reflect dialogue on areas of battlefield specialty, but these are more accurately described as "procedures" or "tactics." They assist in the quest for interoperability and for this reason merit attention and retention. They do not, however, settle any fundamental interservice disputes and they do not force the services to work together. The causal chain begins with a requirement to work together, by either an external organization or a large threat that one service cannot or does not believe it can confront on its own. The services then develop methods and procedures to facilitate this kind of cooperation, because the threat of failure or ineffectiveness seems unacceptable. This is the kind of jointness observed in the COIN air power case study and is not to be dismissed, but it is a lagging phenomenon that barely outlasts the directive or conflict that gave it birth; it is not a compelling, long-term impetus for cooperation.

The third area (training), however, offers hope for progress. As described in the "Joint Potential" component of jointness, it lies in the contestable middle ground of doctrine and materiel. With reference to organizational behavior, it also facilitates (or forces, as needed) opportunities to communicate and to engage in long-term exchanges about competing ideas. Caution in pursuing better training is warranted, because tight fiscal eras are those most prone to cutting opportunities for training that might threaten service-specific habits for building proficiency.

There is also an innate desire to ignore the shortfalls that joint training expresses. The fact that there are over four hundred jointly trained military specialties, though, means that in some ways the US defense establishment turned a corner toward jointness in the 1990s from which there is no return, and this is a good thing.

A better way to derive joint doctrine is by closely watching the services as they do their utmost to achieve what they think they do best. Then, an external broker can quantify and weigh on a common scale the capabilities available to the services when they pursue their own specialized doctrines to the utmost extent. For a given security problem, the best option or palette of options would emerge. Joint doctrine would move away from the philosophically unattainable "best" way to do something. It would instead adjudicate among the alternatives available to arrive at an optimized solution for a given problem. This approach would give voice to the services' most radical, highest-payoff ideas in each decision-situation rather than quashing them under a heavy slab of unanimity. It might, in the lofty vision of Hans Morgenthau (1973, 3), "bring order and meaning to a mass of phenomena which without it would remain disconnected and unintelligible," which is a fair description of most joint doctrine and, all too often, joint command-and-control structures that have not been tested and refined by combat.

A Two-Level Problem

Jointness, as with strategy, merits discussions on two different planes. There is a bifurcated conversation about strategy, for which common usage implies charting a path to victory. There is an undeniable human desire to block off an event, a series of events, or an era, and assign a definitive description to the segment of time. If it involved military action, the description must include "victory" or "defeat." Military and political leaders both—since war and politics are inextricably intertwined—must articulate their strategies, and to most audiences, even some who might be expected to know better, these will connect an existing situation to one that is victorious, or closer to victory than is the present state of affairs. Though "time may appear to be a neutral dimension to war and strategy . . . it is equally usable by all belligerents, [so] its meaning will tend to differ for each" (Gray 2007, 72). All the victorious tactical engagements that the US military waged in the Vietnam conflict gave way to a strategic loss, primarily because the North Vietnamese and Viet Cong used the ten years of time more effectively.

There emerges a second level of strategy, one that exists at a more esoteric level. Victory—along with the finality it implies—is not a primary pursuit within this sphere. This side of the study of strategy allows a sublime indeterminateness. Carl von Clausewitz wrote, "In war the result is never final" (1984, 80). Everett Dolman echoed this in saying that a true strategist "seeks instead of *culmination* a favorable *continuation* of events" (2005, 5).[53] Triumph is anathema to this formulation of strategy, because it demands completion, a final summation of events. The very 'strategy' for which the world clamors, however, is merely a way to bring about these favorable, if fleeting, moments of stability in the march of human affairs.

Both sides of the strategic coin must stay visible if the concept of strategy is to hold practical value and stand up under the weight of rigorous scholarship. While Dolman's exhortation that "the first notion the strategist must discard is victory" (2005, 5) remains useful for a student of strategy, a successful strategist will never forget that victories nonetheless keep alive the will to fight, the imagination that an effort can succeed, and the morale to fight beyond oneself. In Napoleon's immortal formulation, in battle "the moral is to the physical as three is to one" (Dean 2014).

So it is with jointness, both its definition and its pursuit in practice matter. Final resolution of the tension between the two is a wicked problem whose solution is as elusive as a search for complete victory. No country with meaningfully differentiated armed services has attained complete jointness, nor can any serious leader of a military service eschew the unique doctrinal stance that imbues his service with its individualistic view. Doctrine evokes specific entailments in each service and defies a universal definition that reaches across multiple audiences, even in a single country. Yet the public notion of unity and striving toward a shared goal remains useful, because often the pendulum of individual-service striving swings too far for battlefield effectiveness to emerge.

A final point is in order about the proposed definition: it will not necessarily be popular with or accepted by a military audience because it does not match up with the self-conception of jointness that the group holds. More precisely, that group is split about what it believes jointness *is*, though less so over what *should cause it*.[54] Despite the lack of consensus, military leaders tend to have passionate views about which elements do and do not belong in the definition.[55]

Conclusion

Using observations gleaned from a theoretical examination of the case studies, this chapter has identified germane aspects of jointness that present themselves from general and military-specific theories. These distill into four key components that can serve as the basis for studying interservice cooperation. They offer a descriptive framework with increased predictive capability as more data from case-study analysis becomes available. This chapter has also offered a normative, prescriptive definition of "jointness," finding the existent descriptive definition of common usage to be overbroad for helpful defense decision-making. It offered a justification for the pursuit of an improved definition as well as the development of the pretheoretical framework for characterizing and studying joint cooperation.

It seems unlikely, given the incoherent mix of what jointness means to various audiences, that a satisfactory general theory might develop. Part of the incoherence is illustrated by the components that describe jointness listed earlier and by the wide range of values they may reflect. That is not of great concern to the isolated practitioner, though, who, through her own familiarity with the situations in which she engages, will recognize the taxonomic characteristics that make more specialized theories useful. For example, one may pursue a theory of joint acquisition—or a theory that works for a specific kind of acquisition—that is distinct from one appropriate for advancing joint operational command and control in a particular conflict. Indeed, the equifinality of jointness is why so many examples of effective cooperation exist despite the number of potential obstacles that special interests and a robust bureaucracy erect in front of it. The trick for the pragmatist—or the heterogeneous engineer, or the service member who *simply wants to get something done*—seems to be recognizing that there is a plethora of ways to be joint and an accompanying large number of ways to effect cooperation.

Conclusion: Jointness Is a Collaborative Fight

> *Jointness means services specialize and rely on each other.*
> *Jointness has been and will be resisted because no commander*
> *wants to rely on somebody else for make-or-break support.*
> —General Merrill McPeak, US Air Force, September 1994

> *Turning and turning in the widening gyre*
> *The falcon cannot hear the falconer;*
> *Things fall apart; the centre cannot hold;*
> *Mere anarchy is loosed upon the world,*
> *The blood-dimmed tide is loosed, and everywhere*
> *The ceremony of innocence is drowned;*
> *The best lack all conviction, while the worst*
> *Are full of passionate intensity.*
> —William Butler Yeats, "The Second Coming,"
> November 1920

Introduction

This final chapter distills some of the key observations of the preceding case studies into actionable conclusions for practitioners in the defense establishment. Focusing on the components of institutional jointness outlined in chapter 7, it offers some first-iteration hypotheses about the conditions in each component necessary for the generation and support of joint cooperation. This chapter also includes some predictive statements about the nature of institutional jointness based on theoretical commonalities in the case studies. While the review of relevant prior research and the case studies allowed us to sketch out our framing theory in chapter 7, the case studies also allow us a useful summary of the commonalities we found that have concrete and real-life implications. This chapter presents these observations and should be of particular interest to our practitioner audience. We conclude the chapter by looking forward. Keeping the practical conclusions about what influences institutional jointness in mind, we revisit the command-and-control structure initiative, JADC2, and assess the potential of this contemporary opportunity for joint collaboration and action. The apparent struggles establishing JADC2 constitutes a powerful reminder of the continuing difficulties in the pursuit of jointness.

Preliminary Conclusions about Institutional Jointness

We analyzed our lengthy case studies with the specific questions we formulated in chapter 2 in mind. These questions about the conditions under which joint cooperation may be more or less likely, what seems to facilitate and hinder cooperation in other organizational settings, as well as patterns of behavior and strategy in crisis cooperation led us to twelve empirically based assertions about joint cooperation in the US military. These assertions are presented in this chapter in no particular order other than one that seemed to makes sense for readability.

"Useful" Jointness Improves Command-and-Control Methods, Interoperability, or Efficiency

This conclusion arises from the study's effort to create a more constrained definition of "jointness" that did not allow obfuscation or semantic free-riding. However, a real plurality of meanings among the practitioners of jointness demands that the term maintain a flexible definition with respect to its essence. There are in fact at least two sides of jointness: efficiency and cost savings vie with better effectiveness on the battlefield in the minds of those who use the vocabulary of joint cooperation. This characterization extends across all the levels of analysis investigated: internally within services, among services, and in relations to external stakeholders. Just like some members of Congress, certain members of the military feel strongly that interoperability and economies of scale are the only realistic contributions that jointness can make. Others, in both groups, believe in the possibility of unified command-and-control measures that will reduce fratricide and inefficiency in combat while providing a broader range of military effects for national security.[1] Their disagreement prohibits a more-refined definition, but we may conclude with some confidence that the type of jointness sought varies in difficulty across a spectrum. It is *much* more difficult to get services to acquiesce on matters of strategic preference and theory of victory—typically the sticking points that block agreement over interservice command-and-control methods—than it is to agree to purchase common or compatible equipment (even though this is itself more difficult than one would hope if they just read glossy brochures from the Joint Staff).

Despite the vexing degree of difficulty involved in achieving the most idealistic forms of jointness, a reappearing theme encountered in this work is that combat quickly removes barriers to joint cooperation. If basic interoperability thresholds permit safe integration, services will rely on the assistance their counterparts can provide regardless of preconflict ideology. To be flippant, everyone is a doctrinal agnostic in a foxhole. A more serious implication of this trend is that better jointness—and better assessments of what joint cooperation requires—would come from combined-service training that closely approximates combat conditions. For a variety of reasons, such conditions are impossible to attain and difficult to simulate with anything approaching perfect realism. However, the defense establishment must do all it can to resist internal barriers to achieving an ideal. As Monte Cannon writes, "joint training venues and curricula are steps in the right direction" (2012, 299) toward the construction of jointness he advocates. Since true competition and objective assessment of available joint options can improve the effectiveness and efficiency of military forces, accounts of war games that place arbitrary limitations on the mock enemy force to elicit predetermined outcomes are particularly troubling (Van Riper 2013; US Marine Corps History Division 2013).[2]

The case studies examined in this volume, as well as the language of international relations, organization theory, and organizational psychology, all hearken back to a common idea about jointness—it is a complex construct that exists mostly in the eye of the beholder. Jointness can be coincidental and unspoken; it can be intentional and born out of a desire to help; and it can be induced by implicit, explicit, or perceived organizational threats. Where one person sees jointness, another may vociferously deny its existence or even appeal for its removal. This characterization of joint action suggests two competing ontological views. Pessimistically, chance or nefarious purposes are so often at work in thwarting interservice cooperation that even if she so desires, a leader cannot conjure jointness ex nihilo. From an optimist's view, there are so many ways to foster jointness that policy-makers and military innovators who want to encourage joint cooperation have a wide tool kit from which to choose in facilitating their vision. The reality that competition is an integral part of jointness, both in its practice and in debates about its meaning, leads to a corollary that follows from an attempt to define its essence.

Useful Jointness Does Not Assume Common Understanding, Alignment of Interests, or Automatic Cooperation

The published claim that "the Armed Forces of the United States have embraced 'jointness' as their fundamental organizing construct at all echelons" (Joint Chiefs of Staff 2013, 1–2) is a half-truth at best. While this may be the stated goal of legislation, a mandate of DoD policy, and a hope of the Joint

Staff, reality is more reflected by David Deptula's description in a mass communication to Air Force Association members: "Among our four services, a specialized array of capabilities is provided through service or functional components to a joint force commander whose job it is to assemble a plan from this 'menu' of capabilities, applying the most appropriate ones for the contingency at hand" (2014). Far from being organized to fight *together*, an accurate perception is that the services organize, train, and equip forces in the way they *individually* best see fit. In combat, the onus then falls on a joint force commander to assemble an acceptable security solution from the palette of options presented, all in the midst of a crisis-planning environment. In the realm of acquisition, external authority mandates jointness formally; the services use sincere effort (or avoidance mechanisms) to comply with the intended outcome (or to shirk the intent and comply in appearance only). This may be an unpopular truth to voice openly, but unless it is acknowledged, commanders' plans will continue to fail because their expectations do not match the defense establishment's extant structural realities.

Another myth that merits debunking is the idea that the services will get out of one another's way when one branch of the military may be best suited for a mission. It seems more accurate to say that the services want "a fair and stable share of budgets, missions, and commands for each of them" (Sapolsky, Gholz, and Talmadge 2009, 128), and that they will practice "cartelization" to ensure that civilian masters are not able to keep them from this goal. Deptula (2014) believes that jointness "does not mean four separate services deploy to a fight and simply align under a single commander. Nor does jointness mean everybody necessarily gets an equal share of the action. Jointness is not homogeneity—it is not 'going along to get along.' " However, his picture in this respect counters the realists' portrayal of Pentagon politics. Sun-Tzu wrote of the many "estimates" that should presage a military expedition, opining that "with many calculations, one can win; with few one cannot" (Sun-Tzu 2005, 103, section 28). His advice assumes accurate measurements. It is difficult to succeed in an environment where one does not understand the operative terrain and relationships that characterize interservice politics. Jointness would be more successful (and less disappointing) if more practitioners retained an accurate awareness of the political restraints placed on it and did not willfully perpetuate ignorance by repeating comforting platitudes.

Useful Jointness Does Not Provide "A Best-Way to Win All Wars" or "Universal Doctrine"

Helpful joint collaboration may need to productively harness interservice competition while simultaneously resisting pulls toward centralization. In this sense, institutional jointness that emerges from competitive thinking about strategic advantage and what victory might look like and that picks pieces of service competency and capability to make a coherent joint puzzle would be very useful. Joint collaboration could be stronger and more successful if built with an appreciation for the strategic advantage that diversity of thought brings to problem solving, rather than as a zero-sum game or a false front for centralization of decision-making. Furthermore, institutional joint collaboration may not be a helpful way to satisfy expressed desires for more uniform doctrine and strategic vision at the head of the US military as an institution.

The competitive nature that inheres in the military services should be acknowledged, not stifled behind a wall of denial.[3] Military advice to defense principals should include the operative theory of victory upon which it rests. For a given conflict, military services and the planners they send to support a joint effort will arrive equipped with some theory of victory—perhaps explicit, perhaps tacit—of how their expertise can help 'win' the conflict at hand.[4] There is no reason to keep this a secret—it may well be the best idea available. However, if it does not fully address the problem at hand or its implementation defies common sense, a competing theory merits a chance to persuade decision makers.

One may view the practice of using services to organize, train, and equip fielded forces as a stroke of genius or a submission to harsh reality. Either one believes each armed service does best in its respective area of expertise, or else he thinks that a fundamental restructuring of the military establishment is impossible. Barring the imagined cataclysmic event that could drive such a change, this study recognizes that to be joint acknowledges the existence of separate services, and that those services have an imperative to best understand how to exploit the advantages of their particular domain- and specialty-specific operations.[5] What it insists on adding is acknowledgment of the deep-rooted and often contentious competitive spirit that emerges among the services as a result. Deptula phrased it succinctly: "Articulating the virtues and values of your service is being joint" (2014), although he himself might not top a list of defense-establishment figures known best for embracing jointness.

Sapolsky, Gholz, and Talmadge offered a nuanced understanding of the basic philosophical question that underpins the question of joint cooperation. In describing the modern US defense establishment as consisting of two centers of power, the civilian DoD and the military Joint Staff, they made three cogent observations: (1) there is an incessant call for more centralization when there are budget or military crises; (2) playing DoD civilians against the military hierarchies leads to collusion among the military services, which produces politically passable but militarily mediocre short-term solutions; and (3) indecision about the best way to solve security problems is acceptable, because the solutions to emerging and future security problems are by definition unknown entities—a planner's desire to seize upon a "best" course of action for the sake of efficiency risks failure at the hands of an enemy who finds a way around that course of action (Sapolsky, Gholz, and Talmadge 2009, 163–164). They concluded their work on defense politics with a call for continued and spirited multiperspective debate.

Rear Admiral J. C. Wylie acknowledged the fundamental rifts across the services over doctrine, but likewise welcomed the plurality of opinions it provided:

> The basic problem is why they do not agree. Why does the soldier think like a soldier, the sailor like a sailor, and the airman like neither of these but like an airman? Asking why they do not agree is quite a different matter from asserting that they should agree. On the contrary, these differences of judgment, these clashes of ideas, these almost constant pullings and haulings among the services, are the greatest source of military strength that the nation has. We do differ, within and among the services, and may Heaven help us if we ever enter into a period of prevailing sweetness and light and unanimity. Nothing would be more dangerous to our nation than the comfortable and placid acceptance of a single idea, a single and exclusively dominant military pattern of thought. The political parallel is almost too obvious to mention. (1967, 150)

Given the march toward unification and the quelling of service-specific advocacy in the years after he wrote, it is worth spelling out now what Wylie thought was obvious then. The totalitarianism of universal doctrine threatens to suffocate the good ideas that arise in an environment of competing doctrines, and in turn threatens the nation's military potency. The services are positioned to continue debating and to avoid the intellectual tyranny of a universal doctrine. As Eugene Zuckert wrote, "no reason for separate services seems more important than the freedom to apply many years of thinking and experience to operational concepts and weapon requirements," which to him constitute the "big picture of service roles and missions" (1966, 477).

An appreciation of the importance of independent thinking must be accompanied by caution in divining meaning. To wit, "jointness" often becomes the word behind which political interests rally when they are really looking for increased centralization, synonymous with consolidating offices or eliminating a service's capability to do a certain mission. Among the limited resources of the real world, finding efficiencies is a necessity, but today's efficiency can be tomorrow's loss of capability against an unforeseen threat. For this reason, the advocacy for a degree of mission duplication and doctrinal uncertainty among the services that Sapolsky, Gholz, and Talmadge

(2009) advocated is well warranted.[6] There are expenses associated with this pluralistic pursuit of national defense, to be sure, and the degree to which chasing it is feasible reflects the priorities and fiscal health of the nation. To completely eliminate it, though, is to invite collusion among special interests and a sure loss of capability against foes not yet identified.

A curious counterpoint to the theme of pluralistic striving advocated by many authors is the belief that a single, unifying vision can and will emerge within the defense establishment.[7] An overwhelming trend this study uncovered is a desire expressed by military leaders for coherent, singular strategic thought leadership, with relatively fewer voicing a need to welcome and foster enduring competition.[8]

Even Henry Kissinger (1957, 237–238), a legendary practitioner of realpolitik, exhibited this form of unitary optimism in looking to the leadership of a single military leader, specifically the chairman of the Joint Chiefs of Staff, who would participate in the formulation of national strategic doctrine along with the National Security Council.[9] Such hopes are best described as the "strategic monism" against which Samuel Huntington (1961, 264) cautioned. Whether in operations or strategy, pretending there is only one best way to success is a tendency with which the defense establishment must continually struggle.[10] J. C. Wylie cautioned, "Planning for certitude is the greatest of all military mistakes" (1967, 72). In confronting itself with multiple puzzles to solve, the defense establishment moves away from the dangers of monism, but rapidly confronts the problem of resource constraint, which introduces the underlying cause of the next conclusion.

Jointness Does Not Exist as an Interservice State of Nature[11]

Normal interservice dynamics will not result in joint cooperation. Each of the case-study examples of interservice cooperative behavior demonstrated a requirement for a confluence of forces, acting above and beyond the normal interservice bureaucratic exchanges, to come together to effect significant joint behavior, a joint outcome, or an increase in joint capability. The internal structure and processes of the defense establishment provide the most obvious explanation for why this is so. The defense enterprise is a layered, multicomponent bureaucracy. The first level of analysis is the services themselves. They are responsible for organizing, training, and equipping forces, generally specialized to have the most influence within a specific fighting domain, to meet the requirements of commanders who must in accordance with national guidance exercise the military component of the nation's power. They are equipped to be autonomous, and they believe—probably rightly—that they are best equipped to conduct warfare in their domains of specialization.

Therefore, they fight, in the way that all professions do, to advance their particular method of employing military power as the preeminent means of achieving victory in any conflict.

It is the services, however, spurred by institutional preferences and budgetary pressures to vie for their own individual agendas, from which the groups that coordinate among services draw their leadership and talent. The commanders to whom the services provide forces are labeled "joint" commanders, but they pass through the ranks of individual services to attain that level. While Goldwater-Nichols has certainly encouraged many more than would have been in past eras to pursue "joint" assignments, the preponderance of influence (enculturation, promotion, education, and the like) on each potential commander still lies with his or her individual service.

Not only the specified combatant commanders but each and every staff officer in a joint headquarters arrive imbued with a view of the natural order of things, and that view has been influenced a great deal by the service that officer calls "home." The concept of an ideal joint officer—someone able to set aside a career's worth of service loyalty, subconscious prejudices, and the knowledge that the next promotion depends on decisions made by an individual service's corporate process—is a pleasant idea, but one unlikely to be replicated in all but a few individuals who may or may not then subsequently attain high rank. A more realistic outcome of joint service is that a few perceptions of colleagues in other services will be favorably changed (or negative stereotypes calcified) with advocacy for the service in which one holds position and rank proceeding apace. The journey to a truly joint viewpoint happens only with time but usually begins late in a career; only a small percentage of gifted individuals will be able to navigate the difficult straits of articulating positions of compromise without alienating themselves from the service that provides them succor throughout a career.

The possibility of changing service tendencies and preferences toward more institutional jointness over time seems unlikely. The theory of adaptively rational systems holds that if a decision rule used in the past has led to a preferred state for the organization, that decision rule is more likely to be used in the future (Cyert and March 1963 [1992], 117–118). Viewed from this perspective, the Air Force's very existence is the result of prevailing in an institutional argument that air power should be distinct from land power and controlled by individuals not beholden to the will of a ground commander. The organization therefore reinforces decision rules that promote independence. Similarly, the extreme dominance by the Air Force of the DoD budget during the dawning nuclear age witnessed the institution reach a state that was favorable by most measures of bureaucratic success. Because this preferred state occurred in the context of debates about strategic air power, from this theoretical lens one would expect the Air Force to continue to emphasize its contribution to strategic national security aims over those of a tactical nature. All of the other services have similarly deep-seated views that have been reinforced over even longer periods of time.

All Joint Initiatives and Successes Will Fade Away over Time

Though each of the case-study examples in this work demonstrated some degree of interservice cooperative behavior, the ending demonstrated or suggested that the cooperation was short-lived. The same dynamics that make cooperation difficult in the first place remain fixed, eroding the engineered institutional jointness quickly or over a long time. Two illustrative examples come from the aftermath of AirLand Battle. With respect to the theory-of-victory and command-and-control facets of jointness, the doctrine envisioned a big enough challenge to US combat capability that it accommodated an acceptable interservice compromise in both areas. When these concepts were employed on battlefields in Kuwait and Iraq, in a conflict much smaller than the original concept envisioned, the temptation to negate the compromises proved overwhelming. In the aftermath, Deptula's (n.d.) memo characterizing AirLand Battle as doctrinal near-heresy and Scales's (1997) unilateral history of the Army's victory demonstrate how both services snapped back to their preexisting parochial interests.

AirLand Battle also provided an example from the fiscal realm that suggests jointness is the first victim of peacetime budget constraints. The halcyon days of generous Reagan Administration defense budgets during the AirLand Battled years allowed regular liaison of Army and Air Force officers to each other's budget shops. The days of routine Army–Air Force coordination are now a distant memory from the 1980s, though. According to one action officer in what is known informally in the Pentagon as the Air Force's "Engine Room," the planning organization in charge of making service budget recommendations, there is very little joint interaction going on at the action-officer level and no formal exchange of liaisons (Babcock 2014). The reasons for the demise of each joint endeavor will differ in details, but all will relate to service preferences in general. In short, the characteristics that define and portend bureaucratic inertia in other organizations assuredly apply to the military establishment, and they will be overcome only with exceptional circumstances or efforts.

External Influences Are Best Equipped to Drive Cooperative Joint Behavior

Put simply, it seems that fear of losing on the battlefield and a shaming before Congress are the only two things that can coerce the military services to cooperate with one another. Theoretical interpretations rooted in a structural-realist approach to international relations underpin an explanation for this trend.[12] The preponderance of observed forces affecting institutional jointness in this study's cases were external. All observed instances of successful joint cooperation required the attention of organizations existing at levels or hierarchy above the military services. The suggestion is that good intentions alone will not yield jointness; deliberate partnership with higher echelons is a necessity.[13]

Wartime crises drive military organizations to pursue helping behaviors over time, but the cooperation attained is less than ideal. This is an unsurprising, intuitive conclusion, backed by Svedin's empirical research showing that "[a] majority of the interactions between organizations in crises, both in decision-making situations and over the course of a crisis, are conflictual. That is, organizations interacting in crises tend to disagree, fight, and engage in other competitive behavior to a greater extent than they agree, are honest toward each other, or help each other" (2009, 125). While such fighting paradoxically does build trust—probably via a mechanism of repeated engagement—among the organizations, the quality and endurance of joint initiatives suffer, both because the urgency of combat does not allow a great deal of intellectual rigor and because the conditions under which jointness is pursued are exhausting and subject to even greater ongoing change than are routine interservice interactions.

Air Support to Ground Forces Is an Area of Perennial Weakness; Therefore, It Merits Recurring External Attention and Remediation

A collateral conclusion comes from this study's observation of the implementation and demise of AirLand Battle (chapter 4) as well as background research for one of the key observations of the Iraq-Afghanistan close-support case (chapter 5). On the basis of a long history of interservice angst in peacetime and false starts in combat, it is apparent that Army-Air Force cooperation on CAS, ISR, and other disciplines required for air-ground integration are an area of chronic concern.

Absent the involvement of a third-party external forcing function (i.e., congressional attention or service "decapitation" by the Secretary of Defense) or the threat of failure in combat, the two services have a history of letting the command-and-control structures, resources, and training for passably effective mutual support languish when combat is absent and then fighting about them when armed conflict inevitably reappears. An interested secretary of defense or congressional caucus may want to take note of this and provide the master plan that the services have managed to eschew while left to their own devices on the matter. Repeated combat successes that follow initial failures suggest that the services know instinctively what helpful jointness looks like, yet the services refuse to pursue it institutionally absent the threat of exposure to outside scrutiny. The issue thus merits continual peacetime focus, not solely following poor combat performances. Unless someone holds the services' collective feet to the fire on collaboration, they quickly and comprehensively part ways with respect to air support.

Strong Leadership Is Essential to Successful Joint Initiatives

The joint initiatives examined in this study proved to be, in part, efforts to construct stable systems from a wide variety of components. It helps to have strong leaders to act as heterogeneous engineers in pursuing institutional jointness. Theoretical constructs like public-goods theory, organization theory, theory of professions, and some elements of military-innovation theory provide an understanding of why it is difficult to attain joint cooperation and lead to the first conclusion of this chapter. Consideration of this myriad of forces, as well as the wonder that anyone might be able to tame them in creating "self-sustaining networks that are . . . able to resist dissociation," suggests a reason why a singular leader or series of leaders appear in all of the successful case-study examples of joint cooperation (Law 1989, 114).[14] The strong leader with a commensurately forceful personality must initiate movement toward partnership. This was evident in the leadership of General Starry with respect to AirLand Battle (chapter 3), it was present in a series of leaders in the Air Force with respect to the JPATS (chapter 4), and it flickered in and out of view in the case of OEF-OIF air-ground cooperation (chapter 5). Strong leadership has up to this point been largely absent from the development of JADC2, allowing service branches to fall back on primarily developing system solutions within their individual branches, despite the joint purpose and intent of JADC2. (It is curious, but probably not surprising, that four-star leaders in the services have limited their leadership on the topic to summit-type discussions rather than directive memos that commit their names and reputations to the effort.)

The case studies suggest that leaders who drive jointness must be able to conduct the following activities, often simultaneously: (1) articulate a common vision or strategy; (2) engage multiple levels of the defense bureaucracy and find shared interests among them; (3) build coalitions that adopt and advance the central idea; (4) remain in office sufficiently long to realize the attainment of a goal (or impart the vision to a successor); (5) engage in argument and debate to spur dialogue without isolating potential partners and supporters; and (6) do all of the above with sufficient humility and willingness to compromise that personal and service preferences do not preclude cooperation. Individuals able to do all of these things must also be competent and well respected in their fields; they need to accumulate the political capital they will expend during the process.

Finding an individual with all those characteristics appears daunting, yet they did appear in the case studies. In the chapter the AirLand Battle (chapter 3), General DePuy saw a need to change the Army dramatically after its experience in Vietnam; he was also able to pass on his vision to General Starry, who adopted much of it as his own. Generals Meyer, Gabriel, and Wickham, perhaps motivated by individual friendships, yielded their personal as well as service interests to seek compromise and continued cooperation. To the requirement for humility, add bravery. A leader pushing for jointness cannot be deterred by what appear to be mistakes. DePuy took as much internal Army criticism as one might expect in his TRADOC position, but his bold moves opened channels that kept doctrinal development and interservice dialogue going.

Using the framework from chapter 7 as a point of reference, the leadership component of jointness is one that underlies the other three. Without leadership, there is no causal influence, no one available to identify joint potential, and no agent who acts to make multilevel interests coalesce. The role of leaders as indispensable integrators appears over and again in the case-study histories. The singular strength of this component of jointness helps explain why some personalities can be effective even when they do not hold offices at the top of the bureaucratic structure.[15] It also partially explains the transience of joint cooperation—the leaders who assemble the needed components invariably move on to new challenges in different areas of the establishment.

Joint Initiatives Succeed When They Lead to Coalescence of Interests across Levels of the Defense Hierarchy

Successful joint endeavors require a chain of coalescence involving all three levels of analysis: within services, groups coordinating among services, and external stakeholder relations. There is no way of determining which level is most important. Attaining multilevel agreement is analogous to forging both sides of a coin or clapping with two hands; a failure to involve one level portends failure.

The clarity of AirLand Battle's how-to-fight manuals and the tactics derived from the idea unified both the Army and Air Force from the level of the soldier or airman up to the largest fighting units. The concept provided a way for the Joint Staff and DoD to articulate the nation's premier existential military threat, from which it could then define ends, ways, and means to ameliorate that threat. Among external stakeholders, the concept captured the attention and admiration of a sufficient number of members of the executive and legislative branches that it became an accepted reality of the foreign and military-policy environment, even though a Soviet invasion of Europe was not a foregone conclusion. Indeed, nothing like what AirLand Battle envisioned as its central organizing concept even occurred. However, the need for its attendant military acquisition made it interesting to Congress, and the coherent stick it offered in the context of foreign policy made it useful to the executive branch, allowing alliances to form that sustained AirLand Battle through its tenure.

In the instance of the JPATS trainer, the instructor pilots who were training future military aviators had requirements that could be articulated clearly and met with a common platform, but this in itself would not have driven cooperation had that community not been able to find common interests in Congress. Congressional stakeholders were looking for ideas to streamline defense acquisition programs and were at the time particularly enamored with anything that could be labeled "joint." COIN air power (chapter 5) offers the examples of tactical operators who saw the need for improved CAS and ISR capability in the Afghan and Iraqi wars. The sense of urgency they felt carried hard-learned lessons throughout the joint force. Joint commanders, at least during some parts of the conflict, felt spurred to cooperate lest they fail those they led and the expectations of a judging public. The external influence in this case was spotty and inconsistent (albeit *very* forceful at one critical juncture), though, as was the joint cooperation.

The idea that multilevel engagement is a necessary condition for success in a cooperative effort is not unique to the military. In analyzing a factor that came up repeatedly while we researched the case studies, it is clear that a successful effort to force interest groups to coalesce was part of the recipe for gaining passage of the Goldwater-Nichols Act. Anne Marie Getz wrote that a senior staffer, Jim Locher, realized via business theory that "any attempt at reform would have to address the entire organization, no piecemeal approach could be effective" (1998, 143).[16]

The Idea of Jointness Retains Value throughout the Defense Establishment

The same structural realities that require multilevel buy-in to ensure the success of joint cooperation also mean that opportunities exist for a joint advocate, operating at any level, to see an initiative through. Since jointness has cachet everywhere, anyone can use it to an advantage. During times of armed conflict, the effect is magnified, since the visible presence of crisis tends to elicit expectations of cooperation from even the general public, which is otherwise indifferent to military cooperation.[17] From the case studies, the advocacy of low-ranking tacticians who wanted to see jointness increased with technological tools like the Rover video feed presents itself as a prime example. Pragmatism is the close companion of opportunism, though, so the possibility for abuse of the term exists, as evidenced by the hijacking of joint wartime acquisition processes by service interests.

Fighting Is Helpful in the Pursuit of Joint Cooperation

Organizational crisis cooperation literature provides a helpful, nonintuitive approach for the pursuit of jointness. Trust is a precondition for jointness. David Johnson (2014) said, "I don't think we [the US military services] are joint, personally, and it's not because of redundancy. It's because of an absence of trust." While paradoxical at first glance, fighting of a certain kind can aid in building this type of trust. A sense of urgency about a common problem, particularly of the kind inspired by wartime stresses, may best provide this impetus. Although the post-Anaconda dialogue between the Army and Air Force began with incendiary intent, it eventually brought two service chiefs together, and left one willing to go on the record that his counterpart, regardless of the other clamoring voices in both services, "was someone we could work with" (Jumper 2013).[18] Increasing uncertainty about the definition of or effective responses to a crisis increases the likelihood of organizational cooperation, another apparent paradox that explains why both real war and interorganizational conflict, if leveraged with the right intentions, can lead to better cooperation over time.[19] If nothing else, they get organizations talking with one another, which is a necessary precursor to any type of helping behavior.

The Defense Establishment Has a Muddled Conception of Jointness

Birch conducted an internal survey of selected military leaders and a broad sample of people involved in joint military ventures (2014) in 2013–2014.[20] Several trends from the information the survey provided stand out; the passionately held and competing opinions about the essence of jointness discussed earlier reinforce the diversity of meaning mentioned in chapter 1. There are likewise some noticeable differences in what these defense insiders thought of the levels of joint cooperation, whether they were about right or too low. Few thought extant levels to be worryingly high.[21] Moreover, the

ultimate source of jointness is open to wide interpretation and strong debate
—implied wishes for enlightened monism contend with sober expectations
of ongoing competition and argument.

The most pessimistic descriptions—which were expressed in jarringly
similar language—proposed that jointness, such as it exists in the current
national defense establishment, works this way: (1) self-interested advocates
for materiel, weapons systems, and programs (which are merely incremental
improvements of previous versions) drive military acquisitions; (2) military
services decide the tactics, techniques, and procedures they will use to employ these acquired entities; (3) the services determine those areas that are
most critical for interoperability and make partial, after-the-fact efforts to
integrate the entities that they take into areas of conflict; and (4) the services
devise concepts of operations and strategy that show how all of the previous
steps can defeat an anticipated enemy's aggression.

Bleak as these cart-before-the-horse assessments are, opinions of those
involved in the defense establishment are relevant to any discussion of jointness, because these are the people who will or will not make the outsized
effort it seems to require. Their belief in or cynicism toward the concept is a
key predictor of future success in joint endeavors.

Do the differing opinions threaten military effectiveness? Locher, acting
in his role to help pass the Goldwater-Nichols Act, determined prior to its
passage that the essentially separate status of the services within DoD's organizational structure causes its difficulties in pursuing jointness to mount—
he believed that MacArthur's exhortation to "Duty, Honor, Country" he
learned at West Point had given way to "Turf, Power, Service" in Pentagon
politics (2002, 10).

Asserting that DoD's size and challenges are so great as to keep it from
making an effective internal response, he argued for making the change that
the US government has evaded or avoided since a need for better interservice
coordination came to light in 1898. Locher advocated for establishing standing joint task-force headquarters in each regional unified command to imbue
the organization with budgetary authority to buy systems unique to joint operations and to unify the defense support agencies under a single executive.[22]

Such moves would constitute a transition to a new degree of centralization, one that fundamentally gives the joint and umbrella defense organizations power on par with, or exceeding, that of the individual services (Locher
2001, 112–113). Locher recognized that such a change would not likely originate within DoD, and so called on external actors to help it "find that balance
between loyalty to service and devotion to the larger needs of the nation"
(2001, 113). In contrast to Locher, the current ambiguity over essence, level,
and sources of joint cooperation may still have advantages to the black-and-
white certainty of a centrally planned approach. The inflexibility and blind

spots associated with other types of government-led planning may be un-avoidable in a stronger centralized joint task-force structure.

William McNeill, in his magisterial summary of military innovation over the most recent millennium, wrote, "In a given time and place, where alter-nate social structures are in competition, conscious choice and emotional conviction can make the difference in determining which pattern will pre-vail" (1982, 21). He contrasted the market-driven military innovations that led "private and small-group initiatives and self-interest to play a quite ex-ceptional, transitional role in day-to-day behavior" with those enabled by command economies and the "quick halt to breakneck technical change" (1982, 385–386) they wrought.[23] Congressional interest is essential in pursu-ing jointness, but the form matters. Legislative oversight ought to strengthen the invisible hand that promotes continual competition for doctrinal and ma-teriel advances, not merely provide epochal, one-off nudges toward greater bureaucratic centralization. The latter approach drives reliance on assump-tions about defense needs formulated the way Marxism plans a national economy; the likely result is stultifying effects on military capability of the type that have accrued to centrally planned economies.

From the Past the Future: The Prospects for JADC2

Drawing on the preliminary conclusions about jointness just outlined, we look to the future of JADC2 and future of successful joint collaboration. We have concluded that useful jointness is that which the services internally are likely to sustain over time and which improves command-and-control, interoperability, and efficiency. This kind of useful jointness is the hardest to attain but is precisely what JADC2 should embody. The record to date also suggests that JADC2 in its current, separate-and-not-equal configurations has a long way to go to do or be any of these things. However, while the goal of JADC2 should be improved command and control, better interoperability, and increased efficiency, it is not necessarily the case that the way to get there has to be a joint process. Nor do the cases we have examined here suggest that the parties collaborating need to have a common understanding of what it is they are creating. In fact, our preliminary conclusions suggest that interests among the parties that create the public good do not need to be perfectly aligned, and the collaboration does not have to be automatic or a given at the start. Plenty of case evidence suggests that JADC2 has a good chance of utilizing joint collaboration as a process for creating whatever JADC2 ulti-mately turns out to be, and that the useful jointness can emerge even though why and why the parties finally get together may be less than smooth.

Two preliminary conclusions about useful joint collaboration provide spe-cific cautions that are worth highlighting at this early stage of the JADC2's development phase. The first is that, while at times characterized as or envisioned as an 'Internet of all things' for battle, even successful institutional jointness does not provide one best way to win all wars or a doctrine that will cover all services' needs in terms doctrine for fighting wars. With the troubles the services have exhibited in defining what JADC2 is, the excitement among all services about what it *could* be, and the sheer scope that the concept has covered so far, even successful joint ventures beg some caution.

The second caution to heed at the developmental stage is that institutional jointness is not the natural way of doing things among the armed services. The Pentagon's coordinating joint groups (Joint Chiefs, Joint staff, and combatant commanders), try as they might, will need to be much more effective at pursuing the "J" in JADC2 if jointness is to actually follow. The historical record suggests that this organizational level will not be able to create joint collaboration that the services will not resist. The best it can do is to align the interests of the services, even subgroups within the services, to build grassroots support for and advocacy for JADC2 as a function and then joint collaboration as the way to make it work. This could happen by displaying and applying JADC2 capabilities in exercise or in a smaller field test to give the services a sense of what tactical capability they tangibly could gain. However, to do any of these things you need budgetary priority, the preponderance of which still lies with the services themselves and seems marred by bureaucratic politics rather than actual procurement of tangible products or applications.

One of the conclusions from the past is that institutional joint collaboration is unlikely to emerge and lead to results unless championed by strong and persistent leaders. It is still unclear if the current leadership in the defense establishment, such as Chief of the Air Force Brown or current Secretary of Defense retired General Lloyd Austin, will make JADC2 a primary issue. Our conclusion based on the other cases in this study suggests that external influences are best equipped to drive joint cooperative behavior among the services. This suggests that insightful leaders as already mentioned, actors like Congress (currently too poorly informed to push the issue), or a change in the security environment (such as the emergence of a new conflict or crisis) are going to be needed to drive JADC2 to be a truly joint endeavor.

Even with the kind of external pressure that we just talked about, history suggests that joint collaboration will not stand the test of time. The successful joint endeavors and collaborations we have examined, even those where the services agreed that direct service benefits were created, have diminished over time or were abandoned (in favor of service-specific ways of doing things) when the external pressure and attention decreased. If JADC2 were to gain a truly joint collaborative form, it is unlikely that the collaboration would have longer staying power than the direct supervision and accountability that an external stakeholder is ready to apply. Based on the conclusion that the enabling functions, like TACP in the COIN case study, are a perennial weakness in joint collaborations, we can predict that the components currently being explored by the services separately may not hang together for very long if budgets are under threat or there is not an imminent need in a battlefield for JADC2. The glue meant to keep the component parts of the network and the capability together are likely the first to go.

At a time when the US security environment is shifting and the economic constraints on the services are fluctuating, there is likely going to be infighting for resources and the lead on whatever command-and-control system is developed. This may not be altogether bad news, however, as fighting in discrete instances is a predictor of more cooperative strategies across the span of an evolving crisis. Increased pressure may also help the defense establishment clarify the muddled conception it currently has of JADC2 and the concept of jointness.

Conclusion: Lamentations and Optimism for Jointness

The Record: A Collaborator's Lament

The motivation for this study came from a desire to understand the conditions that lead to successful joint cooperation. Realizing that a definitive theory of joint cooperation cannot be built on the small-n sample of cases examined here, we have identified cases that suggest that joint cooperation indeed occurs—sometimes against expectations—and that suggest some of the common features that make a routine appearance when jointness happens. These common features may be helpful, both theoretically and practically, to those who study jointness and who attempt to put it to work for national defense.

The most striking unexpected finding in our study, however, was the visible, repeated demise of jointness once attained. In each of the three historical cases studied, a hard-fought and deliberate struggle to build an interservice something resulted in better cooperation, closer relationships, programmatic efficiency, or superior results on the battlefield. In its frequent praise of jointness, these are the positive outcomes that the defense establishment touts. Many members of the establishment, including almost every person interviewed for this study, readily concede that jointness does in fact have positive outcomes that make a significant contribution to improved national security.

With dismay, however, we note that all of the examples of joint cooperation in this study either ended or exhibited diminishing cooperation. In the case of AirLand Battle, the era of jointness ended rather abruptly, with external factors changing the nature of defense debates in such a way that rendered the existing joint construct impotent. With regard to the JPATS trainer and joint aviation schools, the fast-moving model of jointness that had overcome so much internal friction early on bled out slowly as the faces changed in the offices that had originally enabled such a program. A thousand parochial cuts removed the heart of jointness, replacing it with independent single-service programs.

OEF and OIF are still too contemporary to be evaluated from a fixed historical perspective, and too many of the documents that will provide a clear picture of how battlefield jointness rose and fell remain classified. The sense we have based on the study, though, is that a fear of failure and pressure to demonstrate cooperative competence quickly led to vast improvements in CAS and ISR coordination under the watch of one set of leaders. A pendulum swung during the tenures of another set, though, with enduring improvement unlikely to result. Rather than pursue jointness for the sake of battlefield efficiencies, those running the wars after they had become routine allowed pursuit of parochial doctrinal preferences to not only mar the finish of jointness polished by war, but also threaten to dismantle its underlying structure. An injection of new leadership—one equipped to see the value of jointness—may have saved some of the progress at the end, but only time will reveal whether the Army and Air Force drift apart in the area of air-ground support as they have after every preceding major conflict. If the jointness achieved under the threat of combat unraveled to a large degree, it will, unfortunately, have matched a trend consistent with other observations in this study.

Why this should be troubling merits some discussion. There are a few good reasons to guard against unbridled jointness, not least because of the competition-quashing characteristics attendant with its most negative manifestations. The other argument, one that ties the unifying views of jointness to strategic monism and its eggs-in-one-basket risk is another strong point in any antijointness debate. This study proposes, however, that debate-stifling universal doctrine is the weakest, least useful manifestation of jointness, and that it rarely (if at all) fosters the effectiveness or efficiency toward which it strives. Further, a unifying vision that inspires jointness need not adopt a head-in-the-sand perspective blind to the range of likely threats—there is room for multiple voices in formulating the vision, even if it is likely to be sold by a singularly strong proponent. Once these suspicions about jointness are adequately addressed, many would agree that jointness could and does enable effectiveness on the battlefield and sometimes even efficiency in the halls of the Pentagon. If jointness has uncontested benefits, the question of why it seems always to fall apart after a season of success should be an area of prime concern to practitioners in the defense establishment.

A definitive reason for why successful joint collaboration falls apart is an area for further study. The case studies covered here do not allow for sweeping claims, but they do reveal three threats that the cases had in common. The first is that the pursuit of jointness in every case required leaders to expend a certain amount of energy to divert the normal streams of bureaucratic routine. The motivation for such efforts may have been more or less obviously internal or external, but one or more interested actors did something—going beyond what their military duties called for on the surface—to pursue jointness. Because military service is particularly transient—people hold individual offices for only a few years at most, and often for very short periods of time as they advance in rank—the likelihood that bureaucratic friction will defang a joint effort is high unless it has built up so much momentum early on as to be unstoppable.

The second reality that confronts jointness in this study is the changing nature of the external forces that drive it in the first place. Structural norms of the world shift, both abruptly and gradually, with no predictable pattern. The threats of combat, often the most effective drivers of jointness, over time lose their ability to drive the services together. Even if the war is not over, the battles may become so customary that services feel comfortable enough to slip into advocacy for their parochial preferences rather than compromise in pursuit of joint solutions.

Likewise, if the inspiration for institutional jointness is a domestic-external factor—Congress's focused interest, say, as in the case of the JPATS trainer or the national security establishment's particular concern about the Soviet threat, which set the conditions for AirLand Battle's success—these factors also diminish over time. The Iron Curtain came down. The Gulf War opened up new debates about the primacy of one service over another. Congress always moves on to the next hot political topic. If the world exhibits a structure that shapes major debates about the best way to pursue security, that structure and the debates about it invariably change over time. Where jointness benefits from these debates in one era, it can become a victim in the next.

Finally, the last common threat to jointness that emerged from this study is one of human pride. There is a spectrum to this pride and the velocity with which it unravels jointness. Pride sometimes appears as parochialism, as when the Marine Corps in the Gulf War refused to adopt certain language about the air component commander and referred to him in ways that emphasized their independence. It can be more blatant, as in the Air Force's arrogance following the Gulf War and the Balkan campaigns that effectively shelved a high degree of joint thinking bundled up in AirLand Battle. Pride can even be manifested in ways so innocuous that the services might justifiably refer to them as matters of esprit de corps, such as giving the JPATS trainer aircraft different paint schemes for Navy and Air Force versions, or building modern facilities for Air Force training at Pensacola. After all, these are merely ways to instill service identity and "take care of our people," to use a popular phrase in a commander's vocabulary. Yet items like these seem to have become the first of many small cuts that bled out the jointness in a major initiative that for a time exposed the services' young aviation candidates to genuine jointness, a pursuit that may have paid dividends in later years.

"Hubris" becomes a more appropriate label for this pride when it manifests itself in open interservice scorn among military figures, as seems apparent during some eras of OEF and OIF. The setbacks of blatant personality conflicts and the bickering they induce can be overcome by cooler heads and rational thought in their wake, but the wrenches they throw into command-and-control schemes are perhaps dwarfed by the threat of future interservice mistrust and a tendency to snap back to comfortable, parochial preferences. Once bitten, flag officers are twice shy about losing service prestige and autonomy in the arena of voluntary jointness. For the good of national defense, joint leaders should set aside pride and personal attacks while they make the best case for the theory of victory in which they believe. Because men are not "angels," however, it will require engaged leadership from an informed and involved external source to rescue the services from the morass of pretend jointness in which they currently exist.[24]

The Way Forward: The Case for Optimism

This investigation has resulted in a clear-eyed picture that the favorable pictures of institutional jointness promulgated by joint publications are mirages. However, the study has also showed that there are mechanisms by which interested parties at any level of the defense establishment can foster helpful joint cooperation.

Research on civil-military relations over the past decades has generated a skepticism of individual service's intent, has suggested that senior military leaders ought not be trusted, and has frequently advocated for more civilian control of the military. These sentiments have been reflected in legislation and policy changes, but in practice this has led to more centralization and niggling interference that increases neither efficiency nor effectiveness. Andrew Bacevich is one of few scholars who have called for "untying the hands of senior commanders" (2008, 137) and we think it may be necessary for successful joint collaboration. The scope of American military options is simply too broad to be grasped by itinerant outsiders. If the nation is to have good military advice, it will come from bright, motivated, and empowered joint military leaders. Unfortunately, these leaders will not be able to emerge from the bureaucratic structures of the Pentagon without outside assistance. It will take continual prodding and encouragement from external actors, in cither the executive or the legislative branches, to see where good ideas lie and to ensure they receive proper airing.

There is no compelling, a priori reason to think that this approach to building jointness cannot work, unless one's pessimism about the obstacles is so severe as to cause complete loss of hope for military success. Understanding the correct tension here is paramount. Congress need not micromanage defense policy, and the executive does not have to return to picking targets during lunch.[25] Nor does this study advocate turning over affairs of state to military leaders or giving them leeway to formulate military strategy unchecked by civilian review. There is an imperative for oversight, because it alone can ensure that the best military advice gets a fair hearing. The objective civilian control of the military that exists today we argue would be enhanced, not supplanted, by a competitive process.

There are imperatives at other levels of the defense establishment as well, faced by service chiefs, joint force commanders, and ground-level service members alike. It is the responsibility of truly joint-minded military members to reach out to external organizations, to advocate for better efficiency, and to expose internal service obstacles to cooperation. The idea is not new. Huntington (1957), long the chief advocate of objective control, recognized decades ago the value of strategic pluralism. He described in detail how the American separation of powers leads to an airing of many voices that collectively provide solutions to security problems, but as his illustrative example shows, this debate does not happen without encouragement from the services internally and without protection from smothering interservice politics.[26]

Even an ideal balance of trust, communication, enlightened military strategy, and an responsive industrial sector does not guarantee military-strategic success—jointness is no panacea. "Friction . . . makes the apparently easy so difficult," Clausewitz (1984, 121) reminds; "there is not and cannot be any science of war, and . . . therefore there can be no such thing as a military genius," Tolstoy (2009 [1869], 1243) intones. The humility and sense of striving these sober realities should engender in good civilian and military leadership are cause for optimism, though. Only after realizing the monumental task implicit in crafting national strategy can those charged with its formation assign it the attention and gravity it deserves.

> *I am tempted to declare dogmatically that whatever doctrine the Armed Forces are working on now, they have got it wrong. I am also tempted to declare that it does not matter that they have got it wrong. What does matter is their capacity to get it right quickly when the moment arrives.*
> Sir Michael Howard
> "Military Science in the Age of Peace," 1974

Appendix: Research Method and Case-Study Selection Criteria

Our study investigates cooperation among separate US military services. The motivating desire is to draw predictive conclusions about conditions that lead to convergent, cooperative, and helping behaviors when the services have opportunities to interact to solve emerging security problems. The nature of the question is complex, which makes it unlikely that a single-variable relationship will show itself to be the significant explanatory factor. For this reason, the research method applied in this study should be capable of identifying the interaction of several relevant independent-variable case-study research efforts. These in turn should lead to clusters of contributing independent variables; thus the approach is an appropriate choice of empirical test in this investigation and perhaps the best chance for success given the complexity of the topic (George and Bennett 2005, 26).[1]

Case-study research can help identify multiple and complex combinations of pertinent variables that lead to observed outcomes, but it has inherent limitations. The applicability of the theories it generates may not have broad applicability to all other cases. While able to show the sufficiency of certain variables for a given outcome, case-study research is often unable to prove necessity with certainty beyond a very narrowly defined range of cases. Finally, case-study research is ill equipped to make definite determinations about the relative contribution of a variable to an observed outcome.

Though a certain dependent variable may appear in all cases, small-n samples prove neither that variable's necessity nor the amount of dependence the variable of interest might have on the outcome. Therefore, case-study selection should attempt to strike a balance between the trade-offs this research method imposes and the complex and in-depth understanding of individual cases it makes possible (George and Bennett 2005, 27; Hall 2003).

Only two major works come close to offering theories of jointness. Kenneth Allard made a masterful summary of the command-and-control difficulties inherent in modern joint operations, concluding that a *"baseline of interoperability"* (1996, 257) is a prerequisite for successful joint operations. One of the most relevant insights from this work borrows Wylie's assessment that "there is as yet no accepted and recognized general theory of strategy"

(1967, 67). To Allard, "the absence of a more general strategic paradigm also helps explain why *military organization* has been such a persistent problem in the postwar world" (1996, 262, emphasis in original). Allard deals with essential variables like service culture, joint doctrine, external levels of military control, and the role of defense acquisitions in enabling jointness. Though he did not call it a theory of jointness per se, his summary that "*wise technological choices and tough organizational decisions*" (Allard 1996, 271–272) enable effective command and control is a well-reasoned argument upon which this work builds. Because many of Allard's recommendations seem couched toward the defense acquisition process, including his very useful case-study analysis of data-link networks, a pretheory of jointness requires us to go further.

The second work that merits mention on this short list is Huntington's *The Common Defense*. As with Allard's later work, Huntington's focus was not a discussion of jointness for its own sake, though he did describe interservice rivalry and the effect it has on increasing external civilian control of the military (Snyder 1962, 372). In addition, he named other factors that bear on this study, including the relationship among the hierarchical levels of the defense establishment, the importance of defense acquisitions, and pressure toward unanimity among the Joint Chiefs of Staff. Huntington concluded that Congress had lost its power over the military "not to the President but to the executive branch" (1961, 127), an observation with which this study concurs. Huntington's foundational work, if dated because it was published twenty-five years prior to the Goldwater-Nichols Act, remains relevant even though the structure of US defense institutions has changed. One of the distinctions of this work is its level of analysis: Huntington was primarily concerned with the balance between high-level domestic politics and international relations; his concern for jointness is a by-product of that discussion.

The remaining relevant literature does not advance theories of jointness, instead broadly dealing with American civil-military relations; doctrinal and weapons system innovation; independent service cultures' effect on policy development and implementation; and the prosecution of battles, campaigns, and wars (e.g., Bijker, Carlson, and Pinch 2000; Builder 1989; Huntington 1957; Janowitz 1960; Mahnken 2008; MacKenzie 1990; Millett and Maslowski 1984; Posen 1984; Rosen 1991; Weigley 1973). In order to fill the void of theory on interservice cooperation, this study considered and examined hypothesis-generating candidate cases. This characteristic validates—in the face of the common advice to avoid this practice—the selection of cases based on the dependent variable: cooperative behaviors between military services dealing with emergent security problems.[2] Throughout the study, the term "cooperative behavior" refers to actions taken by individual services that further the ends of one or more other military services while meeting a shared security challenge—such interaction need not necessarily be marked by amicable relations.[3]

Another clarification of the dependent variable (i.e., cooperative behavior between military services dealing with emerging problems) is in order as it has direct bearing on our case selection. Cooperation mandated at the tactical level by one commander who can in a very direct way force service components, troops, and effects to jointly carry out a specific task or achieve a operational objective live in a theatre of war or security concern is a very specific, limited form of jointness. The combatant commanders have considerable control over the forces in the field that are actively carrying out the mission. This type of cooperation can be characterized as "operational jointness," exemplified best by the employment of deployed forces from different services together in war or other operations. While the precise command relationships between the service component commands and the joint force command bodies can vary widely, it is generally speaking a clearer and more forceful direct command relationship. This type of "operational jointness" is not the main focus of this study, even though we acknowledge its significance and that operational jointness is connected to "institutional jointness," which *is* the primary focus of this study. Institutional jointness takes place at the meso-organizational level and is concerned with the building of forces and the capacity to deliver joint effects. In institutional jointness, the units of analysis are the military departments and services or their major force development components. This type of joint cooperation is characterized by a high degree of political input and considerations and longer time horizons than an active deployed mission, and usually carries implications for services' ability to act autonomously in the future. Achieving jointness in this realm is quite challenging but also critical to strategic decisions for the direction in which the US military is developing and its warfighting capability in the present and future. The type of interservice cooperation and infighting that emerge around issues that are suggested for "institutional jointness" is what this study will focus its light on.

Returning to the nature of what we seek to study, we argue that interservice cooperation occurs in an atmosphere characterized by an admixture of "conflicting and complementary interests" (Axelrod and Keohane 1985, 226–227).[4] Therefore, the cooperative behaviors described span a spectrum from what would be called "cooperative" by most casual observers—i.e., noncompetitive, helping, and even sometimes selfless—to competitive or even conflictual in nature (Svedin 2009, 25). In taking this approach, the study considers a wide range of behaviors, and allows for the possibility that cooperation is possible with or without explicit agreement between parties— it may be tacit, negotiated, or imposed from without.[5]

Drawing from Stephen van Evera's and George and Bennett's

recommendations for case-study selection criteria, this study selects a set of historical cases and researches them with the objective of drawing conclusions from past military cooperation opportunities. As such, eligible cases should exhibit (individually or as a collection) a set of six characteristics:[6]

1. provide clear examples of cooperative service behavior;[7]
2. sample all of the US military services;
3. involve all organizational relations levels (within services, among services, and external to the services) that bear on service behavior;
4. describe issues that exhibited considerable bearing on the services' institutional attention and resources;
5. analyze instances of both "most likely" (cooperation where cultures and past history are conducive to cooperation) and "most difficult" (cooperation among services who are historically and culturally prone to fighting) instances of cooperation; and
6. occur in the contemporary US national military decision-making establishment.[8]

Why did we choose this set? The first requirement ensures that each case captures the dependent variable of interest. This specification assumes that the services' default behavior on nontrivial matters is often divergent—they do not cooperate by default unless the threat of failure looms large.[9] Drawing on the documented history of the armed services and assertions in public organization theory, our book assumes that true interservice cooperation is rare; studying instances where it has occurred (as opposed to the plethora of examples where it did not happen) is more likely to reveal useful conclusions about causality, i.e., what supports and what hinders interservice cooperation.[10]

The second requirement assumes that a service's organizational culture is likely an important contributing variable to the cooperative tendencies of the organization.[11] It also ensures selection of cases that subscribe to van Evera's recommendation to choose "cases about which competing theories make opposite predictions" (1997, 83). The willful independence of the Navy toward civilian authorities and other services is well documented with historical evidence, which suggests a lesser proclivity for cooperation than other services (e.g., Barlow 1995; Donnithorne 2013, 381–382; Ehrhard 2000).[12] Using cases from all services where cooperation is in evidence would guarantee at least one outlier case, another desirable characteristic for hypothesis-generating investigations (van Evera 1997, 86).

Common purpose relates the third and fourth criteria; they drive selection of cases of sufficient scope. The interactions of service subcultures, separate services, and the organizations that oversee them (chiefly the Department of

Defense and Congress) have influence on cooperative behavior. However, for all levels of the bureaucracy to become involved, the security problem in question must involve a sufficient degree of force allocation, budgetary share, and service-cultural interest to overcome normal multilevel bureaucratic inertia.[13] The interaction of an Army battalion and Air Force squadron, for example, are not predictors of interservice cooperation unless they reflect larger organizational dynamics. The two criteria in concert—because the time required for an issue to interest multiple bureaucratic levels is often long—also maximize chances that cases selected will exhibit the "large within-case variance in the value on the independent variable, dependent variable or condition variable across time or space" that van Evera (1997, 82) recommends.

The fifth criterion makes it likely that cases will exhibit "extreme values of the independent variable, dependent variable or condition variables" (van Evera 1997, 83). The same criterion also makes it likely that cases will be comparable via the method of differences, exhibiting similar organizational characteristics and differing values of observed variables (van Evera 1997, 84).

The sixth criterion aims to address three separate desirable characteristics of hypothesis-generating case studies. The first is a desire for a sufficient level of documentation that permits thorough research. Specific records and evidence of interservice cooperation and divergence are more accessible in modern archival records. Another characteristic the criterion drives is similarity to contemporary problems. Working with cases in the modern military era also gives them resemblance to current policy issues. Finally, this criterion guarantees that candidate cases will have a measure of shared characteristics While less important than resemblance to current policy issues for hypothesis-generating work, the criterion also makes any hypotheses generated more likely to be widely applicable (van Evera 1997, 84).

Because we are trying to establish a pretheory of a complex dependent variable with a small-n case-study sample, we have adopted the case-study method of process tracing described by George and others (George and Bennett 2005, chapter 10). Specifically, the chapters will reflect *detailed narrative* and *general explanation* to identify the pertinent pretheoretical elements of jointness (George and Bennet 2005, 210–212). The work anticipates varying audience-specific definitions of jointness along with the multitude of applicable theories listed in this appendix. These combinations allow for multiple completing explanations of jointness. This work will not fight the possibility of multiple valid explanations (i.e., equifinality) in explaining jointness. Indeed, process tracing makes it amenable to identifying equifinality. The good news is that if there are many ways to grow the phenomenon of jointness, there may be a corresponding number of means available to those who would attempt to sow the fields of national defense with more of it. In plain English, we think you can take exception to our methods and conclusions, but we hope to provide enough history, analysis, and food for thought that our general, academic, and practitioner audiences will find something worthwhile in these pages. With the rationale for case selection criteria established, an investigation of candidate cases follows.

Case Selection

CASE POPULATION

A host of existing case studies from scholarship about the US military could contribute to this investigation. Though these cases concern military topics other than interservice cooperation, their data may offer evidence about the conditions required for cooperation. Table 2.3 presents a relevant, albeit nonexhaustive, list of candidate cases. This pool of cases intentionally excludes case studies treated by Posen (1984; German *blitzkrieg*, British air defenses, the Maginot Line), Rosen (1991; twenty-one US and UK military innovations), Coté (1996; US Navy and Air Force development of strategic ballistic missiles), Ehrhard (2000; UAV technology deployment), and Donnithorne (2013; Army/Navy response to Goldwater-Nichols Act and Army/Marine Corps during Rapid Deployment Joint Task Force development). The European examples were not selected for this study because of resource limitations and the desire to answer the cooperation question from a US-specific perspective. However, many of these cases would be suitable to further develop the pretheory suggested by this study. The US cases directly test the validity of this study's conclusions, and the European examples provide an opportunity to examine the topic of joint cooperation from a multinational perspective.

SELECTED CASE STUDIES

A shorter list of suitable candidate cases presents itself after applying the first evaluation criterion, which specifies strong evidence of interservice cooperation. These include AirLand Battle, the Joint Primary Aircraft Training System (JPATS) acquisition, and the development of COIN-specific air power capabilities to support operations in Afghanistan and Iraq. There is a modicum of inter-service cooperation demonstrated in Korean tactical-air operations, the Vietnam-era route-pack system, and the Balkan air campaigns. However, the interservice cooperation observed was weak, and the resources required to gain the outcome do not rise above the level of theater commanders or theater component commanders.[14] Several of the US case studies by Posen (1984), Coté (1996), and Donnithorne (2013) do meet the criterion for strong interservice cooperation, and will serve as reserves for preliminary intracase hypothesis testing, comparison, and subsequent theory testing.

Table A.1: Candidate Case Studies

Title	Description	Era	Services	Cooperative Outcome
War Plan Orange	Navy's plan to counter Japanese expansion & aggression in the Pacific	1897–1945	Navy plan; Army & AAF in war	Island-hopping plan; model for ops executed during World War II
Korean Combined Arms	Control of 7th Fleet, 5th Air Force aircraft; CAS coordination	1950–1953	Air Force, Navy, Army	Tactical control/coordination never settled
Vietnam Route Pack Structure	AF & Navy deconfliction of tactical air routes	1966–1972	Air Force & Navy	Deconfliction only; rare coordination between AF/Navy airframes
AirLand Battle	Army & Air Force plan for tactical integration in war with Soviet Union	1973–1991	Army & Air Force	Major policy, tactical & acquisition coordination; created command/control systems used in Op Desert Storm; strategic air power dominance reduced vis-à-vis tactical cooperation w/Army
AEF Structure	AF reorganization; model for joint force presentation to combatant commanders	1990–2002	Air Force (several AF subgroups)	TAC, MAC, SAC dissolved as primary AF sub-commands; ACC became dominant warfighting command
JPATS Acquisition and Development	AF/Navy procure a joint primary aviation trainer, related training systems	1988–1995	Air Force and Navy	Joint aircraft acquisition and joint training programs
Balkan Air Campaigns	Interdiction and CAS during Bosnia and Kosovo conflicts	1995–1999	Air Force and Navy	Dialogue and increased integration of joint air assets via ATO process
Afghanistan and Iraq COIN Air Power	Rapid expansion of CAS and ISR	2001–2012	Air Force, Army, Navy	20-fold increase in UAV orbits; several technological and command-and-control CAS and ISR innovations
JADC2	Contemporary effort to develop joint, all-domain command-and-control capability among all services and weapons systems	2019–	Air Force, Army, Navy, Space Force	Create joint, flexible communications, command, and targeting system to maximize warfighting capability and weapons effects through the use and application of several previously independently controlled weapons and effects

Table A.2 lists the four main candidate cases selected for this study, including expanded descriptions of how they adhere to the first criterion as well as examples of how they meet the third and fourth selection criteria. The table omits the second criterion ("samples all services"); taken as a group, the case selections sample three of the six services—the next section addresses the omission of the Marine Corps and Coast Guard. Similarly, the group of cases in toto meets the fifth criterion ("most likely/most difficult examples"). They demonstrate examples of cooperation between the Air Force independently with the Army (historically a partner with whom the Air Force frequently exhibits an allied bureaucratic relationship) and the Navy (which since the inception of the Air Force has been bureaucratically and operationally aloof and at times fiercely adversarial). Finally, all the cases represent US examples from the current era of military bureaucracy (the sixth criterion), and take place during or after the Goldwater-Nichols Act adjustments to the defense establishment. This aids in the overarching quest to discover conclusions that are relevant to emerging policy questions.

WEAKNESSES

While meeting most of the requirements for the selection of case studies outlined earlier, there are some shortcomings in our sample. The selected cases adequately meet the first criterion and provide evidence of cooperation, but they fall short of the second criterion's specification to sample all military services. The case studies selected do sample from all the military departments, but they do not specifically address the Marine Corps. Though in the Department of the Navy, the Marines are a separate service with a distinct culture, distinct values, and a distinct approach to cooperative matters.[15] Donnithorne (2013, 6n) and Ehrhard (2000, 329) have both highlighted the value of including the Marine Corps independently when evaluating variables affected by service culture. (The Space Force is too new to have any relevant history, and the Coast Guard, while undeniably a military force, exists in substantially a different tactical, operational, strategic, and fiscal ecosystem than the other five.)

The cases selected meet the third criterion, but the JPATS case is perhaps lacking with regard to the fourth criterion. While primary aviation training is foundational to the air arm of both the Air Force and the Navy, the amount of budget it consumes and its impact on cooperation in other arenas are small. Another concern is that all other case studies deal with combat applications; training programs are distinct from military operations pursued in solving immediate security problems. Nevertheless, pilot training is a foundational competence for all six services and this unique instance of Air Force-Navy cooperation and its satisfactory comparison with all other criteria beg the reader to suspend judgment of case selection until the analysis is presented.

Table A.2: Selected Cases

CASE	Criterion #1: Evidence of cooperation	Criterion #3: Involves all bureaucratic levels	Criterion #4: Considerable bearing on institutional resources and culture
AirLand Battle	31 initiatives (Desert Storm equipment, command and control structures, effective joint operations in Iraq)	Cooperation between operations directorates for Army and Air Force Drove major acquisitions and training plans Effort endorsed by DoD; subtext of GNA'86 influenced initiatives	Drove organize/train/equip for two services Provided major equipment, manning, and command/control structure for next major war (Desert Storm)
Joint Primary Aircraft Training System (JPATS)	JPATS acquisition awarded in 1995 (system is still in place and driving primary aviation training for two services)	Training subgroups in both services allied Senior leaders in both services resisted initiative DoD and congressional pressure brought to bear to force innovation	Initiative controls primary aviation training systems for both services Singular example of Air Force/Navy cooperation with respect to air power
Afghan/Iraqi COIN Air Power	Major increase in support to COIN fight from air assets (20-fold increase in UAV ISR orbits; major changes to fighter CAS training and integration with Army after 2005)	Air-ground system participants and integrators recognized need for better capability AF senior leadership resisted change; Army pushed DoD direct involvement to change AF behavior	Major changes to training, unit organization, and acquisition programs Resulted in forced change of senior AF leadership Continues to be a major focus of AF budget debates and force structure planning
JADC2	Not yet observed, but hypothetically might include effective joint development of a system, compatible systems, or joint acquisition	Experimentation, definition of the concept, and attempts to allocate resources have all occurred at service, interservice, and meso-organizational levels of analysis; the effort will succeed or fall into obscurity at an interservice and Defense-Department echelon	Both the title and the aspirations of JADC2 point to a concept that will unify services' ability to function on the battlefield at the cost of individual service autonomy in pursuing proprietary data-collection, information storage/access, targeting, and kinetic or other effects-delivery systems.

The case-study sample meets the fifth and sixth criteria. The sixth limitation, situated in the contemporary era of American military decision-making, is a double-edged sword meriting further comment. Specifying this requirement raises the possibility that the investigation is only applicable within the current era, perhaps neglecting other important enduring characteristics revealed in other time periods. The high probability that the services will address future security challenges from within the confines of the 1947 and 1986 legislation that has defined the boundaries of American defense up to the present makes this an acceptable risk.

Notes

1. The authors describe the conjunctions of variables that are the most useful for showing necessity or sufficiency for given outcome.

2. For a definition of hypothesis-generating case studies and their contribution to theory generation (apart from theory per se), see Levy 2008, 5–6. For discussion of the validity of selection on the dependent variable and small-n case-study selection for theory generation, see Levy 2008, 8.

3. This definition is broad by design and borrows from Milner's (1992) definition of cooperation among international regimes.

4. Axelrod and Keohane (1985) discuss the possibility of cooperation among nations in a context of anarchy. They note findings that "military-security issues display more of the characteristics associated with anarchy than do political-economic ones" (226–227).

5. In comparing theories of international relations cooperation, Milner adopts this broad definition, further asserting that "as long as mutual policy coordination to realize *joint* gains occurs, then it is cooperation by our definition" (1992, 470, emphasis ours).

6. For a list of eleven selection criteria and the types of case studies to which they apply, see van Evera 1997, 77–88; and Milner 1992, 470.

7. A counterargument is that examples of clearly divergent (noncooperative or mutually destructive) behaviors might be instructive in illustrating conditions that are unfavorable to cooperation and thus to be avoided when trying to promote it. Since this is the normal state of bureaucratic politics, there seems to be more value in taking an approach with positive examples.

8. The National Security Act of 1947 and the changes defined by the Goldwater-Nichols Act of 1986 define contemporary military decision-making structures and the context in which military service cultures exist. Prior to 1947, the Department of War and the Department of the Navy existed in a completely different bureaucratic ecosystem than the one that took

shape in the latter half of the twentieth century (e.g., US Congress 1947; US Congress 1986; van Evera 1997, 77–88).

9. This tendency is common to all individuals and organizations (e.g., Halperin 1974, 4–6; Olson 1971, 7–16; Svedin 2009, 7–11). This is a bold, perhaps theoretically falsifiable claim. For example, Jeffrey Vandenbussche surmised that, as a military fight intensifies toward an "existential" level, joint command-and-control issues would fade away along with political sensitivity. One can imagine warfare so intense that services set aside philosophical differences as they make maximum effort. However, since several theaters of war exhibited divergent interservice behavior even in World War II—the most existential conflict yet to arise in US history—this study adopts an assumption that cooperation is elusive enough that it must be pursued. Indeed, comparison of Vandenbussche's central thesis about command and control (briefly, existential conflict drives joint force commanders to decentralized command and control while nonexistential conflict favors centralized command and control) with one of David Johnson's theses about COIN warfare (though nonexistential, it requires decentralized command and control to succeed) shows that there is a tension over strategic preference and the spectrum of war that shows disagreement likely to be just around the corner, no matter how intense a conflict (e.g., Vandenbussche 2007, 68; Johnson 2007, xxiv).

10. This approach follows Rosen (1991), who looked at successful examples of innovation, and differs from Posen (1984) and Coté (1996), who looked at one or more examples of failure to innovate.

11. This study adopts the idea that service cultures and subcultures are both likely to affect cooperation, while agreeing with organizational-culture literature that asserts that cultural phenomena are fluid, not subject to superficial managerial prescription and not yet well understood. See, e.g., J. Steven Ott 2011; Sonja A. Sackman 1991; Lois Recascino Wise 2010. Within the military, Mahnken (2008) argued that service culture generally trumps subcultures—this study reserves judgment on that idea, holding it in tension through the case studies.

12. Since there is no explicit theory of inter-service cooperation, the notion of service cultural contributions to the dependent variable of cooperation is more precisely termed a "pre-theory" (e.g., Martel 2007, 5–6), or "framing theory" as we will call it from this point on.

13. Rosen noted "the relatively minor role civilian political leaders have had in the initiation of and management of military innovation" (1991, 255). Michael Desch (1999, 6) found that civilian control of the military becomes increasingly difficult if the nation is not facing external security concerns against civilian intervention among services that are collaborating (e.g., Coté, 1996, 389).

14. See Ian Horwood (2006, 16–19) for the roots of interservice rivalry over air power in the Korean War. Horwood discusses, at greater length, the "fragmentation of command authority" and "dispersal of responsibility for air power resources" (63). Public disputes between the leading airman (Lieutenant General Short) and the operational commander (General Clark) exemplify the lack of cooperation among the services during Operation Allied Force (in Kosovo 1999); see Johnson (2007, 82–85).

15. Other analyses of service culture make the same omission, notably Builder (1989). Builder's assertion that the Marines do not enter "the defense planning arena as an independent institutional actor with significant voice in the national approach to strategy or military force planning" (9) is quite debatable. When it comes to combined-arms cooperation, our opinion is that the Marines' focus on combined-arms effectiveness allows doctrinal flexibility unadulterated by the lens of service strategic preference. The Marines also enjoy disproportionate public and congressional support compared to their institutional size.

Notes

Chapter 1. The Collaborative Fight

1. Since 1973, Gallup has polled the American public on its confidence in "the military" along with other institutions such as Congress, the Supreme Court, the Presidency, public schools, organized religion, and organized labor (Gallup 2013). In contrast, however, Gallup has also polled since 1949 for opinions about individual services, and the public has no difficulty differentiating distinct preferences for relative prestige and contribution to national defense (Newport 2013).

2. At the direction of the president, two separate orders from the Department of the Navy (July 18, 1903) and the Department of War (July 20, 1903) established the Joint Board. The Joint Board's charter directed it to meet "for the purpose of conferring upon, discussing, and reaching common conclusions regarding all matters calling for the cooperation of the two services." The vague, tautological reference to cooperation inhibited the generals and admirals from forging "a permanent interservice institution" (Godin 2004, 22–23), with personal relationships continuing to facilitate or inhibit joint cooperation as they had prior to the Board's formation.

3. The terms "joint arms" and "combined arms" indicate, respectively, coordinated effort among the services of a single nation and at least one service from two or more nations. Per Colin S. Gray, "Anglo-American usage now agrees that 'joint' operations are those conducted by the forces of more than one armed service, while 'combined' operations are those conducted by more than one country. Until quite recently, 'combined' operations in British usage referred to what now are meant by 'joint' operations" (1999, 240n).

4. Sennacherib's field commander chided Hezekiah, the Judean king, for his reliance on Egypt's alliance, "that splintered reed of a staff, that pierces a man's hand if he leans on it!" See Isaiah 36:6 (New International Version). "There is no place where espionage is not used" (Sun-Tzu 2005, 236) sums up the Chinese philosopher's appreciation of double agents and exploited allegiances. Examples of intrigue and broken alliances are legion in Thucydides (1998, 395, 413, 502, 27), but the havoc wreaked in Peloponnesus by Alcibiades's ever-changing allegiances to Athens, Sparta, and Persia puts a fine point on the practice. Similarly, broken treaties between nearby groups of people who had pledged one another's common defense appear regularly in the history of Rome (e.g., Livy 1982, 156). Machiavelli advised an imagined ruler who has just consolidated power to be "disposed to change according as the winds of fortune and the alterations of circumstance dictate" (Machiavelli 2003).

5. For an outline of the relevant DoD participants and their civil-military relationships, see Samuel P. Huntington (1957, 428–455).

6. Huntington's (1957, 418–422) "interbranch" is equivalent to "interservice" in this work.

7. The quotation describes "strategic monism," and is from Huntington (1957, 418). Owens's (1985) skepticism over JCS reform centered on the idea that the United States had an existing defense structure that favored Huntington's "strategic pluralism," which in Keiser's words "calls for a wide variety of military forces (or services) and weapons to meet a diversity of potential threats" (1982, 121). Both Huntington and Owens viewed pluralism as the more realistic and professional approach to defense. Owens believed that reformers favored strategic monism, which in practice meant withdrawing money from the Navy to build up the Army in Europe. Strategic monism presupposes knowledge of the most likely future enemy and war scenario, or, more cynically, that land power advocates simply want more resources (e.g., Owens 1985, 106–107).

8. The description of the "military spirit" parallels the idea of the public-service ethic that is a core discussion in public administration research. Huntington (1957) explicitly contrasts the differences in outcome wrought by the military public servant with the for-profit businesses at work in downtown Highland Falls, NY, for example. His description of the disorder of the for-profit sphere contrasts unfavorably with the calm order West Point's military academy reservation (465).

9. Clausewitz focused on land armies and offered little discussion of the military capabilities associated with other services today. But his clarion call to "put the largest possible army into the field" (1984, 195), advocacy for "skillful concentration of superior strength" (97), and instruction for the "simultaneous use of all means intended for a given action" (205) stresses the need for unity of command among joint forces, which were in his era different branches of the same service. Gray (1999), noting Clausewitz's neglect of naval matters and the weakness it induced in *On War*, cites the need for contributions by each service in specialized geographic domains in achieving a strategic whole. He argues that the means to achieve sound strategy is not through unification—even though the forces are increasingly interdependent on one another—but rather effective joint employment of forces in the correct balance for an ongoing conflict (212).

10. In an article describing and advocating *jointness*, Lawrence Wilkerson (1997, 66) explicitly disagreed with the assertion that it brings about synergy.

11. For a more detailed distinction of these two concepts, see chapter 2, section "Research Method and Case Study Selection Criteria."

12. While noting that joint cooperation is a necessity for military success, this work does not take the view that it is also *sufficient* for victory. *Blitzkrieg* warfare—a flawless interweaving of joint arms application but absent a coherent operational design—is a potent contrary example (e.g., Naveh 2004, xv). For an opinion that multiservice joint operations were essential for victory in World War II, in a work that examines the conflict from an air power-centric perspective, see Overy 1980, 203.

13. For an Army general exhorting Army personnel to remember the contributions of the Navy and Air Force to the Desert Storm effort, see Schwarzkopf and Petre 1992, 575–576.

14. Psychological research and foreign policy research refer to this phenomenon

as "rallying around the flag" (foreign policy) or in-group cohesion triggered by a perceived external threat (Svedin 2009, 2). William Baker and John Oneal (2001) found that the "rally effect" depends a great deal on the spin that accompanies the presentation of a threat to the public.

15. The USS *Pueblo* was seized by North Korea on January 23, 1968, and its crew was released on December 23, 1968. North Korea still holds the ship (National Security Agency 1992, 1, 134).

16. See, e.g., the Navy's "impassioned testimony" and the accompanying doctrinal disputes with the Air Force aired before the House Armed Services Committee in 1949 described by Jeffrey G. Barlow (2001, 294).

17. This sentence restates Martel's claim about the ambiguity of the term "victory," simply substituting "jointness" where he used "victory" (2007, 3).

18. Huntington describes "objective" civilian control of the military: "A highly professional officer corps stand ready to carry out the wishes of any civilian group which secure legitimate authority within the state" (1957, 84).

19. Because of the inherently unpredictable nature of military threats, all militaries must maintain some slack capacity to succeed when stressed. Indeed, there is something to the Sun-Tzuian argument that the most effective military is one that consists completely of slack capacity (i.e., it never fights because it deters all opponents with 100-percent effectiveness). Thus, a logically appropriate first-order assessment of military capability should always prioritize effectiveness or efficacy above efficiency. This does not mean that inefficiency is favorable to or should be encouraged in a military; to the contrary, unchecked inefficiency can lead to ineffectiveness. The subtle difference here lies in contrasting a self-induced, harmful, and avoidable inefficiency (the lack of ability or desire to cooperate among services) with a natural, unavoidable inefficiency (the requirement for slack capacity).

20. Part of the reason for this appearance is structural: the services themselves do not go to war; they simply organize, train, equip, and present forces to "joint force commanders." These joint commanders are themselves military leaders (having a particular service identity) who are delegated the authority to plan and execute the nation's wars in designated geographic areas or perform functionally specific military tasks in response to the direction of national command authority. The three-department, four-service model contrasts with that of Canada, who maintains a single "Armed Forces" to organize, train, and equip forces to present to combat commanders. Canada, viewing itself as resource-constrained (but desiring to fulfill its commitments to allies such as NORAD and NATO), favors military efficiency and sacrifices some of the independent views and strategic plurality that come with multiple services.

21. Our framing theory of jointness is an example of what other scholars call a "pretheory."

22. Universal consensus about the quality of a given cooperative effort proves elusive, although one of the case studies in this work, AirLand Battle, receives generally good reviews from scholars.

23. To be clear, Rosen's explanatory mechanism for military innovation was intraservice rivalry, specifically the competition of new ideas against established service

bureaucracies. This research borrows from Rosen's approach only in its use of apparently *successful cases*; it does not presume intraservice origins of those cases, nor does it rely solely on the explanatory mechanisms he identified.

24. This tripartite division of the defense establishment considers the backgrounds and perspectives of its constituent parts. Whereas Robert Art considered the services, the Joint Chiefs of Staff, the Secretary of Defense, the Office of the Secretary of Defense, and the defense "superagencies" to be "the Department of Defense," this investigation recognizes descriptive value in finer gradations. The services, our "internal service" level, are heavily steeped in their own unique cultures and dogma. The members of the Joint Chiefs of Staff have a significant time investment in, and are heavily influenced by, their sponsoring services. There is thus a utility in having an "among services" level that includes the Joint Chiefs of Staff, the chairman, and vice chairman, and the joint combatant commanders, who all exhibit these backgrounds and, arguably, dual loyalties (e.g., Art 1968, 161n). On the other hand, leaders of the "external" strata of the defense establishment generally come from the civilian world (or to the degree that they have military service, they take the position after a significant amount of time has transpired since they last served in uniform). They thus more closely resemble the members of the executive or legislative branches, and are better considered as part of this group.

Chapter 2. Factors in Organizational Collaboration and Success Cases

1. Olson's (1971) proof is based on groups being large, though he shows how small groups will tend to further collective interests toward a theoretical maximum if each member of the group has a significant and visible share of the benefit. In this study's sample of *four military services* or the *large military complex*, discussion about the relative size of the group is important, and probably relies in part on whether services are monolithic decision-making organizations or more fractious entities. For now, Olson's nontechnical summary suffices: "In short, the larger the group, the less it will further its interests."

2. A free rider is someone who benefits from a public good without making a contribution to its creation. The imperceptible-contributions problem arises because a lone individual is unable to make a contribution that is significant enough to advance a collective's cause—Olson's (1971, 64) analogy is a man trying to stop a flood with a bucket; such an individual would be deemed insane rather helpful even though his intentions were good.

3. This work uses "coercion" according to Schelling's (1966, 69–72) construct, which defined a negative ("deterrence") element and a positive ("compellence") element. The threat of sanction or criticism would constitute deterrence. The reduction of a service's budget or cancelation of favored programs until it complied with sister-service or DoD priorities would constitute compellence.

4. As the work progressed, and in particular during investigating the second case study about the Joint Primary Aviation Training System (JPATS), the notion of a socially constructed cooperative effort arose, leading to an additional question, "Who works to create stable cooperative structures?" For essays that discuss the relevant form of social constructivism in general, see Wiebe E. Bijker, Thomas P. Hughes, and Trevor Pinch 1987.

5. The agreement between then Army Chief of Staff General Douglas MacArthur and Chief of Naval Operations Admiral William Pratt read, "The Naval Air Force will be based on the fleet and move with it as an important element in solving the primary missions confronting the fleet. The Army Air Forces will be land-based and employed as an essential element to the Army in the performance of its mission to defend the coasts at home and in our overseas possessions, thus assuring the fleet absolute freedom of action without any responsibility for coast defense" (Pratt and MacArthur 1931).

6. Despite the formal independence of the Air Force established in 1947, it has never had complete control of all air assets. Naval aviation has remained independent (albeit with contention about its proper military use), while the Army immediately began developing a helicopter force despite the Air Force's preindependence promises to dedicate air support to the Army (Davis 1987, 92).

7. See their definition of an *agency relationship* as an arrangement "under which one or more persons (the principal[s]) engage another person (the agent) to perform some service on their behalf which involves delegating some decision-making authority to the agent. If both parties to the relationship are utility maximizers, there is good reason to believe that the agent will not always act in the best interests of the principal" (Jensen 1976, 5). See also Jensen and Meckling 1976.

8. Though this work is about military cooperation rather than innovation in military doctrine, it will touch on doctrine to some degree. In doing so, it will rely on a definition shared by the work of Rosen, Posen, and Coté. Military doctrine is the military subcomponent of grand strategy. It answers the "What shall be employed?" and "How shall they be employed?" questions about military means, and the other authors follow suit. Military doctrines set priority among the types of forces available, prescribe organizational structure, determine force-employment guidelines, and specify modes of interservice cooperation. See Posen 1984, 7.

9. "Waltzian" refers to Waltz's (1979, 99) central idea that accurate explanations of international politics center on both the interactions of individual units and the structure of the system in which they operate. Application of Waltz's ideas to systems other than the intercourse of nations retains the idea of individual actors operating in a pseudoanarchic system.

10. In fact, all the authors surveyed for explanations of military innovation assume some level of interservice rivalry to be an integral part of the creative process.

11. The question relevant to this study of jointness is similar to one yielded by consideration of organizational behavior in crises, which links many ideas and demonstrates the utility of a cross-disciplinary approach to the problem. It is particularly useful because the Tjosvold-Rosen construct predicts different results than Svedin's more-specific correlations between crisis type and behavior.

12. This is the author's assertion, and very much subject to debate. The world was not at peace throughout the time periods discussed, but there was little to no threat of thermonuclear war of the kind that ballistic missiles would have enabled. If one accepts the Cold War as "war" per se, the metric of launch-ready thermonuclear missiles then becomes an appropriate strategic metric.

13. The blurring of "wartime" and "peacetime" perceptions that characterized the Cold War is a serious issue that merits further attention. One can argue in favor

of one perception or the other and assert that given metrics of strategic success are or are not important. As the rest of this work will show, confusion or disagreement about the severity of a security situation can have significant impact on jointness.

14. This assertion runs counter joint doctrine. A doctrinaire (pun intended) approach to the way US fighting proceeds might argue that "service" opinions and actions are irrelevant to "jointness" since war is waged by the among services coordinating level of the defense hierarchy (a joint force commander overseeing forces provided to him by the services). This book rejects that view, finding that the service influence over those forces is too high to ignore and that parochialism does not die out even in fairly intense military action. This assertion is in line with the interviews of senior military leaders conducted for this work. It also extends Mahnken's findings about the primacy of service culture over subcultures, and is in agreement with Kenneth Allard's (1996, 4) conclusion that, because of their "organize-train-equip" responsibilities, "service, rather than joint, command structures exercise the dominant influence over" joint forces.

15. This is the central argument of Huntington's *The Soldier and the State* (1957).

16. By "objective control," this work refers to the institutional independence and apolitical obedience of the military organization to congressional and presidential authority. "Subjective control" is the grooming of defense organizations that read and respond to political trends. They have less autonomy about internal structure and policy, but have more participation in political dialogue.

17. The preeminent military strategist Colin Gray frequently cites Leslie White's critique of culture as "not basically anything, . . . a word concept . . . used arbitrarily to designate anything" (White 1975, 4n).

18. Ehrhard takes Builder's seminal effort in the subject of US military service cultures to task as an "uneven, shallow, but occasionally insightful work" (2000, 41n). This work takes the view that, whatever shortcomings exist in a single description, an amalgamation of all four authors presents perhaps a more congruent and detailed view of service cultures.

19. Rosen wrote that peacetime innovation "depends on a senior officer or a group of senior officers who first attract officers with solid traditional credentials to the innovation and then make it possible for younger officers to rise to positions of command while pursuing the innovation" (1991, 96).

20. The acronym later expanded to "PPBES," incorporating "execution."

21. The same description has slightly different connotations for the congressional armed services committees: the Air Force "will lie" or "try to outsmart you," the Navy will "ignore you" or "pretend you don't exist," and the Army "will talk when they should keep their mouths shut" or "try to out-cooperate you" (Scroggs 2000, 57–58).

22. Carl Builder explains the distinctions among service analysis habits and responsiveness to Congress with his "toilet paper parable" (1989, 107–109).

Chapter 3. The Army and Air Force Collaborate on AirLand Battle, 1973–1991

1. The Service chiefs directed "increased joint training" and resolved to settle "any doctrinal and procedural concerns as AirLand Battle doctrine is integrated into joint theater operations"; see Meyer and Gabriel 1983.

2. For Mattis's ban of the term from all joint publications over which he had authority, see Mattis 2008, 1–2, 6.

3. See US Senate 1959, 579.

4. Lind wrote his critique at the behest of a congressional influencer, Sen. Gary Hart, which gave it an external facet this chapter deals with in a later section.

5. For the history and definition of the terms, see Cardwell 1992, 34n.

6. Joint Surveillance and Target Attack Radar System (JSTARS) is an airborne platform optimized to find ground-based enemy armor formations and slower air targets like helicopters. Its survivability and relevance in a modern fight have been severely limited by advanced antiaircraft systems.

7. For a summary of the Army's internal challenges after Vietnam, see Nielsen 2010, 1, 41–42.

Chapter 5. Air Support in Counterinsurgency, 2001–2012

Epigraphs: Creech n.d.; emphasis in original; quotation is from an interview published in McElroy and Slayden Hollis 2002, 7–8.

1. Notably, there were also the tens of thousands of Northern Alliance and Pashtun fighters who had established something of a stalemate with the Taliban at the time the US forces arrived. As such, the US campaign may have added the element that tipped the scales and pushed the Taliban out of power, rather than being the sole explanation for the (temporary) Taliban defeat.

Chapter 6. Joint All-Domain Command and Control

1. "Targets emerge in seconds, incoming enemy fire puts lives at risk and shifting combat dynamics require immediate, on-the-spot decisions in a matter of seconds—all as soldiers navigate the complex web of threats during all-out, high-risk ground-warfare. These kinds of predicaments, which characterize much of what soldiers train to face, are immeasurably improved by emerging applications of AI; artificial intelligence can already gather, fuse, organize and analyze otherwise disparate pools of combat-sensitive data for individual soldiers. Target information from night vision sensors, weapons sights, navigational devices and enemy fire detection systems can increasingly be gathered and organized for individual human soldier decision-makers" (Osborn 2020).

2. "ABMS is the Air Force's version of connecting systems, devices, and platforms—like the internet of things—to make data sharing easier and more usable" (Williams 2020); "Why should our service men and women leave lives where they are connected to everything and come into a military where they are connected to almost nothing" (Roper 2020).

3. All service chiefs tend to roll out defining messages at these service-specific forums.

4. Several sources attribute the aphorism "Amateurs talk strategy. Professionals talk logistics," to Gen. Omar Bradley (e.g., Richards 2018). In the 2019 Joint Warfighting Concept, the Air Force, Navy, and Army were given lead responsibility for, respectively, joint all-domain command and control, global and joint fires, and contested logistics (e.g., Schaus 2021).

5. JSTARS was an airborne system optimized to find and hand off targeting information about Soviet tank columns.

6. One of the most ironic quotes: "'Gen. Brown'—the Air Force Chief of Staff—'and I are both committed to making this happen,' Army Chief of Staff Gen. James McConville tells Breaking Defense. 'It starts at the top'" (Freedberg and Hitchens 2020).

7. "'Right now, the majority of networks have a bureaucratic process that people must go through that makes it very cumbersome for experimentation,' Shell said. 'An experimental network—particularly for capabilities that are at lower maturity levels—if we don't allow them to come on a controlled environment and experiment, then we're not going to learn the lessons needed to enable Multi-Domain Operations as quickly as we need,' Shell said" (Kester 2020).

8. We recognize that the National Defense Strategy theoretically outlines the threat China poses, but we see the question of the United States becoming enmeshed in a real conflict with China and creating a new Pacific theater a less likely prospect.

9. "In our professional community we have built an example of what these future command and control nodes can look like. Called 'all-domain control teams,' we have brought together small groups of people—fewer than 10—from multiple backgrounds, and given them equipment that can communicate with the most commonly used assets in the military. Our teams have experts in air, ground, cyber, and space operations. They are equipped for maneuver. Rather than stay in one place as current command and control nodes mostly do, these teams move, pull-in data, radiate information to others in the network, and repeat. This way of operating increases their survivability. Unlike the hierarchical systems employed in the past, these future command and control nodes would also have the ability to grow, when and where needed, in response to the operational environment" (Cowie et al. 2021). We highlight this passage because it shows how the *operationally successful* model is also likely to be the one that encounters the most bureaucratic resistance.

10. This report likewise underlines the reality that no single service or entity with DoD owns the structure of JADC2.

Chapter 7. An Empirically Informed Theory of Jointness
Epigraphs: Clausewitz 1984, 112; Tolstoy 2009 (1869), 2198–2199.

1. Martel used a multidisciplinary approach to refine the definition of "victory" and to provide a framing theoretical framework for its further study; see chapter 2, pages 15 and 36.

2. Where Martel began his theoretical work with a refined definition, this approach tackles the definition after it outlines the framing theoretical parameters. The reason for swapping the order is the definitional ambiguity encountered in defense-establishment usage, which is itself a function of some of the observed theoretical variables. In short, one's perception of what "jointness" is seems highly dependent on his or her place in the defense establishment. The success or failure of a joint initiative depends in part on a leader's ability to attach meaning that encompasses all of these definitions. Thus, for "jointness," definition *follows* understanding of how the term works in practice.

3. Svedin also observed, "All the strategies contain some mix of cooperative, less cooperative, and competitive behaviors" (2009, 126), which proved true in the case-study observations here as well.

4. The somewhat amorphous and overlapping organizational levels used as levels of analysis—internally within services (level a), among services including coordinating groups like Joint Chiefs, Joint Staff, and combatant commanders (level b), and relations to external stakeholders including the larger DoD, Congress, and industry (level c) proved useful analytically in this study. The scope of influence at each level is qualitatively and quantitatively different. However, there is no reason to hew rigidly to these demarcations; there is room for distinction with levels. For example, while this study has treated DoD, the legislative branch, the executive branch, and the media as "external" members and influencers of the defense establishment, all three case studies suggest that Congress's influence is more extensive than its nominal peers in this category.

5. The literature about organizations cooperating in crises can offer a picture of *whether* organizations are cooperating or not, and if so, to what degree they are, but it does not explain joint behavior. Indeed, crisis cooperation theory makes a fundamental assumption that cooperation in environments marked by threats and uncertainty is a good thing. Proponents of jointness, particularly for what they identify as its most useful manifestation (on the battlefield), would likely agree, but the case studies offered show that this is not always, or even often, the case. Theoretical studies of peacetime military cooperation, such as Coté's (1996, 350–351), argue that cooperation is anathema to achieving the most efficacious slate of military options, even if it leads to short-term efficiency.

6. Barry Posen asserted, "Organizations innovate when they fail, . . . when they are pressured from without, . . . [and] when they wish to expand" (1984, 47). The pressure from without is a dominant theme in his conclusions: "Militaries oppose innovation, but we see some remarkable innovations. . . . Civilians do affect military doctrine. Their intervention is often responsible for the level of innovation and integration achieved in a military doctrine" (227).

7. Rosen, *contra* Posen, stated, "Military organizations do not innovate in peacetime simply in response to defeat or to civilian intervention. Innovation in wartime is not a matter of seeing that existing methods do not work and then correcting them" (1991, 52).

8. When concern about the Warsaw Pact faded in the West, it did not present a challenge only to US strategic thinking. In writing about its relationship with the Russian Federation, Zbigniew Brzezinski asserted that NATO had "to define for itself a historically and geopolitically relevant long-term strategic goal" or risk irrelevance (Brzezinski 2009, 10).

9. Nunn (1985) continued, "The burden of the annual authorization and appropriation process has produced two specific problems. It has led to the trivialization of Congress' responsibilities for oversight and has led to excessive micromanagement. . . . In the defense arena, Congress was to set priorities for programs, not to execute them. Congress' role as the board of directors is eroding; rather, Senators and Representatives and their staffs are acting more and more like national program managers. . . . We have not had a serious debate about the important relationship between our national objectives, our military strategy, our capabilities, and the resources to support that strategy. We all know that there are serious gaps in these important links. . . . These are precisely the questions that Congress is supposed to

consider: Do we have a strategy that achieves our national goals and objectives? Do we have the resources to meet these commitments and support the strategy? What alternative approaches might we adopt for overcoming the strategy-forces mismatch? Those are the questions that Congress should focus on. Instead, we are preoccupied with trivia. . . . Our preoccupation with trivia is preventing us from carrying out our basic responsibilities for broad oversight" (25350).

10. Locher noted that major defense-reform legislation in the twentieth century came about only in 1903 (following the Spanish-American War) and again in 1947 through 1958 (a spate of four actions following World War II), with no further actions "for almost thirty years after 1958" (2002, 29).

11. De Tocqueville's context was the moral quality of a nation; he found that war sharpened it, that it "almost always enlarges the mind of a people, and raises their character" (1840, bk. 3, chap. 22, para. 9).

12. The characterization of interservice cooperation is from Rew 2014.

13. A strong majority of a large sample of defense-establishment professionals, mostly military officers, reflected a belief that unified command-and-control structures were the essence of jointness, rejecting the idea that materiel factors like commonality were the primary goal (see chapter 8).

14. The characterization of congressional involvement in war is from Sapolsky, Gholz, and Talmadge (2009, 131). The "unequal dialogue" is the give-and-take between military leaders and their civilian masters, "unequal, in that the final authority of the civilian leader [is] ambiguous and unquestioned." As Cohen (2003, 209) coined it, the term refers to executive-military interaction—he did not specify legislative-military exchanges.

15. It is not just Congress that focuses attention on CAS after (or during) a conflict. The Army and Air Force continually challenge each other on the topic (see US Senate 1971; "CENTCOMM Commander Urges Close Air Support Examination" 2004; Burgess 2002; Den Beste and Servold 1995; Dinges and Sinnreich 1980; Hasken 2003; Isby 1990; Kent and Ochmanek 1998; Lindsay and Ripley 1994).

16. The characterization comes from previous Army chief of staff (2011–2015), Gen. Raymond Odierno (see "Military Spending: Don't Save the Warthog" 2014).

17. There is a great deal of material decrying the Air Force's perceived distaste for CAS (see Burgess 2002; Den Beste and Servold 1995; Grant 2003; Hasken 2003; Holley 1990; Johnson 2008; McElroy and Hollis 2002; Neuenswander 2014; Olive 2007). However, the Army does not passionately pursue the capability outside of times of conflict either, preferring to train with its organic artillery capability for fire support. The tendency reflects both the inconvenience of joint training and an inherent desire to be self-sufficient whenever possible.

18. One might expect that disputes of this sort will more frequently arise among the services that usually fight on or over land. Operational control of Marine aircraft will be an enduring issue, as the unified Marine Air-Ground Task Force (MAGTF) concept competes with the Air Force's ideal of theater-wide centralized control. A similar issue will appear when the Navy makes a contribution to land warfare with its air power assets.

19. The approach described here is not easy, popular, or fast. McNamara's attempt to exercise this kind of decision-making with regard to the TFX program

serves as a warning to the responses the services will likely exhibit. It also demonstrates why listening to the services with respect to their equipment requirements is just as important as objective control of the military by the executive branch. The TFX was an example of excellent management principles (full-spectrum analysis) plagued by shortsighted and dogmatic execution (the *diktat* of commonality).

20. The most obvious example uncovered in this investigation was CAS incorporation at the JTC early in OEF and OIF, which led Air Force leaders to complain that Army field commanders left their training evolutions with no appreciation of how air power could contribute to maneuver warfare. The situation improved over the duration of the conflicts, but it is by no means perfected, and will likely get worse as interest in CAS fades.

21. Describing debate about strategies for the invasion of Iraq, James Fallows commented, "Rumsfeld proposed something like 75,000 U.S. troops for the invasion force. The Army had in mind something closer to 400,000." Rumsfeld's relentless insistence on a small force gained the upper hand, but back-and-forth negotiations modified his position as well, resulting in a 140,000-strong invasion force and "hybrid" war plan (see Kirk 2008).

22. An alternative explanation, borrowed from Erik Voeten's (2004) work in international relations theory, is simply that dominant regimes invite resistance. He described how the United States is "increasingly isolated in multilateral organizations" (Voeten 2004, 729) to an extent not fully explained by differences in national preference. To extend the analogy, the Air Force may have resisted the dominant COIN framework that shaped OEF and OIF not simply because it has a preference for strategic effect and centralized control, but rather because it did not originate the concept and witnessed it become the dominant idea for more than a decade of conflict.

23. This was the context senior military leaders provided to Army and Air Force AirLand Battle planners (Davis 1987, 3). For the same sentiment applied to all three services in the context of World War II, see Builder 1989, 132–133.

24. Contrast the consistent drive to acquire fifth-generation fighters with the foot-dragging exhibited for COIN air power and the case of "a plane that can provide effective precision close air support and JTAC training, and costs about $1,000 per flight hour instead of the $15,000+ they're paying now." Such a need was "obvious around October 2001," but "took until 2008 for this understanding to even gain momentum within the Pentagon. A series of intra-service, political, and legal fights have ensured that these capabilities won't arrive before 2015 at the earliest, and won't arrive for the USAF at all" ("LAS In, LAS Out" 2013).

25. Sixty-seven percent of respondents to this study's survey (sample size = 54; response rate = 65 percent) felt that "a compelling, shared strategic vision that motivates services to pursue a common goal" should be the "primary reason for joint cooperation." An additional 5 percent cited it as partial motivation for jointness, though this subset of respondents refused to commit to a single primary cause (Birch 2014b).

26. Cannon (2012) by no means dismissed either monitoring or sanction, arguing that the former should be sufficient to give the principal (which is, in his formulation, the joint force commander) an accurate picture of what is happening (without crossing into micromanagement) and that the latter should provide a real means of

punishing or relieving wayward component leaders (without being too draconian) (Cannon 2012, 297–298).

27. "Put simply, the structural, normative approach has run its course and likely taken jointness as far as it can, perhaps in the wrong direction. The alternative is a constructivist approach" (Cannon 2012, 298).

28. In this formulation, the organizational slack must be of a sufficient amount that it allows completion of the task without bankrupting the reserve of intangibles that keep the "coalition" together (see Cyert and March 1963 [1992], 42).

29. Andrew Abbott's (1988) argument about the theory of professions predicts the same challenge through a different means of reasoning, one that relies on the institutional credibility of each service, which have all staked their identity on a given theory of victory and preferred means of operating (see Abbott 1988, 62–65). This is why AirLand Battle writ large was an easier sell than improved CAS and ISR coordination mechanisms: the unconstrained battlefield envisioned in the former did not constrain service preferences or judge one theory of victory inferior to another in the way that air-ground support compromises inevitably do, at least in the eyes of arrantly parochial advocates.

30. This is consistent with Jeffrey Polzer's (2004) finding that subgroup members "with strong organizational loyalties may undermine the [collective] endeavor in the name of protecting organizational interests, ultimately hurting the organization in the process," assuming that, as Thomas Mahnken found, service loyalties tend to trump cross-service subgroup loyalties such as pilot, infantryman, and the like (see Polzer 2004, 93).

31. Recall that a crisis is a "serious threat to the basic structures of the fundamental values and norms of a social system, which—under time pressure and highly uncertain circumstances—necessitates critical decisions" (Rosenthal, Charles, and 't Hart 1989, 3). While the definition is somewhat extreme, it is difficult to conceive that a programmatic or battlefield decision in a military organization would not be couched in such dramatic terms.

32. Behavioral science characterization of aggression, competitive behavior, and cooperation among humans is a segment of the field that defies succinct summary, but a few observations are helpful in discussing the reasons why aggression supplants cooperation. A central theme that has seen its way into social-science research via both philosophy and evolutionary biology is the Lorenzian theory (after Konrad Z. Lorenz) that aggression is an "ineradicably instinctive behavior," regardless of social mores and context; for a description and rebuttal, see Kim 1976, 254. One dominant service influencing national strategy for a protracted time may inhibit cooperation. If interservice relations can be modeled as economic interactions, research would suggest that a chronically wounded sense of pride can be deleterious to cooperation, because "when a person's sense of individual or group honor is repeatedly violated in economic interaction, the reaction may include the release of aggression to repair damaged honor and establish self respect" (Baxter and Margavio 2000, 399). The notion that cooperation "has not evolved in some animal species because of cognitive constraints" may be an epithet one is tempted to hurl in frustration at a recalcitrant interservice counterpart, but the finding is derived from the study of lesser animals and probably does not merit extrapolation to humans (see Stevens, Cushman, and Hauser 2005, 499).

33. Among those interviewed about jointness for this work, most cited trust as a necessary condition for joint cooperation. Reasoning by Freudian analogy, Edward Hoedemaker argued that aggressive individuals are hobbled by distrust and display a need to "take over" others, even though the others may be trustworthy and capable of mutually beneficial helping behaviors. This unfortunate outcome can extend to organizations, and cannot be overcome by "persuasion or argument" (Hoedemaker 1968, 71, 74). Compounding the difficulty is research that suggests that participants in a competitive game attribute the presence of conflict in a game to their counterparts, even if they themselves induce it, and that neither reciprocation nor coaxing produces more helping behavior than simple cooperation, suggesting that even gestures of goodwill among joint rivals will often be scattered on hard ground as a matter of the vagaries of personality (Wilson 1969, 114).

34. Research pertaining to sports teams suggests that "internal rivalry and conflict" in teams can lead to "consistent increments in performance," but that "internal conflict must be kept within limits if it is to contribute to team effectiveness" (Lüschen 1970, 22). This concept of "competitive association" is consistent with case-study conclusions that competitive discussion, even some degree of fighting, fosters joint cooperation more effectively than simply coexisting in relative silence alongside a service or component that one views unfavorably.

35. Threats or fear that lead to communication can be more deleterious than open conflict, rendering "the determinants of action as much a consequence of fantasy and fear as of an appraisal of reality. Corrective information about motives, tools, or plans are left to speculation and the machinations of fear and anxiety" (McNeil 1961, 284).

36. Antonia Chayes (2008) described analogous effects from US behavior with respect to international treaties, finding that, because of "American exceptionalism," "few senators will vote to ratify a treaty over the opposition of their constituents." The outsized representation of conservative special interests results in "selective multilateralism" that manifests itself as "uncooperative treaty behavior" and has "undercut essential international cooperation" (Chayes 2008, 49; see also 72, 47, 48, 74).

37. Riker (1986) wrote that the "fundamental heresthetical device" for someone who might otherwise lose a political decision is to "divide the majority with a new alternative, one that he prefers to the alternative previously expected to win" (1).

38. Martel (2007) began his pretheoretical framework with a "systematic search for a definition of 'victory'" (83).

39. Usage is not always consistent, however. For example, Euro-NATO *Joint* Jet Pilot Training is a specialized US Air Force program that incorporates training for students of many foreign, allied air forces. It is therefore *combined*, but not *joint*, according to the US military establishment's own definitions (Joint Chiefs of Staff 2010, "combined," p. 45). Treaty language and tradition will keep ENJJPT's name and familiar acronym unchanged for the foreseeable future. A further exception: a "combined arms team" denotes intra-service participation in a military operation (Joint Chiefs of Staff 2010, "combined arms team," p. 45).

40. The full definition reads as follows: "Connotes activities, operations, organizations, etc., in which elements of two or more Military Departments participate" (Joint Chiefs of Staff 2014a, 139).

41. The Air Force and Navy versions of the T-6 are quite differentiated, even

though they perform arguably the exact same mission. The various service-specific versions of the F-35 are by comparison designed for missions that are quite different. A RAND report offered, "From the Tactical Fighter, Experimental (TFX)/F-111 program in the 1960s through the JSF program today, the attempt to accommodate multiple operating environments, service-specific missions, and differing performance and technology requirements in common joint fighter designs has increased programmatic and technical complexity and risk, thus prolonging RDT&E and driving up joint acquisition costs" (Lorell et al. 2013, xvii).

42. Explaining and reducing shirking tendencies are, of course, also concerns central to agency theory.

43. Martel (2007) discussed the need for and utility of a *general* theory of victory, one that would resonate with the public as well as specialized audiences in the defense establishment. Cannon emphasized that the *military services*, a subset of that establishment, also lack a "coherent joint vision of victory" (2012, 296) among them.

44. Again, Cannon (2012) also realized that the usual, normative ways of dealing with agent-slack issues—increased monitoring and the threat of punishing noncompliant agents by the principal—were unlikely to succeed given the structural power and individual credibility of the services, and that they have in fact had deleterious effect on interservice cooperation (298).

45. "Efficacy," meaning *the capability to produce an effect*, appears deliberately in the definition in lieu of "effect." Joint efforts should be weighed as if the resultant effect were available and used the best possible ends. Outcomes due to bad or prejudiced generalship should not be part of the adjudication, as they are a symptom of a different problem.

46. On the battlefield, efficacy outstrips efficiency as the measure of greatest interest. The relative standings of these metrics are reversed in peacetime. David Johnson (2014) recommended this priority for combat effectiveness.

47. The authors offer a nuanced understanding of the basic philosophical question that underpins the question of joint cooperation. In describing the modern US defense establishment as consisting of two centers of power, the civilian DoD and the military Joint Staff, they make three cogent observations: (1) there is an incessant call for more centralization when there are budget or military crises; (2) playing DoD civilians against the military hierarchies leads to collusion among the military services, which produces politically passable but militarily mediocre short-term solutions; and (3) indecision about the best way to solve security problems is acceptable, because the solutions to emerging and future security problems are by definition unknown entities—a planner's desire to seize upon a "best" course of action for the sake of efficiency risks failure at the hands of an enemy who finds a way around that course of action (Sapolsky, Gholz, and Talmadge 2009, 164).

48. In describing the pursuit of a definition for "victory," Martel (2007) noted that "it is unlikely that we will be able to formulate definitions for victory that transcend differing political ideologies and world-views or that can achieve the standard of value-free inquiry" (91). In the same vein, Felix Oppenheim wrote, "Since different actors, and different ideologies as well, are committed to different moralities and ideologies, it is not possible to come up with definitions embarking such irreconcilable world views" (1981, 195–196).

49. As Easton (1953 [1971]) cautioned, though, theories of this type should avoid the conceit of "methodological rigor and precise formulation" associated with the physical sciences; these ideas should be a part of the overall consideration of the conduct of the art of war (58–59).

50. Crick's description of doctrine is that of a "theory which claims universal validity, because of a belief that all ideas derive from circumstance, but which then also holds that this truth is deliberately obscured by ruling elites, so that the theory only has to be asserted in the form of propaganda to the masses" (1973, 14). This description is uncomfortably close to the way in which services have propagated their unique doctrines, especially fundamental ideas about victory in war.

51. William Lind (2014), for example, advocates that the sine qua non of joint-ness would be "the creation of a full-time—including peacetime—purple general staff, modeled on the Prussians, obviously." He bemoans parochial interests that constrain creative thinkers: "Clausewitz wouldn't last two weeks at West Point. They are shoved out early because they don't fit the cookie-cutter mold or follow the little Mattel soldier mold of 'first this job, then this job, then this, then that' and all of this career-pattern stuff that produces what the second-generation military wants. It says 'Excellence.' It doesn't want excellence; it wants uniform, dependable, mediocrity, and that's what it gets" (Lind 2014).

52. David Johnson (2014) described joint doctrine as "the least common denominator," i.e., including only those items that the services, each wielding line-item veto authority, found sufficiently noncontroversial to allow into joint publications.

53. Even as Dolman acknowledged the importance of this distinction, he still recognized the importance of divining victory on the battlefield, at least as a preferable alternative to defeat in advancing the interests of a state (2005, 6).

54. Respondents answered three questions, two of which are reported here: The essence question read, "'Jointness' in the U.S. military SHOULD consist primarily of (choose one): A) inter-operability of equipment on the battlefield (compatible radios, computers, situational-awareness systems, e.g.); B) unified command and control under a properly appointed joint commander, or C) efficiency in acquisition and logistics by increased economy of scale?" The cause question read, "Who or what should be the primary influencer, driver, or reason for jointness? A) A compelling, shared strategic vision that motivates services to pursue a common goal; B) external direction from the President, Congress, the Department of Defense, or another entity that oversees the defense establishment; or C) the need for greater fiscal or combat efficiency due to constrained resources?" In all cases, other answers reflect a respondent who chose more than one multiple-choice answer.

55. A survey of fifty-four people interviewed for this project, administered after all interviews had been completed, yielded a response rate of 69 percent.

Conclusion: Jointness Is a Collaborative Fight

Epigraphs: McPeak 1994, slide 205; Yeats 1970 (1920).

1. In an internal survey study of selected military leaders, 42 percent felt that Jointness in the US military *should* consist of "unified command and control under a properly appointed joint commander," while 16 percent believed it consists of "inter-operability of equipment on the battlefield" (Birch 2014). Of the remainder, 3

percent believed it to be "efficiency in acquisition and logistics from increased economies of scale," and 39 percent chose more than one of the above choices. Thirty-eight
of fifty-five leaders solicited responded to the survey (Birch 2014b).

2. Lt. Gen. Van Riper is well known for his unconventional tactics as the leader
of red forces in the *Millennium Challenge 02* war game, which he quit in protest after
"Pentagon officials who managed the game simply disregarded or overruled the [sim-
ulated opposition] militias' most devastating moves" (Kaplan 2003).

3. Charles Hermann et al. described three models of group decision-making
processes that tend to reach "Concurrence (producing a tendency to *avoid* group
conflict); Unanimity (producing a tendency to *resolve* group conflict; and Plurality
(producing a tendency to *accept* group conflict" (C. Hermann et al. 2001, 138, emphasis in original). The problem with military-service logrolling and a unified front
across groups that coordinate among services is that concurrence and unanimity tend
to mask the sources of conflict from external decision makers, sometimes stifling
innovation and operational thinking. While avoidance and resolution are acceptable
in some cases and are frequently necessary timesaving devices for effective defense
operations, the external level of the defense establishment must selectively insert
itself to ensure that enough plurality remains intact. Absent this necessary oversight,
defense decisions may become moribund as differences are papered-over or some
viable options never see the light of day outside the Pentagon.

4. Desert Storm provides a ready and recent example. The Air Force posited a
vociferous argument that carefully crafted strikes using stealth and precision could
throw into confusion Saddam Hussein's regime, making the ground fight in Kuwait
easier and perhaps even hastening the dictator's departure. The Army and Marines
believed that closing with and destroying the forces of the Iraqi Army was required
to eject the Iraqi presence in Kuwait.

5. The conclusion that the services are not prone to voluntary cooperation without significant outside motivation seems to be backed by both ample evidence as well
as a preponderance of social science theory. If bureaucratic barriers in place are so
massive as to threaten the delivery of national security, prowess on the battlefield, or
any other common-good commodity that one could reasonably expect to derive from
investing in a military force, a military paradigm-shifting "crisis" or "pronounced
professional insecurity" that leads to the destructive-constructive cycle of paradigm
shift could result (Kuhn 1970, 66–68). Given the recent record of the US armed
forces, however, such a crisis does not appear imminent.

6. "It seems strange to hail confusion and indecision, but we must. . . . Wise leaders . . . will search out other planners and analysts for confirmation or, more importantly, for dissent" (Sapolsky, Gholz, and Talmadge 2009, 164).

7. Among those interviewed for this work, General Jumper was probably the most
forceful and articulate holder of such an opinion. He wrote,

> What would really drive jointness is the development of Joint Concepts of Operation
> [CONOPS] that would analyze how we plan to engage jointly before we run out and
> start buying the equipment we will use to fight with. A true CONOPS would guide
> interoperability, C2 [command and control], and acquisition.

The JROC [Joint Requirements Oversight Council] should be reformed to produce CONOPS instead of requirements. . . . The glue that we don't have, and really need, is the Joint CONOPS that forces the services, along with each COCOM, to develop a plan for how they plan to fight, or carry out the strategic direction. They would be forced to answer questions about shared responsibility, shared capability, efficient joint logistics, etc." (Jumper 2013).

8. Several interviewees for this study articulated the opinion that a unifying strategic vision would emerge from a joint commander strong enough to gain unity of command and quell interservice competition during a conflict (Deptula 2013; Jumper 2013; Kem 2014; Kwast 2013; Peters 2013; Van Riper 2013). A competing viewpoint is summarized in the words of David Johnson (2014), who argued that there is no "joint theory of victory" that competes with existing theories debated for land, sea, and air domains and that, therefore, there could be no dominant corresponding joint doctrine.

9. Kissinger believed that strategic doctrine "enables society to deal with most problems as a matter of routine and reserves *creative* thought for unusual or unexpected situations," (1957, 403–404).

10. Johnson's prescription to avoid strategic stasis was to confront multiple simulated security crises with planning and exercises: "I think the ISCs, the Defense Planning Scenarios, give you problems that are kind of real-world but also are not present." He went on to describe two different scenarios that involved different combinations of likely combat intensity, threat, geography, and regional participants—a challenging and varied spectrum of military planning problems. "How do I solve those problems?" (Johnson 2014).

11. Given their remarkable autonomy and the agency problems others have described, it seems most apt to describe the services as existing in a constrained Hobbesian environment, with regard to both joint-force commanders (among services level of analysis) and their would-be external masters like the DoD, the executive, and Congress (external relations level of analysis). Interservice rivalry is as potent as one can imagine, just as long as one party does not drive another to seek relief outside the Pentagon. For Hobbes's description of the "state of nature" as "the ungoverned passions of individuals," which serve as "an analogue for the state of anarchy to which civil society is too easily returned," see Springborg 2007, 9. Sixty-nine percent of those surveyed identified a "compelling, shared strategic vision" as the primary cause that should drive jointness. An additional 5 percent selected it as one of multiple reasons (Birch 2014b).

12. A similar basis surfaces in Posen's (1984) explanation for military innovation. Recall his posited causes of and barriers to innovation: "Organizations innovate when they fail . . . when they are pressured from without . . . [and] when they wish to expand" (1984, 47). "Militaries oppose innovation, but we see some remarkable innovations. . . . Civilians do affect military doctrine. Their intervention is often responsible for the level of innovation and integration achieved in a military doctrine" (227).

13. If the services pursue the coalescence required to bring a joint initiative to fruition (as in seeking congressional support for the JPATS), it seems to counter Huntington's idea of objective military control. Such control lets military organizations

pursue defense matters according to priorities that they determine internally. As a stipulation for this freedom, a social contract levies a requirement that military organizations will not politicize themselves by reaching out for influence within the legitimate power-holding organizations like the executive and congressional branches that exert objective control over them.

14. Strictly speaking, in Law's vocabulary of network coalescence, he asserts that "the social should not be privileged" (1989, 113). Where jointness is primarily a social, not technological, construct, there is a case for dropping the qualifying statement about social influence if one extends the discussion to jointness.

15. Chiabotti (2014) noted that "a spirit of compromise and humility" among action officers who worked on the JPATS over many years was able to keep the cooperative vision of the program intact across a wide array of senior officers who rotated more frequently through Air Training Command. The desire to achieve a successful joint outcome was sufficient to reconcile the program's enduring themes with the "different agenda and . . . different take on just what it meant to be 'joint'" each senior leader brought to his term in office.

16. In fact, Locher's approach involved triangulation among elements within Congress, the Defense Department, and the greater executive branch, all which contained significant antireform factions, not least of whom was the secretary of defense himself, Caspar Weinberger.

17. "First, the public often expects organizations, in a time of threat, to rally around the flag in order to limit the uncertainty about the threat and to maximize the use of resources to meet a common challenge" (Svedin 2009, 2). Baker and Oneal (2001) summarized a correlation between the use of military force and public interest that resulted in greater presidential support when the United States initiated hostilities, and attempted to "effect some change in the international or regional geopolitical environment" (678), and the act received both media attention and aggressive promotion by the White House. All characteristics listed apply to the COIN air power observations in OEF and OIF.

18. General Jumper made the remark about General Shinseki (Jumper 2013).

19. The more uncertain the organizations are about how to define the crisis, or about how to respond to the crisis, the more likely they are to cooperate. They are more likely to agree in decision situations and tend, as an overarching strategy, to pursue concurrence seeking across the crisis (Svedin 2009, 133).

20. The sampled cross-section included active-duty and retired military personnel, those who have studied defense as an academic discipline, historians who specialize in aspects of military cooperation or strategy, and people who have worked for members of the external defense establishment. Several people fall into more than one category. While the goal of this study was not to make a scientific sample of the opinions of this large, diverse group, one of the entailments of the research was an opportunity to ask each of them, absent a preconceived opinion or leading questions, what they thought about the topic of jointness, whether it was worth pursuing as an end in itself, and what the most effective means of pursuit might be.

21. Thirty-four percent of respondents agreed that the current level of joint cooperation was "near-optimal," and most of these cited the enduring nature of OEF and OIF, which have brought the services into routine contact with one another.

Sixty-six percent of respondents replied that jointness was "too little (or almost non-existent)." While no one responded that it was "too much" at present, a handful commented that external actors (Congress, DoD) at times sought a level of jointness that was unrealistic or harmful (Birch 2014b).

22. This quest for a "true General Staff" has remained a central root of the Defense Reform movement, according to one of its central architects and thought leaders (Lind 2014). Lind does not believe that the Goldwater-Nichols Act changed the defense establishment in ways that further jointness or increase military effectiveness and efficiency.

23. McNeill (1982, vii) wrestled with the problem of how to move societies back from the precipice of increasingly destructive technologies—military-industrial developments that threatened to wipe out huge segments of the population whenever combat erupted. His observations about the innovative path to that point, though, are instructive to military professionals, whom he labeled the "macroparasites" of the civilization he was interested in saving.

24. The idea borrows from James Madison's idea (writing as "Publius") of the fundamental necessity for checks and balances in government. If men were angels, no government would be necessary. "If angels were to govern men, neither external nor internal controls on government would be necessary. In framing a government which is to be administered by men over men, the great difficulty lies in this: you must first enable the government to control the governed; and in the next place oblige it to control itself. A dependence on the people is, no doubt, the primary control on the government; but experience has taught mankind the necessity of auxiliary precautions" (Madison 1788).

25. President Johnson's "Tuesday Lunch Group," convened for management of the Vietnam War, offered him a chance to meet with close advisers in a forum smaller than the National Security Council, which assisted in his continual battle against press leaks. The series of meetings, though, gave a "widespread impression that Johnson micro-managed the bombing of North Vietnam," which was "dramatically confirmed" (Barrett 1991, 678) in archival evidence, and stands as an example of a low point in executive trust of the military services to achieve national objectives.

26. Huntington (1957) used the B-36 controversy (the "Revolt of the Admirals") and the attendant congressional hearings to show how defense advocates use a "pluralistic strategy" (419) to pursue their relatively weak political position.

Abbott, Andrew D. 1988. *The System of Professions: An Essay on the Division of Expert Labor.* Chicago: University of Chicago Press.

"Air Force, Navy Name Joint Training Aircraft 'Texan II.'" 1997. *Airman* 41 (8): 12.

"Air Force and Navy Agree to Closer Ties." 1983. *Air Force Magazine.* January, 30.

"Air Force Approves Raytheon's T-6A Trainer for Full-Rate Production." 2001. *Defense Daily International* 3 (6): 1.

Alcala, Raoul. 2013. Chief (1987–1991), Doctrine, Concepts, and Systems Integration Division, US Army Deputy Chief of Staff for Operations and Plans. Washington DC, personal interview with Jeffrey Donnithorne, February 13.

Allard, Kenneth. 1996. *Command, Control, and the Common Defense.* New Haven, CT: Yale University Press.

Allison, Graham T., and Philip Zelikow. 1999. *Essence of Decision: Explaining the Cuban Missile Crisis.* 2nd ed. New York: Longman.

Allvin, David. 2022. Vice Chief of Staff, US Air Force. Washington DC, online address to US Air Force School of Advanced Air and Space Studies alumni, February 17.

Almquist, Peter. 1987. "Moscow's Conventional Wisdom: Soviet Views of the European Balance." *Arms Control Today* 17 (10): 16–21.

"America's Approach to Command and Control Goes Peer to Peer: Warfare's Worldwide Web." 2021. *Economist,* January 9.

Ancker, Clinton J., III. 2013. "The Evolution of Mission Command in U.S. Army Doctrine, 1905 to the Present." *Military Review* 93 (2): 42–52.

Anderegg, C. R. 2001. *Sierra Hotel: Flying Air Force Fighters in the Decade after Vietnam.* Washington, DC: US GPO.

Aris, Hakki. 2003. "A Programme for Success." *NATO's Nations and Partners for Peace,* Fall, 147–150.

Armstrong, Brian S. 2014. US Air Force instructor pilot (1999–2003), Joint Specialized Undergraduate Pilot Training, Naval Air Station Whiting Field, FL. Washington, DC, personal interview with Paul Birch, April 7.

Armstrong, David A. 1995. "Jointness and the Impact of the War." *Joint Forces Quarterly,* no. 8 (Summer): 36–37.

"Army: Tremendous Demand for UAVs in Afghanistan." 2010. *Army Times,* December 16.

Art, Robert J. 1968. *The TFX Decision: McNamara and the Military.* Boston: Little, Brown.

Aspin, Les. 1993a. Policy on the Assignment of Women in the Armed Forces.

Secretary of Defense official correspondence. Washington, DC: Department of Defense, April 28.

———. 1993b. Roles, Missions, and Functions of the Armed Forces of the United States. Secretary of Defense official correspondence. Washington, DC: Department of Defense, April 16. Copy reviewed in personal collection of Raymond O'Keefe, Universal City, TX.

Atkinson, Rick. 1993. *Crusade: The Untold Story of the Persian Gulf War*. Boston: Houghton Mifflin.

Avant, Deborah D. 2007. "Political Institutions and Military Effectiveness: Contemporary United States and United Kingdom." In *Creating Military Power: The Sources of Military Effectiveness*, edited by Risa A. Brooks and Elizabeth A. Stanley, 80–105. Stanford: Stanford University Press.

Axe, David. 2021. "Pentagon Scientists Mull a Big Problem—How to Destroy a Whole Lot of Chinese Ships and Russian Tanks, and Fast." *Forbes*, March 9. www.forbes.com/sites/davidaxe/2021/03/09/pentagon-scientists-mull-a-big -problem-how-to-destroy-a-whole-lot-of-chinese-ships-and-russian-tanks/?sh =1c119aa876d5.

Axelrod, Robert, and Robert O. Keohane. 1985. "Achieving Cooperation under Anarchy: Strategies and Institutions." *World Politics* 38 (1): 226–254.

Babcock, Anthony D. 2014. Chief (2013–2014), Exercise Branch, US Air Force Strategic Plans and Programs Division. Washington DC, personal interview with Paul Birch, March 21.

Bacevich, Andrew J. 2008. *The Limits of Power: The End of American Exceptionalism*. New York: Metropolitan.

Baker, William D., and John R. Oneal. 2001. "Patriotism or Opinion Leadership? The Nature and Origins of the 'Rally 'Round the Flag Effect.'" *Journal of Conflict Resolution* 45 (5): 661–687.

Banks, Howard, and Thomas Jaffe. 1994. "Pat's Fit Snit." *Forbes* 153 (3): 20.

Barajas, Ashley. 2022. "Bell's High-Speed Vertical Take-Off and Landing Hits a Milestone." *Flying*, February 23.

Barlow, Jeffrey G. 1995. *Revolt of the Admirals: The Fight for Naval Aviation, 1945–1950*, 2001 ed. Washington, DC: Ross & Perry.

Barnes, Julian E. 2007. "He Helped Clear the Fog of War." *Los Angeles Times*, September 13. articles.latimes.com/2007/sep/13/nation/na-rover13.

Barnett, Jackson. 2021. "Air Force Inks New ABMS Concept Document." *Fed Scoop*. www.fedscoop.com/air-force-abms-concept-document-signed-jadc2/.

Barrett, David M. 1991. "Doing 'Tuesday Lunch' at Lyndon Johnson's White House: New Archival Evidence on Vietnam Decisionmaking." *Political Science and Politics* 24 (4): 676–679.

Bartholomew, Derek. 2014 US Air Force instructor pilot (1999–2003), Joint Specialized Undergraduate Pilot Training, Naval Air Station Whiting Field, FL. Washington, DC, telephonic interview with Paul Birch, May 2.

Bates, Matthew. 2009. "Sniper Pod Improves Capabilities, Lethality of B-1." US Air Force press release. San Antonio, TX: Defense Media Activity, March 20. www.af .mil/News/Article-Display/Article/120867/sniper-pod-improves-capabilities -lethality-of-b-1/.

Baxter, Vern, and A. V. Margavio. 2000. "Honor, Status, and Aggression in Economic Exchange." *Sociological Theory* 18 (3): 399–416.

Beare, Stuart A. 2014. Commander (2011–2014), Canadian Joint Operations Command. Ottawa, ON, telephonic interview with Paul Birch, February 27.

Beebe, Kenneth. 2006. "The Air Force's Missing Doctrine: How the US Air Force Ignores Counterinsurgency." *Air & Space Power Journal* 20 (1): 27–34.

Belasco, Amy. 2009. *Troop Levels in the Afghan and Iraq Wars, FY2001–FY2012: Cost and Other Potential Issues*. Rpt. R40682. Washington, DC: Congressional Research Service.

Bell, Isaac. 2014. Aide-de-Camp (2012–2014) to the Chief of Staff of the US Air Force, social media post, March 18.

Berman, Paul. 1978. *The Study of Macro and Micro Implementation of Social Policy*. Santa Monica, CA: RAND Corporation.

Biass, Eric. H., and Wesley Fox. 2012. "Baseline Aircraft on Steroids." *Armada International* 36 (4): 22–32.

Bijker, Wiebe E., Thomas P. Hughes, and Trevor Pinch. 1987. *The Social Construction of Technological Systems*. Cambridge MA: MIT Press.

Bijker, Wiebe E., W. Bernard Carlson, and Trevor Pinch. 2000. *Inside Technology*. Cambridge, MA: MIT Press.

Birch, Paul. 2014a. "Purple Reign: Ascent and Decline of Joint Behavior in the U.S. Military." PhD diss., School of Advanced Air and Space Studies, Air University.

———. 2014b. "Survey of Views on Joint Cooperation." Unpublished survey of military leaders conducted for dissertation research. Montgomery, AL, July.

Birch, Paul, Ray Reeves, and Brad DeWees. 2020. "Build ABMS from the Bottom Up, for the Joint Force." *Breaking Defense*. breakingdefense.com/2020/05/build-abms-from-bottom-up-for-the-joint-force/.

Bochain, Louis G. 2014. Commander (July 2001–July 2003), US Air Force 20th Air Support Operations Squadron. Hampton VA, personal interview with Paul Birch, January 24.

Boot, Max. 2002. *The Savage Wars of Peace: Small Wars and the Rise of American Power*. New York: Basic.

Boyington, Ed. 2002. "CNATRA: Situation Report." *Wings of Gold* 27 (3): 23.

Boyle, Edmund J., Mark M. Higgins, and Ghon S. Rhee. 1997. "Stock Market Reaction to Ethical Initiatives of Defense Contractors: Theory and Evidence." *Critical Perspectives on Accounting* 8 (6): 541–561.

Boyne, Walter J. 2009. "How the Predator Grew Teeth." *Air Force Magazine* 92 (7): 42–45.

Braybrook, Roy. 2013. "Trainers, Aviation's Sine Qua Non." *Armada International* 37 (3): 42–52.

Brinkley, Douglas. 1998. *Unfinished Presidency: Jimmy Carter's Journey to the Nobel Peace Prize*. New York: Penguin.

Broder, John M. 1990. "US War Plan in Iraq: 'Decapitate' Leadership." *Los Angeles Times*, September 16.

Brose, Christian. 2020. *The Kill Chain: Defending America in the Future of High-Tech Warfare*. New York: Hachette.

Brownlee, Romie L., and William J. Mullen, III. 1979. *Changing an Army: An Oral*

History of General William E. DePuy, USA Retired. Washington DC: US Army Center of Military History.

Brzezinski, Zbigniew. 1979. "NSC Weekly Report #92." Weekly Reports (to the President) 91–101 (3/79–6/79); Container 42; Donated Historical Material; Zbigniew Brzezinski Collection. Atlanta, GA: Jimmy Carter Library, March 30.

———. 2009. "An Agenda for NATO: Toward a Global Security Web." *Foreign Affairs* 88 (5): 2–20.

Buede, Dennis M., and Terry A. Bresnick. 2007. "Applications of Decision Analysis to the Military Systems Acquisition Process." In *Advances in Decision Analysis: From Foundations to Applications*, edited by Ralph F. Miles Ward Edwards, Jr., and Detlof von Winterfeldt, 539–563. Cambridge: Cambridge University Press.

Builder, Carl H. 1989. *The Masks of War: American Military Styles in Strategy and Analysis.* Baltimore: Johns Hopkins University Press.

———. 2002. *The Icarus Syndrome: The Role of Air Power Theory in the Evolution and Fate of the US Air Force.* New York: Transaction.

Burgess, Mark. 2002. *Killing Your Own: The Problem of Friendly Fire during the Afghan Campaign.* Washington, DC: Center for Defense Information.

Call, Steve. 2007. *Danger Close: Tactical Air Controllers in Afghanistan and Iraq.* College Station: Texas A & M University Press.

Callahan, Bryan. 2013. "The Limits of Airpower in Information-Dominant Warfare." MA thesis, School of Advanced Air and Space Studies, Air University.

Campo, Joseph L. 2013. Commander (2012–2014), US Air Force 26th Weapons Squadron. Las Vegas NV, telephonic interview with Paul Birch, November 19.

Canan, James W. 1984. "NATO on the Upbeat." *Air Force Magazine*, September 1. www.airandspaceforces.com/article/0984nato/.

Cannon, Monte R. 2012. "Cleaning Up the Joint: Command, Control, and Agency in American War Fighting." PhD diss., Air University.

Cardwell, Thomas A., III. 1992. *Airland Combat: An Organization for Joint Warfare.* Maxwell Air Force Base, AL: Air University Press.

Carroll, Matt. 1995. "Raytheon Wins Contract for Planes." *Boston Globe*, June 23:85.

Carver, Michael. 1986. "Conventional Warfare in the Nuclear Age." In *Makers of Modern Strategy from Machiavelli to the Nuclear Age*, edited by Peter Paret, 779–814. Princeton, NJ: Princeton University Press.

Castelli, Christopher J. 2010. "Report: Navy Wasting Hundreds of Millions on Unneeded Aircraft." *Inside the Pentagon's Inside the Navy* 23 (48), December 6.

———. 2011. "Navy Slashes Training Aircraft Procurement by Twenty Planes, Shifts Funds." *Inside the Pentagon* 27 (1), January 6.

"CENTCOMM Commander Urges Close Air Support Examination." 2004. *Jane's IHS: Jane's Defense & Security*, April 2.

Chai, Sun-Ki. 1997. "Rational Choice and Culture: Clashing Perspective or Complementary Modes of Analysis?" In *Culture Matters: Essays in Honor of Aaron Wildavsky*, edited by Richard Ellis and Michael Thompson, 45–56. Boulder, CO: Westview.

Chapman, Anne W., et al. 1998. *Prepare the Army for War: A Historical Overview of the Army Training and Doctrine Command, 1973–1998* Fort Monroe, VA: TRADOC Military History Office.

Charette, Robert N. 2008. "What's Wrong with Weapons Acquisitions." *IEEE Spectrum*, November 1.

Charleston News and Courier. 1988. "Navy Accused of 'Stonewalling' on Base Closure Issue." December 30.

Chayes, Antonia. 2008. "How American Treaty Behavior Threatens National Security." *International Security* 33 (1): 45–81.

Chiabotti, Stephen D. 2008. "'Heterogenius' Engineering and JPATS: Leadership, Logic, and Acquisition Requirements." In *Concepts for Air Force Leadership*. Maxwell Air Force Base, AL: Air University Press.

———. 2014. Chief (1987–1992), Aircraft Program Management, US Air Force Air Training Command. Maxwell Air Force Base, AL, e-mail exchange with Paul Birch, June 6.

Chomsky, A. Noam, and David Barsamian. 2013. *Power Systems: Conversations on Global Democratic Uprisings and the New Challenges to US Empire*. New York: Henry Holt.

Church, Aaron M. U. 2014. "Are RPA Pilots the New Normal?" *Air Force Magazine*, April, 34–38. www.airandspaceforces.com/article/0414rpa/.

Cichowski, Kurt A. 1993. *Doctrine Matures through a Storm: An Analysis of the New Air Force Manual 1–1*. Maxwell Air Force Base, AL: Air University Press.

Clancy, Tom, and Chuck Horner. 1999. *Every Man a Tiger*. New York: G. P. Putnam & Sons.

Clark, Stephen. 2013. Staff officer, US Air Force Strategy, Concepts and Wargaming Division. Washington, DC, personal interviews with Paul Birch, November 17 and 27.

Clausewitz, Carl von. 1984. *On War*. Translated by Michael Eliot Howard and Peter Paret. Edited by Michael Eliot Howard and Peter Paret. Rev. ed. Princeton, NJ: Princeton University Press.

Cohen, Eliot A., 2003 ed. *Supreme Command: Soldiers, Statesmen, and Leadership in Wartime*. New York: Anchor.

Colarusso, Laura M. 2002. "Raytheon Receives $30 Million to Produce T-6A Trainers for the Navy." *Inside the Air Force* 13 (17): 11–12. www.jstor.org/stable /e24789562.

Coleman, James S. 1988. "Free Riders and Zealots: The Role of Social Networks." *Sociological Theory* 6 (1): 52–57.

Congressional Budget Office. 1977. *Assessing the NATO/Warsaw Pact Military Balance*. Washington, DC: Congressional Budget Office.

Conway, John L., III. 2007. "New USAF Doctrine Publication: Air Force Doctrine Document 2–10, Homeland Operations." *Air & Space Power Journal* 21 (1): 23.

Cooling, Benjamin Franklin. 1990. "Introduction." In *Case Studies in the Development of Close Air Support*, edited by Benjamin Franklin Cooling, 1–12. Washington, DC: US Government Printing Office.

Cope, Jay. 2009. "The Future Has Arrived—First T-6Bs Land at Whiting Field to Fanfare." *Whiting Tower* 65 (35): 1–5. ufdcimages.uflib.ufl.edu/UF/00/09/86 /19/00034/00009-02-2009.pdf.

Corum, James S., and Wray R. Johnson. 2003. *Airpower in Small Wars: Fighting Insurgents and Terrorists*. Lawrence: University Press of Kansas.

Costello, Peter A., III. 1997. "A Matter of Trust: Close Air Support Apportionment and Allocation for Operational Level Effects." MA thesis, Air Command and Staff College, Air University.

Coté, Owen R. 1996. "The Politics of Innovative Military Doctrine: The U.S. Navy and Fleet Ballistic Missiles." PhD diss., Massachusetts Institute of Technology. hdl.handle.net/1721.1/11217.

Coulam, Robert F. 1977. *Illusions of Choice: The F-111 and the Problem of Weapons Acquisition Reform.* Princeton, NJ: Princeton University Press.

Cowie, Leland, Todd Graff, Craig Cude, and Brad DeWees. 2021. "To Build Joint Command and Control, First Break Joint Command and Control." *War on the Rocks,* July 2. warontherocks.com/2021/07/to-build-joint-command-and -control-first-break-joint-command-and-control/.

Creech, Wilbur L. 1981. "Air-Land Battle of the Future; Personal Notes for a Speech to the Association of the U.S. Army." IRIS No. 01126108. Maxwell Air Force Base, AL: Air Force Historical Research Agency, October 21.

Crick, Bernard. 1973. *Political Theory and Practice.* New York: Basic.

Cyert, Richard M., and James G. March. 1963. *A Behavioral Theory of the Firm.* 1992 ed. Hoboken, NJ: Wiley-Blackwell.

Dabros, Michael R. 2014. Commander (2011–2013), Canadian Defense Liaison Staff. London, ON, telephonic interview with Paul Birch, February 7.

Davis, Mark G. 2004. "Operation Anaconda: Command and Confusion in Joint Warfare." MA degree, School of Advanced Air and Space Studies, Air University.

———. 2005. *Operation ANACONDA: An Air Power Perspective.* Office of Air Force Lessons Learned (AF/XOL), Headquarters United States Air Force (The Pentagon).

Davis, Richard G. 1987. *The 31 Initiatives: A Study in Air Force-Army Cooperation.* Washington, DC: Office of Air Force History.

Dawes, Robyn M., Alphons J. C. Van de Kragt, and John M. Orbell. 1990. "Cooperation for the Benefit of Us—Not Me, or My Conscience." In *Beyond Self-Interest,* edited by J. J. Mansbridge, 97–110. Chicago: University of Chicago Press.

Dean, Peter J. 2005. "Napoleon as a Military Commander: The Limitations of Genius." *The Napoleon Series.* www.napoleon-series.org/research/napoleon /c_genius.html.

DeConcini, Dennis W. 1992. Additional Statements. 102nd Cong., 2nd sess., *Congressional Record* 138 (102): S9956.

DeGarmo, James. 2014. Manager, Raytheon and Hawker Beechcraft, various T-6 and AT-6 programs; former JPATS program manager (July 1989–April 1992). San Antonio, TX, personal interview with Paul Birch, April 28.

Demarest, Geoff. 2010. "Let's Take the French Experience of Algeria out of U.S. Counterinsurgency Doctrine." *Military Review* 90 (4): 19–24.

Dempsey, Martin E. 2012. *America's Military—A Profession of Arms.* Chairmans of the Joint Chiefs of Staff white paper. Washington, DC: The Joint Staff.

Den Beste, Edwin J., and Gary M. Servold. 1995. *CAS Integration Lessons.* Rpt. 95-1. Fort Leavenworth, KS: Center for Army Lessons Learned.

Deptula, David A. 1991. "Trends in Joint, Army, and USAF Doctrine Development." K239.0472-1; IRIS No. 0876156, August 27. Maxwell Air Force Base, AL: Air Force Historical Research Agency.

―――. 2013. Director (2001), Combined Air Operations Center, Operation Enduring Freedom; Air Force Deputy Chief of Staff (2006–2010) for Intelligence, Surveillance, and Reconnaissance. Arlington, VA, personal interview with Paul Birch, December 5.

―――. 2014. Electronic mailing distribution to Air Force Association members, June 5.

―――. 2021. "Happy Billy Mitchell Day!" Twitter Post, @Deptula_David, December 29.

DePuy, William. 1994a. " 'Pot of Soup' Letter to TRADOC Subordinate Commanders." Folder O; Box: Personal Files 1974–1975. In *Selected Papers of General William E. DePuy*, edited by Richard M. Swain. Carlisle Barracks, PA: US Army Military History Institute.

―――. 1994b. "Letter to General Frederick C. Weyand." Folder: Field Manuals 100-5, 1974–1977; Box: Transcripts and Diplomas. In *Selected Papers of General William E. DePuy*, edited by Richard M. Swain. Carlisle Barracks, PA: US Army Military History Institute.

―――. 1994c. "Letter to General Creighton W. Abrams, Jr. (analyzing the Arab-Israeli War, 1973);" Box: Deputy CG TRADOC; The Orwin C. Talbott Papers. In *Selected Papers of General William E. DePuy*, edited by Richard M. Swain, 69–74. Fort Leavenworth, KS: Combat Studies Institute.

―――. 1994d. "Presentation to the TRADOC Commanders' Vision '91 Conference." Folder: Army of the Future; Box 2: Official Papers-CG, TRADOC; etc.; The Maxwell R. Thurman Papers. In *Selected Papers of General William E. DePuy*, edited by Richard M. Swain. Carlisle Barracks, PA: US Army Military History Institute.

―――. 1994e. "Letter to General Frederick J. Kroesen." Folder: Miscellaneous Correspondence. Carlisle Barracks, PA: US Army Military History Institute.

―――. 1994f. "Letter to Robert W. Komer." Folder: Correspondence 1973–1977; The William E. DePuy Papers." Carlisle Barracks, PA: US Army Military History Institute.

―――. 1994g. "Letter to Major General Gordon Sumner, Jr." Folder: R-S-T; Box 8: TRADOC Commander's Conference, 1975, The William E. DePuy Papers. In *Selected Papers of General William E. DePuy*, edited by Richard M. Swain. Carlisle Barracks, PA: US Army Military History Institute.

―――. 1994h. "Talking Paper on Field Manual 100-5, Operations." In *Selected Papers of General William E. DePuy*, edited by Richard M. Swain. Carlisle Barracks, PA: US Army Military History Institute.

―――. 1994i. "Memorandum (draft) for the U.S. Army Chief of Staff; Subject: How to Determine Requirements for the Army's Weapons; Field Manuals 100-5, 1974–1977." Box: Transcripts and Diplomas. In *Selected Papers of General William E. DePuy*, edited by Richard M. Swain. Carlisle Barracks, PA: US Army Military History Institute.

Desch, Michael C. 1999. *Civilian Control of the Military: The Changing Security Environment*. Baltimore: Johns Hopkins University Press.

De Tocqueville, Alexis. 1835. *Democracy in America*, vol. 1. Translated by Henry Reeve. Apple Books ed.

————. 1840. *Democracy in America*, vol. 2. Translated by Henry Reeve. Apple Books ed.

Deutch, John M. 1994. Consolidation of Fixed-Wing Flight Training. Deputy Secretary of Defense official correspondence. Washington, DC: Department of Defense, October 24. Copy in personal collection of Raymond O'Keefe, Universal City, TX.

Dinges, Edward A., and Richard H. Sinnreich. 1980. "Battlefield Interdiction: Old Term, New Problem." *Field Artillery Journal* 48 (1): 14–20. permanent.fdlp.gov /lps13201/1980/JAN_FEB_1980_FULL_EDITION.pdf.

Dixon, Robert. 1976. "Letter to General William E. DePuy." K168.03-2342 Pt. 3; IRIS No. 1137059, July 13. Maxwell Air Force Base, AL: Air Force Historical Research Agency.

Dixon, Robert, and William DePuy. 1975. *Air Force Policy Letter for Commanders.* Washington, DC: Department of the Air Force.

"DoD Says Acquisition Reform Pilot Programs Making Progress." 1996. *Defense Daily*, March 19.

Dolman, Everett Carl. 2005. *Pure Strategy: Power and Principle in the Space and Information Age*. New York: Frank Cass.

Donnelly, Peter A. 2014. Former Assistant Director of Operations, US Air Force 20th Air Support Operations Squadron. Hampton VA, personal interview with Paul Birch, January 24.

Donnithorne, Jeffrey W. 2013. "Principled Agents: Service Culture, Bargaining, and Agency in American Civil-Military Relations." PhD diss., Georgetown University. repository.library.georgetown.edu/bitstream/handle/10822/559479 /Donnithorne_georgetown_0076D_12323.pdf.

Doolittle, James H., and Carol V. Glines. 1991. *I Could Never Be So Lucky Again.* 1992 ed. New York: Bantam.

Dorell, Oren. 2013. "Pentagon Moves Naval Forces Closer to Syria." *USA Today*, August 25. www.usatoday.com/story/news/world/2013/08/23/syria-chemical -attack/2690757/.

Dorr, Robert F. 2012. "History Mystery." *Air Power History* 59 (4): 64.

Dougherty, Tim. 2002. "Jumper: Ingenuity Is Key to Transformation." *Air Force Print News*, March 21.

Dunlap, Charles J., Jr. 2007. "Air-Minded Considerations for Joint Counterinsurgency Doctrine." *Air & Space Power Journal* 21 (4): 63–74. www.e-ir.info/2012 /01/25/air-minded-considerations-for-joint-counterinsurgency-doctrine/.

Dunnigan, Jim, ed. 2012. "Warplanes: UAV Growth Continues." *StrategyPage*. October 30. strategypage.com/htmw/htairfo/articles/20121030.aspx.

Easterly, William. 2008. "Institutions: Top Down or Bottom Up?" *American Economic Review* 98 (2): 95–99.

Easton, David. 1953. *The Political System: An Inquiry into the State of Political Science*. 1971 ed. New York: Knopf.

Eddins, Joseph. 2021. "Valenzia: ABMS Will Deliver the 'Decision Advantage.'" *Airman Magazine*. May 26.

Edwards, John R. 2014. B-52 instructor weapons systems officer (2003-2006), US

Air Force Weapons School, Washington, DC, personal interview with Paul Birch, February 5.

Ehrhard, Thomas P. 2000. "Unmanned Aerial Vehicles in the United States Armed Services: A Comparative Study of Weapon System Innovation." PhD diss., Johns Hopkins University.

Eisenhardt, Kathleen M. 1989. "Agency Theory: An Assessment and Review." *Academy of Management Review* 14 (1): 57–74.

Eisenhower, Dwight D. 1958. Special Message to the Congress on Reorganization of the Defense Establishment. The American Presidency Project, by Gerhard Peters and John T. Woolley, April 3. www.presidency.ucsb.edu/node/234642.

Emerson, Richard M. 1962. "Power-Dependence Relations." *American Sociological Review* 27 (1): 31–41.

Emme, Eugene M. 1979. "The American Dimension." In *Air Power and Warfare: The Proceedings of the 8th Military History Symposium*, edited by Alfred F. Hurley and Robert C. Ehrhart, 56–82. Washington, DC: Office of Air Force History.

Emmons, Richard H. 1991. *Specialized Undergraduate Pilot Training and the Tanker-Transport Training System*. Air Training Command official history. Randolph Air Force Base, TX: Air Education and Training Command Office of History and Research.

Erlanger, Steven. 1993. "Russia's Workers Pay Price as Military Industries Fade." *New York Times*, December 3, 1993.

Eversden, Andrew. 2020. "US Army, Air Force Sign Agreement to Develop Joint All-Domain Concept." *C4ISRNet*, October 5. www.c4isrnet.com/battlefield-tech/it-networks/2020/10/05/us-army-air-force-sign-agreement-to-develop-joint-all-domain-concept/.

"F-15E Debuts with Targeting Pod, Bomb." 2005. *Jane's Defense Weekly*, January 14.

Feaver, Peter. 1999. "Civil-Military Relations." *Annual Review of Political Science* 2:211–241.

Foglesong, Robert H. 2004. "Springboard for Airpower: Remarks by General Robert H. Foglesong." *Air Force Magazine*, March 1. www.airforcemag.com/MagazineArchive/Pages/2004/March2004/0304airpower.aspx.

Foote, Sheila. 1999. "Senate Appropriators Add Money for LHD-1 Ship, NMD." *Defense Daily* 202 (39): 1.

Freedberg, Sydney J., Jr., and Theresa Hitchens. 2020. "Army, Air Force Get Serious on JADC2: Joint Exercises in 2021." *Breaking Defense*, October 9. breakingdefense.com/2020/10/army-air-force-get-serious-on-jadc2-joint-exercises-in-2021/.

Frieden, Jeffry A. 1999. "Actors and Preferences in International Relations." In *Strategic Choice and International Relations*, edited by David A. Lake and Robert Powell, 39–76. Princeton, NJ: Princeton University Press.

Friscolanti, Michael. 2005. *Friendly Fire: The Untold Story of the U.S. Bombing That Killed Four Canadian Soldiers in Afghanistan*. Mississauga, Ontario: John Wiley & Sons Canada.

Fulghum, David A. 1991. "Lack of Firm Requirements, Force Reductions Raise JPATS Questions." *Aviation Week & Space Technology* 135 (9): 66–67.

————. 1995. "Ranger 2000 Tops in JPATS Test." *Aviation Week & Space Technology* 142 (23): 24–25.

Fulghum, David A., and John D. Morrocco. 1994. "Pentagon Battles over Raiding JPATS." *Aviation Week & Space Technology* 141 (8): 23.

Futrell, Robert Frank. 1989a. *Ideas, Concepts, Doctrine: Basic Thinking in the United States Air Force 1907–1960.* Vol. 1. Maxwell Air Force Base, AL: Air University Press.

————. 1989b. *Ideas, Concepts, Doctrine: Basic Thinking in the United States Air Force, 1961–1984.* Vol. 2. Maxwell Air Force Base, AL: Air University Press.

Gallup, Inc. 2013. *Confidence in Institutions.* www.gallup.com/poll/1597/confidence-institutions.aspx.

Gambold, Keven. 2013. Chief Operations Officer, Unmanned Experts LLC. Washington DC, telephonic interview with Paul Birch, November 12.

"GAO Rules in Favor of Raytheon in JPATS Protest." 1996. *Defense Daily*, February 6.

Gates, Robert M. 2014. *Duty: Memoirs of a Secretary at War.* Kindle ed. New York: Alfred A. Knopf.

Gear, James. 2013. Vice President, Strategic Business Development, L-3 Unmanned Systems. Washington, DC, telephonic interview with Paul Birch, November 18.

Geibel, Adam. 2002. "Operation Anaconda, Shah-i-Khot Valley, Afghanistan, 2–10 March 2002." *Military Review* May–June, 72–77. www.govinfo.gov/content/pkg/GOVPUB-D110-c8aaef82e26966248e7c3ab0267dc0a9/pdf/GOVPUB-D110-c8aaef82e26966248e7c3ab0267dc0a9.pdf.

Gelb, Leslie H. 1975. "Ford Fires Schlesinger, Colby." *Pittsburgh Post-Gazette*, November 3:1–2.

George, Alexander L., and Andrew Bennett. 2005. *Case Studies and Theory Development in the Social Sciences.* Cambridge, MA: MIT Press.

Gersten, Peter E. 2013. Deputy Director (2013–2014), Joint Staff Politico-Military Affairs (Western Hemisphere). Washington, DC, personal interview with Paul Birch, October 22.

Gething, Michael J. 2010. "USAF Launches Sensor Upgrade for Sniper and Litening Pods." *Jane's International Defense Review*, October 28.

Gething, Michael J., Scott R. Gourley, and Joris Janssen Lok. 2006. "Imagery Intelligence—Boom Time for Image Intelligence as Digital Exploitation Burgeons." *Jane's International Defense Review*, November 7.

Getz, Anne Marie. 1998. "Congressional Policy Making: The Goldwater-Nichols Defense Reorganization Act of 1986." PhD diss., Yale University.

Gilbert, M. Thomas P., Andrew Rambaut, Gabriela Wlasiuk, Thomas J. Spira, Arthur E. Pitchenik, and Michael Worobey. 2007. "The Emergence of HIV/AIDS in the Americas and Beyond." *Proceedings of the National Academy of Sciences of the United States of America* 104 (47): 18566–18570. pubmed.ncbi.nlm.nih.gov/17978186/.

Gillespie, Paul G. 2006. *Weapons of Choice: The Development of Precision-Guided Munitions.* Tuscaloosa: University of Alabama Press.

Gilmore, Gerry J. 2002. "Anaconda Is Success; Enemy Killed Unknown, Say Officials." American Forces Press Service, March 15. www.defense.gov/News/NewsArticle.aspx?ID=44244.

Godin, Jason Robert. 2004. "Coordinating Rooks and Bishops: An Institutional History of the Joint Army and Navy Board." MA thesis, Texas A & M University.

Goodnight, G. Thomas. 1986. "Ronald Reagan's Re-Formulation of the Rhetoric of War: Analysis of the 'Zero Option,' 'Evil Empire,' and 'Star Wars' Addresses." *Quarterly Journal of Speech* 72 (4): 390–414.

Gordon, Michael R. 1983. "The Army's Air-Land Battle Doctrine Worries Allies, Upsets the Air Force." *National Journal* 15 (25): 1274–1277.

———. 1991. "Ground Strategy: Focus on Rear Line." *New York Times*, February 17, 1991.

Gordon, Michael R., and Bernard E. Trainor. 2006. *Cobra II: The Inside Story of the Invasion and Occupation of Iraq.* New York: Pantheon.

Goyer, Robert. 1999a. "T-6A Texan II Gets FAA Certification." *Flying* 126 (11): 44.

———. 1999b. "Wagstaff to Demo JPATS." *Flying* 126 (5): 36.

Grant, Rebecca. 2003. "The Clash about CAS." *Air Force Magazine*, January, 54–59.

———. 2013. "The ROVER." *Air Force Magazine*, August, 38–42. www.airforcemag.com/PDF/MagazineArchive/Documents/2013/August%202013/0813rover.pdf.

Gray, Colin S. 1978. "Force Planning, Political Guidance, and the Decision to Fight." *Military Review* 58 (4): 5–16.

———. 1999. *Modern Strategy.* Oxford: Oxford University Press.

———. 2007. *Fighting Talk: Forty Maxims on War, Peace, and Strategy.* Westport, CT: Praeger Security International.

Greenert, Jonathan W. 2014. "The Navy's Rebalance to Asia: Challenges and Opportunities." Washignton, DC, remarks at The Center for Strategic and International Studies, May 19. www.csis.org/events/navys-rebalance-asia-challenges-and-opportunities-featuring-admiral-jonathan-greenert.

Groendyk, Dave. 2007. "T-6A Texan II Reaches Half-Million Flight Hours." US Air Force press release, August 15. www.wpafb.af.mil/News/Article-Display/Article/401221/t-6a-texan-ii-reaches-half-million-flight-hours/.

Guest, Robert. 2013. "The 'Good War' Winds Down." *Economist: The World in 2014*, October 30:62.

Hacker, Jacob S., and P. Pierson. 2010. *Winner-Takes-All Politics: How Washington Made the Rich Richer and Turned Its Back on the Middle Class.* New York: Simon & Schuster.

Hadley, Greg. 2021a. "New NDAA Takes Aim at F-35 Sustainment Costs, Joint Program Office." *Air Force Magazine*, December 10. www.airforcemag.com/new-ndaa-f-35-joint-program-office-sustainment/.

———. 2021b. "Lack of JADC2 Coordination across Services Is 'Recipe for Disaster,' Analyst Warns." *Air Force Magazine*, August 6. www.airandspaceforces.com/lack-of-jadc2-coordination-across-services-is-recipe-for-disaster-analyst-warns/.

Hall, Michael D., Peter H. Gressy, and Charles G. Boyd. 1988. "Memorandum of Understanding: USAF and USN Concerning Development and Acquisition of Trainer Aircraft." SD III-15; ATC Periodic History 1988; Vol. X; (6 December). Randolph Air Force Base, TX: Air Education and Training Command History Office.

Hall, Peter A. 2003. "Aligning Ontology and Methodology in Comparative Politics." In *Comparative Historical Analysis in the Social Sciences*, edited by James Mahoney and Dietrich Rueschemeyer, 373–406. Cambridge: Cambridge University Press.

Hall, Thomas. 2014. Lieutenant Commander, US Navy. Washington, DC, telephonic interview with Paul Birch, March 31.

Halperin, Morton H. 1974. *Bureaucratic Politics and Foreign Policy*. Washington, DC: Brookings Institution.

Hampson, Fen Osler. 1999. "Barriers to Negotiation and Requisites for Success." In *Multilateral Negotiations: Lessons from Arms Control, Trade, and the Environment*, 23–51. Baltimore: Johns Hopkins University Press.

Harrison, Todd. 2021. "Battle Networks and the Future Force." Washington, DC: Center for Strategic and International Studies, August 5. www.csis.org/analysis/battle-networks-and-future-force.

Harvey, William. 2014. Major, US Marine Corps; Joint action officer, Air Land Sea Applications Center. Washington, DC: telephonic interview with Paul Birch, January 22.

"HASC: Air Force May Need More B-2s for Global Strike Task Force," 2001. *Defense Daily International* 2 (30): 1–2.

Hasken, Scott A. 2003. "A Historical Look at Close Air Support." MA thesis, US Army Command and General Staff College.

Hayden, Teresa N. 2006. "Back When IBM Had Balls." Blog post, August 20. http://nielsenhayden.com/makinglight/archives/007893.html.

Hayward, Thomas. 1982. Reorganization Proposals for the Joint Chiefs of Staff, Admiral Hayward's testimony.

Hechter, Michael. 1990. "The Emergence of Cooperative Social Institutions." In *Social Institutions: Their Emergence, Maintenance and Effects*, edited by Michael Hechter, Karl-Dieter Opp, and Reinhard Wippler, 13–33. Berlin and New York: Walter de Gruyter.

Heckathorn, Douglas D. 1990. "Collective Sanctions and Compliance Norms: A Formal Theory of Group-Mediated Social Control." *American Sociological Review* 55 (3): 366–384.

Heines, Vivienne. 2004. "Full-Blooded Texan T-6A Puts Muscle into Primary Flight Training." *Training & Simulation Journal*, October, 40.

Heller, Charles E., and William A. Stofft, eds. 1986. *America's First Battles, 1776–1965*. Lawrence: University Press of Kansas.

Henke, Glen A. 2008. "Planning Full Spectrum Operations: Implications of FM 3-0 on Planning Doctrine." *Military Review* 88 (6): 97101.

Herbert, Paul H. 1988. "Deciding What Has to Be Done: General William E. DePuy and the 1976 Edition of FM 10-5, Operations." *Leavenworth Papers* 16. Fort Leavenworth, KS: Combat Studies Institute.

Hermann, Charles F., et al. 2001. "Resolve, Accept, or Avoid: Effects of Group Conflict on Foreign Policy Decisions." *International Studies Review* 3 (2): 133–168.

Hermann, Margaret G., et al. 2001. "Who Leads Matters: The Effects of Powerful Individuals." *International Studies Review* 3 (2): 83–131.

Hewish, Mark. 2001. "Switchboards in the Sky." *Jane's International Defense Review*. June 20.

Hewson, Robert. 2008. "On Call: Honing Weapon Effects for the Close Air Support Role." *Jane's International Defense Review*, July 21.

Hiatt, Fred. 1984. "Army and Air Force Chiefs Vow to Cooperate on Arms, Tactics." *Washington Post*. May 23, A3.

Hinote, Clinton S. 2021. "Project Convergence 2021 Backbrief." Online presentation to US Air Force Futures directorate. Washington DC, November 12.

Hitchens, Theresa. 2019. "OSD & Joint Staff Grapple With Joint All-Domain Command." *Breaking Defense*. November 14. breakingdefense.com/2019/11/osd-joint-staff-grapple-with-joint-all-domain-command/.

———. 2020. "MDO Exclusive: Air Force Targets Primary Role in Joint C2." *Breaking Defense*, January 21. breakingdefense.com/2020/01/mdo-exclusive-air-force-targets-primary-role-in-joint-c2/.

———. 2021. "The Joint Warfighting Concept Failed, until It Focused on Space and Cyber." *Breaking Defense*. July 26. breakingdefense.com/2021/07/the-joint-warfighting-concept-failed-until-it-focused-on-space-and-cyber/.

Hoedemaker, Edward D. 1968. "Distrust and Aggression: An Interpersonal-International Analogy." *Journal of Conflict Resolution* 12 (1): 69–81.

Holley, I. B., Jr. 1990. "A Retrospect on Close Air Support." In *Case Studies in the Development of Close Air Support*, edited by Benjamin Franklin Cooling, 535–555. Washington, DC: Office of Air Force History.

Horwood, Ian. 2006. *Interservice Rivalry and Airpower in the Vietnam War*. Fort Leavenworth, KS: Combat Studies Institute Press.

Howard, Michael. 1974. "Chesney Memorial Gold Medal Lecture: Military Science in the Age of Peace." *RUSI Journal* 119 (1): 3–9.

———. 1976. *War in European History*. 2001 ed. New York: Oxford University Press.

Hrebenar, Ronald J., and Ruth K. Scott. 1997. *Interest Group Politics in America*. Armonk, NY: ME Sharpe.

Hughes, David. 1994. "USAF, Navy Enter Joint Training Era." *Aviation Week & Space Technology* 141 (8): 40–42.

Hughes, Thomas A. 1995. *Over Lord: General Pete Quesada and the Triumph of Tactical Air Power in World War II*. New York: Free Press.

Hughes, Thomas P. 1987. "The Evolution of Large Technological Systems." In *The Social Construction of Technological Systems*, edited by Wiebe E. Bijker, Thomas P. Hughes, and Trevor J. Pinch, 51–82. Cambridge MA: MIT Press.

Huntington, Samuel P. 1957. *The Soldier and the State: The Theory and Politics of Civil–Military Relations*. Cambridge, MA: Harvard University Press.

———. 1961. *The Common Defense: Strategic Programs in National Politics*. New York: Columbia University Press.

Hurley, Alfred F. 1975. *Billy Mitchell: Crusader for Airpower*. Bloomington: Indiana University Press.

Hurley, Alfred F., and William C. Heimdahl. 1997. "The Roots of US Military Aviation." In *Winged Shield, Winged Sword: A History of the United States Air Force*, edited by Bernard C. Nalty, 3–34. Washington, DC: Air Force History and Museums Program.

Insinna, Valerie. 2022. "With a New Office, Air Force Revamps ABMS Acquisition . . . Again." *Breaking Defense*, September 19. breakingdefense.com/2022/09/with-a-new-office-air-force-revamps-abms-acquisition-again/.

Isby, David. 1990. *Written Statement on Close Air Support.* Submitted to US House of Representatives, Committee on Armed Services, Hearing on Roles and Missions of Close Air Support Investigation Subcommittee. Washington, DC: US Government Printing Office, September 27.

Jacobs, Jody, David E. Johnson, Katherine Comanor, Lewis Jamison, Leland Joe, and David Vaughan. 2009. *Enhancing Fires and Maneuver Capability Through Greater Air-Ground Joint Interdependence.* Santa Monica, CA: RAND Corporation. www .rand.org/content/dam/rand/pubs/monographs/2009/RAND_MG793.sum .pdf.

Jaffe, Greg. 2003. "Divisions over Iraq Strain NATO-US Plan for War with Iraq Hinges on Close Air Support." *Wall Street Journal*, February 11.

Janis, Irving L. 1982. *Groupthink: Psychological Studies of Policy Decisions and Fiascoes.* 2nd ed. Boston: Houghton Mifflin.

Janowitz, Morris. 1960. *The Professional Soldier: A Social and Political Portrait.* Glencoe, IL: Free Press.

Jensen, Michael C. 1976. "Theory of the Firm: Managerial Behavior, Agency Costs and Ownership Structure." *Journal of Financial Economics* 3 (4): 305–360.

Jensen, Michael C., and William H. Meckling. 1976. "The Theory of the Firm: Managerial Behavior, Agency Costs, and Ownership Structure." In *The Theory of the Firm: Critical Perspectives on Business and Management*, edited by Nicolai J. Foss, 248–306. New York: Routledge.

Jervis, Robert. 1976. *Perception and Misperception in International Politics.* Princeton, NJ: Princeton University Press.

Jinnette, James. 2014. Division Chief (2011–2014), Headquarters US Air Force Combat Air Forces. Washington DC, telephonic interview with Paul Birch, January 7.

Johns, Lionel S., Peter Sharfman, Alan Shaw, Stephen Budiansky, Michael Callahan, Allen Greenburg, Peter Lert, and Nancy Lubin. 1987. "Technology Issues: Reconnaissance, Surveillance, and Target Acquisition to Support Follow-On Forces Attack." In *New Technology for NATO: Implementing Follow-On Forces Attack*, 143–167. Washington, DC: Office of Technology Assessment.

Johnson, David E. 1998. *Fast Tanks and Heavy Bombers: Innovation in the U.S. Army 1917–1945.* Ithaca: Cornell University Press.

———. 2007. *Learning Large Lessons: The Evolving Roles of Ground Power and Air Power in the Post-Cold War Era.* Santa Monica, CA: RAND Corporation.

———. 2011. *Hard Fighting: Israel in Lebanon and Gaza.* Santa Monica, CA: RAND Corporation.

———. 2014. Director, Chief of Staff of the Army Strategic Studies Group. Washington DC, personal interview with Paul Birch, February 24.

Johnson, David T., and Barry R. Schneider. 1991. "Current Issues in U.S. Defense Policy." Washington, DC: Center for Defense Information.

Johnson, Ken. 1995. "Raytheon in Bay State Slumps amid Records." *Eagle-Tribune*, July 14:21.

Johnson, Michael H. 2008. "Cleared to Engage: Improving the Effectiveness of Joint Close Air Support." *Air & Space Power Journal* 22 (2): 71–81.

Joint Chiefs of Staff. 1984a. *Department of Defense Dictionary of Military and Associated Terms.* Washington, DC: The Joint Staff.

———. 1984b. *Subject: Improving the Quality and Timeliness of JCS Advice/Responses*. Washington DC: The Joint Staff. Archival copy retained at Washington Navy Yard; Naval History and Heritage Command Archive.

———. 1989. *Department of Defense Dictionary of Military and Associated Terms*. Joint Publication 1-02. Washington, DC: The Joint Staff.

———. 1995. *Joint Tactics, Techniques, and Procedures for Joint Suppression of Enemy Air Defenses*. Joint Publication 3-01.4. Washington, DC: The Joint Staff.

———. 2009. *Close Air Support*. Joint Publication 3-09.3. Washington DC: The Joint Staff, July 8. www.bits.de/NRANEU/others/jp-doctrine/jp3_09_3%2809c%29.pdf.

———. 2010. *Department of Defense Dictionary of Military and Associated Terms*. Joint Publication 1-02. Washington, DC: The Joint Staff.

———. 2013. *Doctrine for the Armed Forces of the United States*. Joint Publication 1. Washington, DC: The Joint Staff.

———. 2014a. *Department of Defense Dictionary for Military and Associated Terms*. Joint Publication 1-02. Washington, DC: The Joint Staff.

———. 2014b. *Command and Control of Joint Air Operations*. Joint Publication 3-30. Washington, DC: The Joint Staff.

———. 2017. *Doctrine for the Armed Forces of the United States*. Joint Publication 1, Change 1. Washington, DC: The Joint Staff. irp.fas.org/doddir/dod/jp1.pdf.

Jones, Archer. 1978. "The New FM 100-5: A View from the Ivory Tower." *Military Review* 58 (2): 27–36.

Jones, David C. 1982. "Why the Joint Chiefs of Staff Must Change." *Presidential Studies Quarterly* 12 (2): 138–149.

Jones, Lawrence R., and Jerry L. McCaffery. 2008. *Budgeting, Financial Management, and Acquisition Reform in the U.S. Department of Defense: Research in Public Management*. Charlotte, NC: Information Age.

Jordan, Andrew Q. 2014. Major, US Army; Current Operations, Arabian Peninsula Working Group; US Special Operations Command Central. Washington, DC: telephonic interview with Paul Birch, February 3.

Joseph, Suzanne. 2000. "Anthropological Evolutionary Ecology: A Critique." *Journal of Ecological Anthropology* 4:6–30.

Jumper, John P. 2013. Chief of Staff (2001–2005), US Air Force. Reston VA, personal interview with Paul Birch, December 16.

———. 2014. Washington, DC, email exchange with Paul Birch, June 23.

Kaplan, Fred. 2003. "War-Gamed." *Slate*, March 28. www.slate.com/articles/news _and_politics/war_stories/2003/03/wargamed.single.html.

———. 2008. *Daydream Believers: How a Few Grand Ideas Wrecked American Power*. Hoboken, NJ: John Wiley & Sons.

Karber, Phillip A. 1976. "Dynamic Doctrine for Dynamic Defense." *Armed Forces Journal International* 114: 28–29.

Keiser, Gordon W. 1982. *The U.S. Marine Corps and Defense Unification, 1944–47: The Politics of Survival*. Washington, DC: National Defense University Press.

Keller, John. 2013. "MQ-9 Reaper Hunter-Killer UAV to Receive Electro-Optical Targeting Systems from Raytheon in $50.2 Million Contract." *Military Aerospace Electronics*, October 27. www.militaryaerospace.com/commercial-aerospace

/article/14227350/mq9-reaper-hunterkiller-uavs-to-receive-electrooptical
-targeting-systems-from-raytheon-in-502-million-contract.

Kem, Jack D. 2014. Supervisory Professor, United States Army Command and General Staff College. Fort Leavenworth, KS, telephonic interview with Paul Birch, May 20.

Kent, Glen A., and David A. Ochmanek. 1998. *Defining the Role of Airpower in Joint Missions.* Santa Monica, CA: RAND Corporation.

Keys, Ronald. 2004. "Lt Gen, USAF; Deputy Chief of Staff for Air and Space Operations (A-3), U.S. Air Force." Shephard's Air Power Conference, January.

Kim, Samuel S. 1976. "The Lorenzian Theory of Aggression and Peace Research: A Critique." *Journal of Peace Research* 13 (4): 253–276.

King, Samuel, Jr. 2013. "Air Force Begins Testing New Pod Capability." US Air Force press release, January 18. www.dvidshub.net/news/100728/air-force-begins -testing-new-pod-capability.

Kirk, Michael. 2008. "Bush's War: Part I." *Frontline*, edited by Tim Mangini, March 24. Arlington, VA: Public Broadcasting Service. www.pbs.org/wgbh/frontline /documentary/bushswar/.

Kissinger, Henry A. 1957. *Nuclear Weapons and Foreign Policy.* 3rd ed. New York: Council on Foreign Relations.

———. 1976. "Building an Enduring Foreign Policy: Creative Leadership in a Moment of Uncertainty." *Vital Speeches of the Day,* January 1:166–172.

———. 2008. "What Vietnam Teaches Us." *Newsweek,* November 3:44–46.

Koester, Jay. 2020. "JADC2 'Experiment 2' Provides Looking Glass into Future Experimentation." US Army press release, April 23. www.army.mil/article/23 4900/jadc2_experiment_2_provides_looking_glass_into_future_experimen tation.

Kostro, Stephanie S. 2014. "Internal Army Tensions Put National Security at Risk." *FYSA: For Your Situational Awareness,* March, 2014. csis.org/files/publication /140321_ISP_newsletter_FYSA_MARCH_2014.pdf.

Kotter, John P. 1996. *Leading Change.* Boston: Harvard Business School Press.

Kraft, Nelson. 2014. Chief of Tactics, US Army Maneuver Center of Excellence; Commander (2000–2002), Charlie Company, 1st Battalion, 87th Infantry Brigade, 5th Mountain Division. Fort Moore (then Fort Benning) GA, telephonic interview with Paul Birch, January 30.

Kramer, Roderick M., and M. B. Brewer. 1986. "Social Group Identity and the Emergence of Cooperation in Resource Conservation Dilemmas." In *Experimental Social Dilemmas* edited by H. A. M. Wilke, D. M Messick, and C. G. Rutte, 177–203. New York: Peter Lang.

Kristensen, Hans M., and Matt Korda. 2020. "Chinese Nuclear Forces, 2020." *Bulletin of the Atomic Scientists* 76 (6): 443–457.

Krooner, Kevin. 2006. "Staff Sergeant, US Air Force; Joint Terminal Attack Controller (JTAC) Remarks." *Jane's Defense Weekly,* January.

Kross, Walter. 2014. Commander (1996–1998), US Transportation Command and Air Mobility Command; Deputy Chief of Staff (1988–1990), Plans and Requirements, Headquarters Air Training Command. Washington DC, telephonic interview with Paul Birch, 1 May.

Kuhn, Thomas S. 1970. *The Structure of Scientific Revolutions*. 2nd ed. Chicago: University of Chicago Press.

Kwast, Steven L. 2013. Director (2013), Air Force Quadrennial Defense Review. Washington, DC, personal interview with Paul Birch, December 17.

Kyle, Deborah, and Benjamin Schemmer. 1982. "Navy, Marines Adamantly Oppose JCS Reforms Most Others Tell Congress Are Long Overdue." *Armed Forces Journal International* 119 (10): 61–67.

Lacdan, Joe. 2020. "Army, Air Force Form Partnership, Lay Foundation for CJADC2 Interoperability." *Army News Service*, October 2. www.af.mil/News/Article -Display/Article/2369626/army-air-force-form-partnership-lay-foundation -for-cjadc2-interoperability/.

Lambeth, Benjamin S. 2005. *Air Power Against Terror: America's Conduct of Operation Enduring Freedom*. Santa Monica, CA: RAND Corporation.

———. 2007. *Combat Pair: The Evolution of Air Force-Navy Integration in Strike Warfare RAND Project Air Force*. Santa Monica, CA: RAND Corporation.

———. 2014. "AirLand Reversal." *Air Force Magazine*. February, 60–64.

Langer, Emily. 2011. "Four-Star General Developed Cold War Strategy." *Washington Post*, September 2.

"Laser Guided JDAM Debuts in Iraq." 2007. *Defense Update*, August 27. defense -update.com/20070827_laserguidedjdamdebutsiniraq.html.

"LAS In, LAS Out: Counter-Insurgency Planes for the USA and Its Allies." 2013. *Defense Industry Daily*, October 3. www.defenseindustrydaily.com/las-in-las-out -counter-insurgency-planes-for-the-usa-and-its-allies-010548/.

Law, John. 1989. "Technology and Heterogeneous Engineering: The Case of Portuguese Expansion." In *The Social Construction of Technological Systems*, edited by Wiebe E. Bijker, Thomas P. Hughes, and Trevor J. Pinch, 111–134. Cambridge, MA: MIT Press.

Lawton, Alan, Julie Rayner, and Karin Lasthuizen. 2013. *Ethics and Management in the Public Sector*. London: Taylor & Francis.

Laymon, Robert. 2014. Acquistion Logistics Superintendent (1983–1986), T-46 Next Generation Trainer (1983–1986); Acquisition Logistics Superintendent (1986–1990), T-1 Tanker-Transport Training System & T-6 JPATS, Air Training Command Headquarters. San Antonio, TX, personal interview with Paul Birch.

Leach, William D., and Paul A. Sabatier. 2005. "To Trust an Adversary: Integrating Rational and Psychological Models of Collaborative Policymaking." *American Political Science Review* 99 (4): 491–503.

Lee, Caitlin. 2013. "USAF Debates Reduction in UAV Orbits." *Jane's Defense Weekly*, November 13.

Legro, Jeffrey. 1996. "Culture and Preferences in the International Cooperation Two-Step." *American Political Science Review* 90 (1): 118–137.

Lemann, Nicholas. 2002. "The War on What? The White House and the Debate about Whom to Fight Next." *New Yorker*, September 9, 40–41.

Leonhard, Robert R. 1991. *The Art of Maneuver: Maneuver-Warfare Theory and AirLand Battle*. Novato, CA: Presidio.

Levy, Jack S. 2008. "Case Studies: Types, Designs, and Logics of Inference." *Conflict Management and Peace Science* 25 (1): 1–18.

Lewis, Kevin N. 1990. *The U.S. Air Force Budget and Posture over Time.* Santa Monica, CA: RAND Corporation.

Lexington. 2014a. "Boots on the Ground." *Economist*, February 28.

————. 2014b. "Medals for Drone Pilots?" *Economist*, March 28.

Lind, William S. 1977. "Some Doctrinal Questions for the U.S. Army." *Military Review* 57 (3): 54–65.

————. 2014. Legislative aide for military affairs (1977–1986) for Senator Gary Hart. Washington DC, telephonic interview with Paul Birch, July 1.

Lindsay, James M., and Randall B. Ripley. 1994. "How Congress Influences Foreign and Defense Policy." *Bulletin of the American Academy of Arts and Sciences* 47 (6): 7–32.

Linn, Brian McAllister. 2007. *The Echo of Battle: The Army's Way of War.* Cambridge, MA: Harvard University Press.

Livy. 1982. *Rome and Italy.* Translated by Betty Radice. London: Penguin.

Locher, James R. 2001. "Has It Worked? The Goldwater-Nichols Reorganization Act." *Naval War College Review* 54 (4): 95–115.

————. 2002. *Victory on the Potomac: The Goldwater-Nichols Act Unifies the Pentagon.* College Station: Texas A & M University Press.

Lock-Pullan, Richard. 2005. "How to Rethink War: Conceptual Innovation and Air-Land Battle Doctrine." *Journal of Strategic Studies* 28 (4): 679–702. doi.org/10.1080/01402390500301087.

Loeb, Vernon. 2003. "General Defends Tactics in Afghan Battle." *Washington Post*, March 12. www.washingtonpost.com/archive/politics/2003/03/12/general-defends-tactics-in-afghan-battle/8f1862c9-0689-4905-91cc-f0d948396ea1/.

London, Herbert I. 1984. *Military Doctrine and the Amerian Character: Reflections on AirLand Battle.* New Brunswick, NJ: Transaction.

Loomis, Dan G. 1977. "FM 100-5 Operations: A Review." *Military Review* 57 (3): 66–69.

Lorell, Mark A., Michael Kennedy, Robert S. Leonard, Ken Munson, Shmuel Abramzon, David L. An, and Robert A. Guffey. 2013. *Do Joint Fighter Programs Save Money?* Santa Monica, CA: RAND Corporation.

Lorenz, Stephen A. 2013. Commander (2008–2011), Air Education and Training Command. Washington DC, telephonic interview with Paul Birch, December 16.

Losey, Stephen. 2021. "Christine Wormuth Pledges to Defend Army from Navy, Air Force Budget Grabs." *Daily News.* www.military.com/daily-news/2021/05/14/christine-wormuth-pledges-defend-army-navy-air-force-budget-grabs.html.

Lupia, Arthur, and Gisela Sin. 2003. "Which Public Goods Are Endangered? How Evolving Communication Technologies Affect the Logic of Collective Action." *Public Choice* 117 (3): 315–331.

Lüschen, Günther. 1970. "Cooperation, Association, and Contest." *Journal of Conflict Resolution* 14 (1): 21–34.

Machiavelli, Niccolò. 2003. *The Prince.* Translated by Daniel Donno. Bantam Classics ed. New York: Bantam Dell.

MacKenzie, Donald. 1990. *Inventing Accuracy: A Historical Sociology of Nuclear Missile Guidance.* Reprint ed. Cambridge MA: MIT Press.

Madison, James. 1788. "The Federalist, No. 51," February 6. Washington, DC:

National Archives. founders.archives.gov/documents/Madison/01-10-02 -0279.

Magnuson, Stew. 2006. "Revamped Flag Exercises Reflect New Missions." *National Defense*, December, 34–37.

———. 2021. "ANALYSIS: Air Force's Roper Is Gone, but His Vision Lives On." *National Defense*, February 24. www.nationaldefensemagazine.org/articles/2021 /2/24/air-forces-roper-is-gone-but-his-vision-lives-on.

Mahaffey, Fred K., and John T. Chain. 1984. *Memorandum of Understanding, Army/ Air Force Exchange of Staff Officers*, June 1. Washington, DC: Headquarters US Army and US Air Force.

Mahan, Alfred Thayer. 1890. *The Influence of Sea Power upon History, 1660–1783*. Boston: Little, Brown.

Mahnken, Thomas G. 2008. *Technology and the American Way of War since 1945*. New York: Columbia University Press.

Maloney, Matthew. 2013. Lieutenant Commander and F-18 pilot, US Navy. Washington DC, telephonic interview with Paul Birch, December 12.

Manning, Thomas A., et al. 1991. "History of the Air Training Command." Randolph Air Force Base, TX: History and Research Office, Air Training Command.

Marsh, Robert T. 1999. Deputy (1969–1973), US Air Force Systems Command reconnaissance, strike and electronic warfare division. Washington DC, personal interview with Thomas P. Ehrhard, April 19.

Martel, William C. 2007. *Victory in War: Foundations of Modern Military Policy*. Cambridge: Cambridge University Press.

Mason, Joseph, et al. 2005. "History of the Air Education and Training Command 2002–2003." Edited by Thomas A. Manning. Randolph Air Force Base, TX: History and Research Office, AETC.

Mattis, James N. 2008. *Assessment of Effects Based Operations*, August 14. Norfolk, VA: US Joint Forces Command.

May, Ernest R., John D. Steinbruner, and Thomas W. Wolfe. 1981. *History of the Strategic Arms Competition 1945–1972; Top Secret Report (Declassified)*. Archival holding in National Security Archive. Washington, DC: George Washington University. Nsarchive2.gwu.edu/NSAEBB/NSAEBB277/front%20matter %20chs%201-5.pdf.

McAdam, Mario Diani Doug. 2003. *Social Movements and Networks: Relational Approaches to Collective Action*. Oxford: Oxford University Press.

McCoy, Daniel. 2012. "USAF Puts Hold on LAS Contract amid Hawker Protest." *Wichita Business Journal*, January 5. www.bizjournals.com/wichita/news/2012 /01/05/usaf-puts-hold-on-las-contract-amid.html.

McElroy, Robert H., and Patricia Slayden Hollis. 2002. "Afghanistan: Fire Support for Operation Anaconda." *Field Artillery*, September–October, 5–9.

McFadden, Robert D. 2014. "James R. Schlesinger, Willful Aide to Three Presidents, Is Dead at 85." *New York Times*, March 28: A18.

McLeary, Paul. 2020. "Exclusive: Navy Makes Major JADC2 Push, Linking Sensors & Shooters." *Breaking Defense*. https://breakingdefense.com/2020/11/memo -shows-navy-making-major-jadc2-push-linking-sensors-shooters/.

McMillin, Molly. 2013. "GAO Rejects Beechcraft Protest." *Wichita Eagle*, June 14.

McNeil, Elton B. 1961. "Personal Hostility and International Aggression." *Journal of Conflict Resolution* 5 (3): 279–290.

McNeill, William H. 1982. *The Pursuit of Power: Technology, Armed Force, and Society since A.D. 1000.* Chicago: University of Chicago Press.

McPeak, Merrill A. 1994. *Presentation to the Commission on Roles and Missions of the Armed Forces.* Air Force Rpt. 1994 O-382-819 (D 301.2: R 64). Washington, DC: US Air Force.

Mearsheimer, John J. 1982. "Maneuver, Mobile Defense, and the NATO Central Front." *International Security* 6 (3): 104–122.

Meilinger, Phillip S., ed. 1997. *The Paths of Heaven: The Evolution of Airpower Theory.* Maxwell Air Force Base, AL: Air University Press.

Mets, David R. 1998. "A Glider in the Propwash of the Royal Air Force? Gen. Carl A. Spaatz, the RAF, and the Foundations of American Tactical Air Doctrine." In *Airpower and Ground Armies,* edited by Daniel R. Mortensen, 45–92. Maxwell Air Force Base, AL: Air University Press.

———, ed. 1999. *The Air Campaign: John Warden and the Classical Airpower Theorists Revised.* Maxwell Air Force Base, AL: Air University Press.

Meyer, Edward C. 1979. "Letter to General Donn A. Starry; TRADOC, Fort Monroe." June 13. Washington, DC: Headquarters, US Army.

Meyer, Edward C., and Charles A. Gabriel. 1983. "Memorandum of Understanding on Joint USA/USAF Efforts for Enhancement of Joint Employment of the AirLand Battle Doctrine." April 21. Washington, DC: Headquarters US Army and US Air Force.

Middleton, Drew. 1983. "U.S. Developing Flexible Battlefield Strategy." *New York Times,* March 13.

"Military Spending: Don't Save the Warthog." 2014. *Economist,* June 14.

Millett, Allan R., and Peter Maslowski. 1984. *For the Common Defense: A Military History of the United States.* New York: Free Press.

Mills, C. Wright. 1956. *The Power Elite.* Oxford: Oxford University Press.

Milner, Helen. 1992. "International Theories of Cooperation among Nations: Strengths and Weaknesses." *World Politics* 44 (3): 466–496.

Mintz, John. 1996. "Just Plane Too Big? Challenge to Military Trainer's Suitability for Some Female Pilots Stalls Big Contract." *Washington Post,* January 30: E1.

Momyer, William W. 1978. *Airpower in Three Wars.* Washington, DC: US Government Printing Office.

———. 1981. Letter to Colonel Duncan R. McNabb: Background Information on Air Force Perspective for Coherent Plans. In *Doctrine Information Publication No. 10.* Washington, DC: Headquarters US Air Force.

Morgan, Gareth. 2006. "Nature Intervenes: Organizations as Organisms." In *Images of Organization,* edited by Gareth Morgan, 33–69. Thousand Oaks, CA: Sage.

Morgenthau, Hans J. 1973. *Politics among Nations: The Struggle for Power and Peace.* 5th ed. New York: Alfred A. Knopf.

Morton, Louis. 1962. "Interservice Cooperation and Political-Military Collaboration." In *Total War and Cold War: Problems in Civilian Control of the Military,* edited by Harry L. Coles, 132–136. Columbus: Ohio State University Press.

Mulrine, Anna. 2009. "UAV Pilots." *Air Force Magazine*, January. www.airforcemag.com/article/0109uav/.

Nagl, John A. 2002. *Learning to Eat Soup with a Knife: Counterinsurgency Lessons from Malaya and Vietnam.* Chicago: University of Chicago Press.

Najam, Adil. 2001. "Getting Beyond the Lowest Common Denominator: Developing Countries in Global Environmental Negotiations." PhD diss., Massachusetts Institute of Technology.

Naveh, Shimon. 2004. *In Pursuit of Military Excellence: The Evolution of Operational Theory.* London: Frank Cass.

"Navy Awards Boeing $23 Million for Laser JDAM." 2012. *Defense Update*, September 7. defense-update.com/20120907_navy_laser-jdam.html.

"Navy Secretary Strafes Bureaucrats." 1983. *Washington Post*, March 5.

Naylor, Sean. 2005. *Not a Good Day to Die: The Untold Story of Operation Anaconda.* New York: Berkeley.

Neal, Curtis V., Robert B. Green, and Troy Caraway. 2012. "Bridging the Gap from Coordination to Integration." *Joint Forces Quarterly* 67:97–100.

Neuenswander, Matthew D. 2003. "JCAS in Operation Anaconda—It's Not All Bad News." *Field Artillery Journal*, May–June: 2–4.

———. 2014. Commander (2001–2002), US Air Force 332nd Air Expeditionary Group. Washington, DC, telephonic interview with Paul Birch, January 7.

Newport, Frank. 2013. *Americans See Army, Marines as Most Important to Defense.* Gallup, Inc. www.gallup.com/poll/148127/americans-army-marines-important-defense.aspx.

Newton, Robert E. 1992. *The Capture of the USS Pueblo and Its Effect on SIGINT Operations*, Declassified NSA report, doc. ID 3997429, FOIA case #40722. Archival copy in National Security Archive. Washington, DC: George Washington University. nsarchive2.gwu.edu/NSAEBB/NSAEBB278/US_Cryptologic_History—The_Capture_of_the_USS_Pueblo.pdf.

Nielsen, Suzanne C. 2010. *An Army Transformed: The U.S. Army's Post-Vietnam Recovery and the Dynamics of Change in Military Organizations.* Carlisle Barracks, PA: Strategic Studies Institute.

Nielson, Daniel L., Michael J. Tierney, and Catherine E. Weaver. 2006. "Bridging the Rationalist-Constructivist Divide: Re-Engineering the Culture of the World Bank." *Journal of International Relations and Development* 9 (2): 107–139.

Nissenbaum, Dion. 2013. "Pentagon Says Afghanistan Needs U.S. Help." *Wall Street Journal*, July 30.

Nolan, Dave. 1997. "Air Force, Navy Name Joint Training Aircraft 'Texan II.'" *Airman* 41 (8): 12.

North, David M. 1992. "Power Tucano H Handles Well at High Low Altitudes." *Aviation Week & Space Technology* 136 (16): 46–49.

Nunn, Mark, and Tim Arnold. 2006. "Trainers." *Flying Safety* 62 (1 and 2): 38–41.

Nunn, Sam. 1985. "Congressional Oversight of National Defense." 99th Cong., 1st sess., *Congressional Record* 131 (125): S12338.

———. 1991. "Tribute to Gen. Carl E. Vuono." 102nd Cong., 1st sess., *Congressional Record* 137 (121): S11998.

Odierno, Raymond T., James F. Amos, and William H. McRaven. 2013. *Strategic Landpower: Winning the Clash of Wills*. Washington, DC: Strategic Landpower Task Force.

O'Keefe, Raymond. 2014. US Navy JPATS requirements officer. Universal City, TX, telephonic interview with Paul Birch, May 6.

Olive, Steven G. 2007. "Abdicating Close Air Support: How Interservice Rivalry Affects Roles and Missions." MA degree, US Army War College.

Oliver, Pamela. 1980. "Rewards and Punishments As Selective Incentives for Collective Action: Theoretical Investigations." *American Journal of Sociology* 85 (6): 1356–1375.

Oliver, Pamela E., and Gerald Marwell. 1988. "The Paradox of Group Size in Collective Action: A Theory of the Critical Mass. II." *American Sociological Review*: 1–8.

Oliver, Pamela, Gerald Marwell, and Ruy Teixeira. 1985. "A Theory of the Critical Mass. I. Interdependence, Group Heterogeneity, and the Production of Collective Action." *American Journal of Sociology* 91 (3): 522–556.

Olson, Mancur. 1971. *The Logic of Collective Action: Public Goods and the Theory of Groups*. Cambridge, MA: Harvard University Press.

Oppenheim, Felix E. 1981. *Political Concepts: A Reconstruction*. Oxford: Basil Blackwell.

Orchard, John T. 2013. Commander (2012–2013), US Air Force 492d Fighter Squadron. Washington, DC, telephonic interview with Paul Birch, November 24.

Orr, Verne, and John Lehman, Jr. 1982. "Joint USN/USAF Efforts for Enhancement of the Joint Cooperation," October 25. Washington, DC: Department of the Navy and Headquarters US Air Force.

Orton, Megan. 2006. "Air Force T-6A Texan II Flies 250,000th Hour." Press release, US Air Force Air Education and Training Command, August 25. www.aetc.af.mil /News/Article-Display/Article/263550/air-force-t-6a-texan-ii-flies-250000th -hour/.

Osborn, Kris. 2020. "A New Generation of AI." *Warrior Maven*. warriormaven.com /future-weapons/army-research-lab-top-10-2019-future-technologies.

Ostrom, Elinor. 1990. *Governing the Commons: The Evolution of Institutions for Collective Action*. Cambridge: Cambridge University Press.

Ott, J. Steven. 2011. "Understanding Organizational Culture." In *Classics of Public Administration*, 7th ed., edited by J. M. Shafritz and A. C. Hyde, 490–496. Belmont, CA: Wadsworth.

Overy, Richard J. 1980. *The Air War, 1939–1945*. London: Europa.

Owens, MacKubin Thomas. 1985. "The Hollow Promise of JCS Reform." *International Security* 10 (3).

Park, Francis H. J. 2012. "The Unfulfilled Promise: The Development of Operational Art in the U.S. Military, 1973–1997." PhD diss., University of Kansas.

Pavelec, Sterling Michael. 2010. "By Land and Sea: Non-Carrier Naval Aviation." Chap. 14 in *One Hundred Years of US Navy Air Power*, edited by Douglas V. Smith, 301–321. Annapolis: Naval Institute Press.

Peck, Michael. 2013. "Why Can't the U.S. Air Force Find Enough Pilots to Fly Its Drones?" *Forbes*, August 22.

Pede, Charles. 2020. "Address to the Duke University National Security Law Conference." National Security Law Conference, Raleigh, NC, February 28.

Pengelley, Rupert. 2004. "In CAS of Emergency, Contact the Universal Observer." *Jane's International Defense Review*, April 1.

"Pentagon Says JPATS Plan Would Burden Contractors." 1992. *Aviation Week & Space Technology* 137 (16).

Perry, James L., and Lois R. Wise. 1990. "The Motivational Basis of Public Service." *Public Administration Review* 50 (3): 367–373.

Peters, Ralph. 2013. Lieutenant Colonel, US Army, author, and television commentator. Washington, DC, telephonic interview with Paul Birch, December 10.

Phillips, Edward H. 2000. "T-6A to Begin MOT&E Testing." *Aviation Week & Space Technology* 152 (13): 37–38.

Piatt, Walter E. 1999. "What is Operational Art?" MA thesis. Fort Leavenworth, KS: Command and General Staff College, School of Advanced Military Studies. apps.dtic.mil/sti/pdfs/ADA370243.pdf.

Pierson, Paul, and Theda Skocpol, eds. 2007. *The Transformation of American Politics: Activist Government and the Rise of Conservativism*. Princeton, NJ: Princeton University Press.

Pike, John. 2011. "T-46 Eaglet Next Generation Trainer." *Global Security*, July 7. www.globalsecurity.org/military/systems/aircraft/t-46.htm.

Pincus, Walter. 2009. "Airborne Intelligence Playing Greater Role in Irregular Warfare." *Washington Post*. www.washingtonpost.com/wp-dyn/content/article/2009/04/27/AR2009042703672.html.

———. 2012. "Defense Procurement Problems Won't Go Away." *Washington Post*, May 2. www.washingtonpost.com/world/national-security/defense-procurement-problems-wont-go-away/2012/05/02/gIQAyQNvxT_story.html.

Polzer, Jeffrey T. 2004. "How Subgroup Interests and Reputations Moderate the Effect of Organizational Identification on Cooperation." *Journal of Management* 30 (1): 71–96.

Poole, Walter S. 2013. "Adapting to Flexible Response 1960–1968." Vol. 2, *History of Acquisition in the Department of Defense*, edited by Glen R. Asner. Washington, DC: OSD Historical Office.

Posen, Barry R. 1984. *The Sources of Military Doctrine: France, Britain, and Germany between the Wars*. Ithaca, NY: Cornell University Press.

Pratt, Everett, and William Hayden. 1993. "Joint Fixed-Wing Training." Briefing slides from unofficial working papers; SECDEF-Directed Joint Pilot Training Working Papers. Washington, DC: Department of Defense. Archival copy in collection of Raymond O'Keefe, Universal City, TX.

Pratt, William, and Douglas MacArthur. 1931. Agreement between the US Army Chief of Staff and the US Navy Chief of Naval Operations, "MacArthur-Pratt Agreement"; IRIS No. 123080; (January 7, 1931). Air Force Historical Research Agency, Maxwell AFB, Alabama.

Proxmire, E. William, and Peter Almquist. 1987. "No, the Soviet Military Is Not Ten Feet Tall." 100th Cong., 1st sess., *Congressional Record* 133 (206): S18641.

Pursell, Carroll W. 1972. *The Military-Industrial Complex*. New York: Harper & Row.

Pyatt, Everett. 2014. "Save the A-10: Give It to the Army." *Real Clear Defense*, January 21. www.realcleardefense.com/articles/2014/01/22/save_the_a-10__give_it_to_the_army_107047.html.

Rasmussen, Robert D. 1978. "The Central European Battlefield: Doctrinal Implications for Counterair-Interdiction." *Air University Review* 29 (5): 2–20.

"Raytheon Plane Found Unreliable by Pentagon." 2001. *Los Angeles Times*, November 22: C4.

"Raytheon Sees Foreign Sales on the Horizon for JPATS." 1995. *Defense Daily*, July 7.

Rew, William J. 2010. "Operational Flexibility." Presentation to US Air Force Air War College. Maxwell Air Force Base, AL, January 22.

———. 2014. Vice Commander (2009–2013), Air Combat Command; Director of Operations (2003–2004), 9th Air Force and U.S. Central Command Air Forces. Washington, DC, telephonic interview with Paul Birch, January 7.

Rice, Donald B. 1990. *The Air Force and US National Security: Global Reach— Global Power*. US Air Force white paper, June. Washington, DC: Department of the Air Force.

Richards, Brendan. 2018. "Amateurs Talk Strategy: Professionals Talk Logistics." *KPMG Newsroom*, July 11. newsroom.kpmg.com.au/amateurs-talk-strategy -professionals-talk-logistics/.

Richman-Loo, Nina, and Rachel N. Weber. 1996. "Gender and Weapons Design." In *It's Our Military, Too! Women and the U.S. Military*, edited by Judith Hicks Stiehm. Philadelphia: Temple University Press.

Riker, William H. 1986. *The Art of Political Manipulation*. New Haven, CT: Yale University Press.

Roberson, Darryl L. 2014. Vice Director (2012–2014) for Operations, the Joint Staff. Washington, DC, personal interview with Paul Birch, March 21.

Roberts, John W. 1982. Commander (1975–1979), US Air Force Air Training Command. Personal interview with James C. Hasdorf and David W. Shircliffe. Maxwell Air Force Base, AL: Air Force Historical Research Agency, November 2–4.

"Rockwell Protests Air Force JPATS Selection." 1995. *Inside Defense Daily*, July 27.

Romjue, John L. 1984a. "The Evolution of the AirLand Battle Concept." *Air University Review* 35 (4): 6–7.

———. 1984b. *From Active Defense to AirLand Battle: The Development of Army Doctrine 1973–1982*. Edited by Henry O. Malone. TRADOC Historical Monograph Series. Fort Monroe, VA: US Army Training and Doctrine Command.

Rondeau, William. 2007. "561st Joint Tactics Squadron Prepares Force, Captures Today's Tactical Issues." *Military News*, June 8. www.militarynews.com /peninsula-warrior/news/top_stories/561st-joint-tactics-squadron-prepares -force-captures-today-s-tactical-issues/article_3604a678-37ee-5a3f-a1c5-4a762d a7ff84.html.

Roper, Will. 2019. "Official Air Force Biography." US Air Force website. August 2019. www.af.mil/AboutUs/Biographies/Display/Article/1467795/dr-will -roper/.

———. 2020. "Dr. Will Roper ABMS Ask Me Anything." *Youtube*, August 25. www .youtube.com/watch?v=yuNJwyrVJo4.

Rosen, Stephen P. 1991. *Winning the Next War: Innovation and the Modern Military.* Ithaca, NY: Cornell University Press.

Rosenthal, Uriel. 1986. "Crisis Decision Making in the Netherlands." *Netherlands' Journal of Sociology* 22 (2): 103–129.

Rosenthal, Uriel, Michael T. Charles, and Paul 't Hart. 1989. "Introduction: The World of Crises and Crisis Management." In *Coping with Crises: The Management of Disasters, Riots and Terrorism*, edited by Uriel Rosenthal, Michael T. Charles, and Paul 't Hart, 3–33. Springfield: Charles C. Thomas.

Rosser, Sue V. 2001. "Will EC 2000 Make Engineering More Female Friendly?" *Women's Studies Quarterly* 29 (3 and 4): 164–186.

Rotsker, Bernard. 2013. *Right-Sizing the Force: Lessons for the Current Drawdown of American Military Personnel: Active Duty Military Personnel, 1940–2011.* March 30. Washington, DC: Center for a New American Security. www.infoplease.com /ipa/A0004598.html.

Roughead, Gary. 2001. "Capitol Hill: Update." *Wings of Gold* 26 (3): 19.

Roughton, Randy. 2011. "Rise of the Drones: 9/11 and War on Terror Sparked an Explosion in Unmanned Aerial Vehicle Technology." *Airman.* http://airman .dodlive.mil/2011/09/rise-of-the-drones/.

"Rover Develops into All-Purpose Battlefield Comms System." 2010. *Jane's International Defense Review*, November 25.

Sabatier, Paul A. 1986. "Top-Down and Bottom-Up Approaches to Implementation Research: A Critical Analysis and Suggested Synthesis." *Journal of Public Policy* 6 (1): 21–48.

Sabatier, Paul A., and Daniel Mazmanian. 1980. "The Implementation of Public Policy: A Framework of Analysis." *Policy Studies Journal* 8 (4): 538–560.

Sackman, Sonja A. 1991. "Uncovering Culture in Organizations." *Journal of Applied Behavioral Science* 27 (3): 295–317.

Sandler, Todd. 2015. "Collective Action: Fifty Years Later." *Public Choice* 164 (3): 195–216.

Sapolsky, Harvey M., Eugene Gholz, and Caitlin Tarmadge. 2009. *U.S. Defense Politics: The Origins of Security Policy.* New York: Routledge.

Scales, Robert H. 1997. *Certain Victory: The US Army in the Gulf War.* Washington, DC: Brassey's.

Scarborough, Rowan. 2002. " 'Friendly Fire' Judge's Memo Assailed: General Criticized Actions That Led to Deaths of Canadians." *Washington Times*, July 30.

Schaus, John. 2021. "Bad Idea: Overprioritizing 'Jointness' in the Joint Warfighting Concept." *Defense 360*, December 10. Washington, DC: Center for Strategic and International Studies. Defense360.csis.org/overprioritizing-jointness-in-the -joint-warfighting-concept/.

Schein, Edgar H. 2010. "The Concept of Organizational Culture: Why Bother?" In *Classics of Organization Theory*, 7th ed., edited by J. M. Shafritz and J. Steven Ott, 349–360. Belmont, CA: Wadsworth.

Schelling, Thomas C. 1966. *Arms and Influence.* New Haven, CT: Yale University Press.

Schmitt, Eric. 1990. "Confrontation in the Gulf; Air Force Chief Is Dismissed for Remarks on Gulf Plan; Cheney Cites Bad Judgment." *New York Times*, September 18.

Schneider, Greg, and Renae Merle. 2004. "Reagan's Defense Buildup Bridged Military Eras." *Washington Post*, June 9. www.washingtonpost.com/archive/business /2004/06/09/reagans-defense-buildup-bridged-military-eras/ec621466-b78e -4a2e-9f8a-50654e3f95fa/.

Schwarzkopf, H. Norman, and Peter Petre. 1992. *It Doesn't Take a Hero*. New York: Bantam.

Scroggs, Stephen K. 2000. *Army Relations with Congress: Thick Armor, Dull Sword, Slow Horse*. Westport: Praeger.

Seip, Mark. 2020. "Bad Idea: All Sensors, All Shooters, All the Time—a Joint All-Domain Command and Control System That Prioritizes Centralization," *Defense 360*, December 15. Washington, DC: Center for Strategic and International Studies. Defense360.csis.org/bad-idea-all-sensors-all-shooters-all-the -time-a-joint-all-domain-command-and-control-system-that-prioritizes-cen tralization/.

"Senate Appropriators Direct DoD to Buy More JPATS." 1996. *Defense Daily*, June 21, 1.

"Senators Want Training Aircraft to Accommodate Female Pilots." 1993. *Minerva's Bulletin Board* 6 (3), September 30.

Serbu, Jared. 2021. "Why DoD's Fighting Force Is 'Ever Shrinking' despite Robust Budgets." *Federal News Network*, August 1. federalnewsnetwork.com/on-dod /2021/08/why-dods-fighting-force-is-ever-shrinking-despite-robust-budgets/.

Shanker, Thom. 2007. "New Strategy Vindicates Ex-Army Chief Shinseki." *New York Times*, January 12:A13.

———. 2008. "At Odds with Air Force, Army Adds Its Own Aviation Unit." *New York Times*, June 22.

———. 2010. "Pentagon Plans Steps to Reduce Budget and Jobs." *New York Times*, August 10:A10.

Shapley, Deborah. 1982. "The Army's New Fighting Doctrine." *New York Times Sunday Magazine*, November 28:SM36.

Sherlock, Michael J. 2003. "Modify the Goshawk and the Pilot Training Syllabus." *Proceedings* 129 (12): 72–73.

Singer, Peter W. 2012. "Comparing Defense Budgets, Apples to Apples." *Time*, September 25.

Sirak, Michael. 2007. "OSD Defers JASSM Recertification While Reliability Plan Worked." *Defense Daily* 234 (47): 1.

"Sit Tight." 1996. *Aviation Week & Space Technology* 144 (2), January 8:329–334.

Siuru, William D. 1994. "JPATS: Finally a New Primary Trainer." *Marine Corps Gazette* 78 (5): 68–70.

Skinner, Douglas W. 1988. *AirLand Battle Doctrine*. Rpt. No. 463. Alexandria VA: Center for Naval Analyses. Apps.dtic.mil/sti/pdfs/ADA202888.pdf.

Sladek, Rick. 2014. Chief (1992–1997), US Air Force Air Education and Training Command aircraft requirements. San Antonio TX, personal interview with Paul Birch, May 13.

Sladek, Rick, and Bob McGee. 1993. "Anticipated hearing questions and answers: SAF/AQ on JPATS and Other Aviation Training Programs." SD V-29; K220.01

V.17; IRIS No. 01115057, March 25. Maxwell Air Force Base, AL: Air Force Historical Research Agency.

Sligh, Robert. 2003. "History of the Air Education and Training Command 2000–2001." Edited by Thomas A. Manning. Randolph Air Force Base, TX: History and Research Office, Air Education and Training Command.

Sloan, John F. 1977. "Letter to the Editor." *Military Review* 57 (7): 111.

Snider, Don M. 1996. "The U.S. Military in Transition to Jointness: Surmounting Old Notions of Interservice Rivalry." *Airpower Journal* 10 (3): 16–27. www.airuniversity.af.edu/Portals/10/ASPJ/journals/Volume-10_Issue-1-Se/1996_Vol10_No3.pdf.

Snyder, Glenn H. 1962. "The Politics of National Defense: A Review." *Journal of Conflict Resolution* 6 (4): 368–373.

Spires, David. 1998. "Patton and Weyland: A Model for Air-Ground Cooperation." In *Airpower and Ground Armies: Essays on the Evolution of Anglo-American Air Doctrine, 1940–1943*, edited by Vincent Orange et al., 147–163. Maxwell Air Force Base: Air University Press.

Springborg, Patricia. 2007. "General Introduction." In *The Cambridge Companion to Hobbes's* Leviathan, edited by Patricia Springborg, 1–27. Cambridge: Cambridge University Press.

"Spring Takeoff for JPATS Ground System." 1997. *Seapower* 40 (2): 14–15.

Starry, Donn A. 1979. "Letter to LTG William C. Meyer; TRADOC, Fort Monroe." June 26.

———. 1981. "The Air Land Battle." Teletype message distributed to US Army TRADOC subordinate commanders, January 29.

Stevens, Jeffrey R., Fiery A. Cushman, and Marc D. Hauser. 2005. "Evolving the Psychological Mechanisms for Cooperation." *Annual Review of Ecology, Evolution, and Systematics* 36:499–518.

Sun-Tzu. 2005. *The Illustrated Art of War*. Translated by Samuel B. Griffith. Oxford: Oxford University Press.

Svedin, Lina M. 2009. *Organizational Cooperation in Crises*. Farnham, UK: Ashgate.

Swain, Richard M., ed. 1994. *Selected Papers of General William E. DePuy*. Carlisle Barracks, PA: US Army Military History Institute.

Swartz, Peter M., and Karin Duggan. 2011. *U.S. Navy—U.S. Air Force Relationships: 1970–2010*. Rpt. No. 20110725336. Alexandria, VA: Center for Naval Analysis.

Sweetman, Bill. 1997. "Choosing a Primary Air Trainer." *Jane's International Defense Review* 30 (3): 71.

———. 2000. "Northrop Grumman Back from the Brink." *Interavia* 55 (645): 18–19.

Thomas, George M., Henry A. Walker, and Morris Zelditch, Jr. 1986. "Legitimacy and Collective Action." *Social Forces* 65 (2): 378–404. https://doi.org/10.1093/sf/65.2.378.

"Thrown Out." 1995. *Defense Daily*, August 28.

Thucydides, 1998 ed. *The Landmark Thucydides: A Comprehensive Guide to the Peloponnesian War*. Edited by Robert B. Strassler. 1st Touchstone ed. New York: Simon & Schuster.

Tiron, Roxana. 2004. "Air Force Chopper Pilot Training Splits from Army." *National Defense* 89 (613): 38–39.

Tirpak, John A. 2014. "Gates versus the Air Force." *Air Force Magazine*, March 1, 54–57.

Tjosvold, Dean. 1984. "Cooperation Theory and Organizations." *Human Relations* 37 (9): 743–767.

———. 1988. "Effects of Shared Responsibility and Goal Interdependence on Controversy and Decisionmaking between Departments." *Journal of Social Psychology* 128 (1): 7–18.

Tolstoy, Lev Nikolayevich. 2009 (1869). *War and Peace*. Translated by David Widger. Project Gutenberg iBooks ed. Moscow: Russian Messenger.

Toppe, Alfred. 1953. *Night Combat*. Washington, DC: Center of Military History.

TRADOC, Combined Arms Center. 2010. *Army: Profession of Arms 2011—the Profession After 10 Years of Persistent Conflict*. Blackwell, OK: Schatz.

Trest, Warren A. 1998. *Air Force Roles and Missions: A History*. Washington, DC: Air Force History and Museums Program.

Udéhn, Lars. 1993. "Twenty-Five Years with the Logic of Collective Action." *Acta Sociologica* 36 (3): 239–261.

Umansky, Eric. 2001. "Studs and Duds." *Washington Monthly* 33 (12): 15–21.

United Nations. 1955. *Treaty of Friendship, Co-Operation, and Mutual Assistance*. Warsaw: United Nations.

"USAF Orders Sniper Targeting Pods." 2002. *Jane's Defense Weekly*, July 12.

US Air Force. 1953. *Air Force Manual 1-2: United States Air Force Basic Doctrine*. Washington, DC: Department of the Air Force.

———. 1984. *Air Force Manual 1-1: Basic Aerospace Doctrine of the United States Air Force*. Washington, DC: Department of the Air Force.

———. 1988. *Trainer Masterplan*. Washington, DC: Department of the Air Force. Archival copy in SD III-1; ATC Periodic History 1988; Vol. X; Air Education and Training Command History Office, Randolph Air Force Base, TX.

———. 2009. *Unmanned Aircraft Systems Flight Plan, 2009–2047*, May 18. www.fas .org/irp/program/collect/uas_2009.pdf.

———. 2010. CAP Requirements: Single 24/7/365 MQ-1/9. US Air Force press release. Washington, DC: US Air Force, September 11.

———. 2011. *Air Force Doctrine Document 1: Air Force Basic Doctrine, Organization, and Command*. Washington, DC: Department of the Air Force.

US Air Force Air Education and Training Command. 1999. *History of Air Education and Training Command 1 July 1993–31 December 1995. Vol. 1—Narrative*. Randolph Air Force Base, TX: AETC History and Research Office.

"US Air Force Eyes Better Integration with Army." 2004. *Jane's Defense Weekly*, November 5.

US Air Forces Central. 2008. *Airpower Summary for July 17*. US Air Force public affairs release. Al Udeid Air Base, Qatar: July 18. www.airandspaceforces .com/PDF/SiteCollectionDocuments/Reports/2008/July/Day21/Airpower _summary_071708.pdf.

US Army. 1986. *Field Manual 100-5: Operations*. Washington, DC: Headquarters, US Army.

————. 2012. *Army Doctrine Publication 1 (Field Manual 1): The Army*. Washington, DC: Department of the Army.

US Congress. 1947. National Security Act of 1947. Pub. L. 235; 61 Stat. 495; 80th Cong., 1st sess., July 26.

————. 1986. The Goldwater-Nichols Department of Defense Reoganization Act of 1986. Pub. L. 99-433, 100 Stat. 992; 99th Cong., 2nd sess. www.govinfo.gov /content/pkg/STATUTE-100/pdf/STATUTE-100-Pg992.pdf.

————. 1988. National Defense Authorization Act for Fiscal Year 1989. Pub. L. 100-456, 100th Cong., 2nd sess. www.govinfo.gov/app/details/COMPS-634 #:~:text=Title,Forces%2C%20and%20for%20other%20purposes.

————. 1989. National Defense Authorization Act for Fiscal Years 1990 and 1991. Pub. L. 101-189, 103 Stat. 1352; 101st Cong., 1st sess.

US Department of Defense. 1989. "Trainer Aircraft Masterplan." Washington, DC: Department of Defense. Archival copy in SD III-12; ATC Periodic History 1988; Vol. X; Air Education and Training Command History Office, Randolph Air Force Base, TX.

————. 2007. "Department of Defense Releases Selected Acquisition Reports." DoD press release based on Selected Acquisition Reports (SARs) to Congress, April 9.

US Government Accountability Office. 2013. "Defense Acquisitions: Assessment of Selected Weapon Programs." GAO rpt. GAO-13-294SP. Washington, DC: GAO. www.gao.gov/products/gao-13-294sp.

————. 2014. "Defense Acquisitions: Assessment of Selected Weapon Programs." GAO rpt. GAO-14-340SP. Washington, DC: GAO. www.gao.gov/products/gao -14-340sp.

————. 2020. "Action Is Needed to Provide Clarity and Mitigate Risks of the Air Force's Planned Advanced Battle Management System." Washington, DC: GAO. www.gao.gov/assets/710/706165.pdf.

US House of Representatives. 1959. Subcommittee of the Committee on Appropriations. "Department of Defense Appropriations for 1960, Part 1." 86th Cong., 1st sess.

————. 1981. Subcommittee of the Committee on Appropriations. "Department of Defense Appropriations for 1982." 97th Cong., 1st sess.

————. 1982. Investigations Subcommittee of the Committee on Armed Services, "Reorganization Proposals for the Joint Chiefs of Staff." April 21. 97th Cong., 2nd sess.

————. 1984. Committee on Armed Services, "Hearings on H.R. 5167, Department of Defense Authorization of Appropriations for FY85 and Oversight of Previously Authorized Programs." 98th Cong., 2nd sess., H.Rept 98-691.

————. 1986. Debate on National Defense Authorization Act for Fiscal Year 1987. 99th Cong., 2nd sess., *Congressional Record* 132 (110): H5891.

————. 1989. "National Defense Authorization Act 1989." Conference rpt. On Pub. L. 100-456, 100th Cong., 1st sess.

————. 1997. "National Defense Authorization Act for Fiscal Year 1998." Conference rpt. on Pub. L. 105-85, 105th Cong., 1st sess., *Congressional Record* 143 (144).

———. 2009. Committee on Armed Services, Subcommittee on Tactical Air and Land Forces, "Combat Aviation Programs Update." 112th Cong., 1st sess., November 2.

US Marine Corps History Division. 2013. "Biography of Lieutenant General Paul K. Van Riper." US Marine Corps History Division. www.mcu.usmc.mil /historydivision/Pages/Who's%20Who/V-X/van_riper_pk.aspx.

"USMC Orders Litening Pods." 2001. *Jane's Defense Weekly.* August 3.

US Navy. 1994. *Naval Doctrine Publication 1: Naval Warfare.* Washington, DC: Department of the Navy.

US Senate. 1959. Committee on Aeronautical and Space Sciences, Subcommittee on Governmental Organization for Space Activities, "Investigation of Governmental Organization for Space Activities." 86th Cong., 1st sess., S. Hrg. March 24–May 7.

———. 1969. Subcommittee of the Committee on Appropriations, "Department of Defense Authorization for Appropriations for Fiscal Year 1970." 91st Cong., 1st sess. www.congress.gov/91/crecb/1969/12/23/GPO-CRECB-1969-pt31 -1.pdf.

———. 1971. Committee on Armed Services, Preparedness Investigation Subcommittee, Hearing before the Special Subcommittee on Close Air Support. 92nd Cong., 1st sess., October 22–November 8. play.google.com/books/reader ?id=199zy-8T_ukC&pg=GBS.PA6&hl=en.

———. 1976. Committee on Armed Services, "Fiscal Year 1977 Military Procurement Authorizations." 94th Cong., 2nd sess., S. Hrg. 94. www.govinfo.gov/app /details/CHRG-94shrg67524Op3/context.

———. 1985. Committee on Armed Services. "Defense Organization: The Need for Change." 99th Cong., 1st sess., S. Rpt. 99-86, October 16. dair.nps.edu/handle /123456789/3764.

———. 2009. Committee on Armed Services, "The Secretary of Defense's 2010 Budget Recommendations." 111th Cong., 1st Sess., S. Hrg. 111-258, April 30. www.govinfo.gov/content/pkg/CHRG-111shrg54649/html/CHRG-111shr g54649.htm.

———. 2014. Committee on Armed Services, "Carl Levin National Defense Authorization Act for Fiscal Year 2015." 113th Cong., 2nd sess., S. Rept 113-176. www .congress.gov/congressional-report/113th-congress/senate-report/176/1.

US State Department. 2014. "The 1973 Arab-Israeli War." *Milestones in the History of US Foreign Relations.* Washington, DC: US State Department. history.state .gov/milestones/1969-1976/arab-israeli-war-1973.

US War Department. 1943. *Field Manual 100-20: Command and Employment of Air Power.* Washington, DC: War Department.

Vandenbussche, Jeffrey L. 2007. "Centering the Ball: Command and Control in Joint Warfare." MA degree, School of Advanced Air and Space Studies, Air University.

Vandiver, James. 2014. Commander (2011–2014), US Navy Naval Aviation Schools Command. Pensacola FL, telephonic interview with Paul Birch, April 11.

van Evera, Stephen. 1997. *Guide to Methods for Students of Political Science.* Ithaca, NY: Cornell University Press.

Van Riper, Paul K. 2013. Commanding General (1995–1997), US Marine Corps Combat Development Command. Quantico, VA, personal interview with Paul Birch, December 9.

Van Tol, Jan, et al. 2010. *Airsea Battle: A Point-of-Departure Operational Concept.* Washington, DC: Center for Strategic and Budgetary Assessments.

Voeten, Erik. 2004. "Resisting the Lonely Superpower: Responses of States in the United Nations to U.S. Dominance." *Journal of Politics* 66 (3): 747–748.

Wacker, Rudolph F. 1967. "Managing the Infinities of Basic Doctrine." MA thesis, Air Command and Staff College, Air University.

Walters, Tome H., Jr. 2014. Director (2000–2004), Defense Security Cooperation Agency; Director (1989–1991), Air Training Command requirements. Washington DC, telephonic interview with Paul Birch, April 28.

Waltz, Kenneth N. 1979. *Theory of International Politics.* New York: Random House.

Webb, Grant A. 1996. "The 'Plane' Truth about DoD Undergraduate Helicopter Pilot Training Consolidation." MA thesis, United States Marine Corps Command and Staff College.

Weber, Rachel N. 1997. "Manufacturing Gender in Commercial and Military Cockpit Design." *Science, Technology, & Human Values* 22 (2): 235–253.

Weigley, Russell F. 1973. *The American Way of War: A History of United States Military Strategy and Policy.* Bloomington: Indiana University Press.

Weinberger, Caspar W. 1998. Fifteenth Secretary of Defense, 1981–1987. Personal interview with James R. Locher, October 27. Archival copy in Box 63, Locher's personal papers.

"Welcome to the 19th Century." 2014. *Wall Street Journal,* March 17, A14.

Welsh, Mark A. 2014. "Military Strategy Forum: General Mark A. Welsh on the Future of the Air Force." Washington, DC: Center for Strategic and International Studies, March 27. www.youtube.com/watch?v=mxCobhGl-oE.

White, Leslie A. 1975. *The Concept of Cultural Systems: A Key to Understanding Tribes and Nations.* New York: Columbia University Press.

Whitmore, Bishane A. 2021. "Identity Crisis: How Air Force Identity Influences Rejection of Disruptive Innovations." PhD diss., School of Advanced Air and Space Studies, Air University.

Wickham, John A., Jr., and Charles A. Gabriel. 1984. "Memorandum of Agreement on US Army—US Air Force Cross-Service Participation in the POM Development Process." Washington, DC: Department of the Army and Department of the Air Force, November 29.

Wilkerson, Lawrence B. 1997. "What Exactly Is Jointness?" *Joint Forces Quarterly* 16:66–68. ndupress.ndu.edu/portals/68/Documents/jfq/jfq-16.pdf.

Williams, Lauren C. 2020. "Air Force seeks proof of concept in latest JADC2 experiment." *FCW,* August 26. fcw.com/it-modernization/2020/08/air-force-seeks-proof-of-concept-in-latest-jadc2-experiment/258046/.

Williams, Rhys H. 2004. "The Cultural Contexts of Collective Action: Constraints, Opportunities, and the Symbolic Life of Social Movements." In *The Blackwell Companion to Social Movements,* edited by Sarah A. Soule, David A. Snow, and Hanspeter Kriesi, 91–115. Oxford: Blackwell.

Williamson, Murray. 1983. "A Tale of Two Doctrines: The Luftwaffe's Conduct of the Air War and the USAF's Manual 1–1." *Journal of Strategic Studies* 6 (4): 84–93. doi.org/10.1080/01402398308437169.

Willis, Danielle. 2021. Commander (2020–2022), US Air Force 93rd Air Ground Operations Wing. Arlington VA, personal interview with Paul Birch, December 16.

Wilson, James Q. 1974. *Political Organizations*. New York: Basic.

Wilson, Warner. 1969. "Cooperation and the Cooperativeness of the Other Player." *Journal of Conflict Resolution* 13 (1): 110–117.

Winton, Harold R. 1996. "Partnership and Tension: The Army and the Air Force between Vietnam and Desert Shield." *Parameters* 26 (1): 100–119.

Wise, Lois Recascino. 2010. "The Public Service Culture." In *Public Administration Concepts and Cases*, edited by Richard J. Stillman, II, 320–330. Boston: Wadsworth Cengage Learning.

Wolfe, Frank. 2001. "G-Suits for T-6 Pilots to Reduce Loss of Consciousness Incidents during Training." *Defense Daily International* 2 (40): 1.

———. 2002. "HASC Panel Adds $3.2 Billion for Porcurement, Restricts Comanche Funding." *Defense Daily* 214 (23): 1.

Woodward, Bob. 1987. *Veil: The Secret Wars of the CIA, 1981–1987*. New York: Pocket.

Worden, Mike. 1998. *Rise of the Fighter Generals: The Problem of Air Force Leadership*. Maxwell Air Force Base, AL: Air University Press.

Wormuth, Christine. 2021. "Secretary of the Army's Keynote Speech." Transcribed by Dontavian Harrison. Washington, DC: Association of the US Army, October 11. www.army.mil/article/251180/ausa_2021_secretary_of_the_armys_keynote_speech_11_october_2021.

Wylie, J. C. 1967. *Military Strategy: A General Theory of Power Control*. Annapolis, MD: Naval Institute Press.

Yeats, William Butler. 1920. *Michael Robartes and the Dancer*. 1970 Irish University Press ed. Churchtown, Dundrum, Ireland: Cuala Press.

Yoshitani, Gail E. S. 2011. *Reagan on War: A Reappraisal of the Weinberger Doctrine*. College Station, TX: Texas A & M University Press.

Yost, David S. 1998. *NATO Transformed: The Alliance's New Roles in International Security*. Washington, DC: United States Institute of Peace.

Yuchtman, Ephraim, and Stanley E. Seashore. 1967. "A System Resource Approach to Organizational Effectiveness." *American Sociological Review* 32 (6): 891–903.

Zessin, Cynthia G., Michael H. Oelrich, James Simpson, and Greg Hutto. 2008. "Can You Hear Me Now? F-15E Enhanced Radio Test Using DOE." Symposium rpt., June. Eglin Air Force Base, FL: 76th Military Operations Research Symposium.

Zuckert, Eugene M. 1966. "The Service Secretary: Has He a Useful Role?" *Foreign Affairs* 44 (3): 458–479.

Index

Printed in the USA
CPSIA information can be obtained
at www.ICGtesting.com
LVHW042356110124
768731LV00014B/286/J